Euclidean Shortest Paths

"Beauty on the Path", a digital painting by Stephen Li (Auckland, New Zealand), September 2011, provided as a gift for this book.

Fajie Li · Reinhard Klette

Euclidean Shortest Paths

Exact or Approximate Algorithms

 Springer

Fajie Li
School of Information Science
and Technology
Huaqiao University
P.O. Box 800
Xiamen Fujian
People's Republic of China
li.fajie@yahoo.com

Reinhard Klette
Dept. Computer Science
University of Auckland
P.O. Box 92019
Auckland 1142
New Zealand
r.klette@auckland.ac.nz

ISBN 978-1-4471-6064-9   ISBN 978-1-4471-2256-2 (eBook)
DOI 10.1007/978-1-4471-2256-2
Springer London Dordrecht Heidelberg New York

British Library Cataloguing in Publication Data
A catalogue record for this book is available from the British Library

Cover design: VTeX UAB, Lithuania

Printed on acid-free paper

Springer is part of Springer Science+Business Media (www.springer.com)

To Zhixing Li, and to the two youngest in the Klette family in New Zealand

Foreword

The world is continuous the mind is discrete.

David Mumford (born 1937)

Recently, I was confronted with the problem of planning my travel from Israel to New Zealand, home of the two authors of this book. When taking two antipodal points on the globe, like Haifa and Queenstown, there is an infinite number of shortest paths connecting these points. Still, due to constraints like reachable airports and airlines, finding the optimal solution was almost immediate.

Throughout the long history of geometry sciences, the problem of finding the shortest path in various scenarios occupied the minds of researchers in many fields. Even in Euclidean spaces, which are considered simple, the introduction of obstacles leads to challenging problems for which efficient computational solvers are hard to find. The optimal path in 3D space with polyhedral obstacles was among the first geometric problems proven to be, at least formally, computationally hard to solve. It took almost 20 years for a team of 5 programming experts to eventually implement a method approximating the continuous Dijkstra algorithm that is reviewed in this book. Exact problems are hard to solve, and approximations are obviously required.

My personal line of work when dealing with geometric problems somewhat differs from the school of thought promoted by this book. A numerical approximation in my vocabulary involves the notion of accuracy that depends on an underlying grid resolution. This grid is defined by sampling the domain of the problem and leads to the field of numerical geometry in which efficient solvers are simple to design.

The alternative computational geometry school of thought describes obstacles as polyhedral structures that allegedly define the "exact" problem. The resulting challenges under this setting are extremely difficult to overcome. Still, the unifying bridge between these two philosophical branches is defined by the geometric problems. Without being familiar with the difficulty involved in designing a path between points in a weighted domain, one could not appreciate the conceptual simplicity of numerical Eikonal solvers.

This book addresses the type of hard problems in the computational geometry flavor while inventing constraints that allow for efficient solvers to be designed. For example, the creative rubberband methods explored in this book restrict the optimal paths to bands of bounded width, thereby redefining problems and simplifying the challenges, proving yet again Aleksandr Pushkin's observation that *"inspiration is needed in geometry, just as much as in poetry."* I hope that, like me, the reader would find the geometrical challenges introduced in this book fascinating and also appreciate the elegance of the proposed solutions.

Haifa, Israel   Ron Kimmel

Preface

A Euclidean shortest path connects a source with a destination, avoids some places (called *obstacles*), visits some places (called *attractions*), possibly in a defined order, and is of minimum length. Euclidean shortest-path problems are defined in the Euclidean plane or in Euclidean 3-dimensional space. The calculation of a convex hull in the plane is an example for finding a shortest path (around the given set of planar obstacles). Polyhedral obstacles and polyhedral attractions, a start and an endpoint define a general Euclidean shortest-path problem in 3-dimensional space.

The book presents selected algorithms (i.e., not aiming at a general overview) for the exact or approximate solution of shortest-path problems. Subjects in the first chapters of the book also include fundamental algorithms. Graph theory offers shortest-path algorithms for discrete problems. Convex hulls (and to a lesser extent also constrained convex hulls) have been discussed in computational geometry. Seidel's triangulation and Chazelle's triangulation method for a simple polygon, and Mitchell's solution of the continuous Dijkstra problem have also been selected for a detailed presentation, just to name three examples of important work in the area.

The book also covers a class of algorithms (called *rubberband algorithms*), which originated from a proposal for calculating minimum-length polygonal curves in cube-curves; Thomas Bülow was a co-author of the initiating publication, and he coined the name 'rubberband algorithm' in 2000 for the first time for this approach.

Subsequent work between 2000 and now shows that the basic ideas of this algorithm generalised for solving a range of problems. In a sequence of publications between 2003 and 2010, we, the authors of this book, describe a class of rubberband algorithms with proofs of their correctness and time-efficiency. Those algorithms can be used to solve different Euclidean shortest-path (ESP) problems, such as calculating the ESP inside of a simple cube-arc (the initial problem), inside of a simple polygon, on the surface of a convex polytope, or inside of a simple polyhedron, but also ESP problems such as touring a finite sequence of polygons, cutting parts, or the safari, zookeeper, or watchman route problems.

We aimed at writing a book that might be useful for a second or third-year algorithms course at the university level. It should also contain sufficient details for students and researchers in the field who are keen to understand the correctness proofs, the analysis of time complexities and related topics, and not just the algorithms and their pseudocodes. The book discusses selected subjects and algorithms at some depth, including mathematical proofs for most of the given statements. (This is different from books which aim at a representative coverage of areas in algorithm design.)

Each chapter closes with theoretical or programming exercises, giving students various opportunities to learn the subject by solving problems or doing their own experiments. Tasks are (intentionally) only sketched in the given programming exercises, not described exactly in all their details (say, as it is typically when a costumer specifies a problem to an IT consultant), and identical solutions to such vaguely described projects do not exist, leaving space for the creativity of the student.

The audience for the book could be students in computer science, IT, mathematics, or engineering at a university, or academics being involved in research or teaching of efficient algorithms. The book could also be useful for programmers, mathematicians, or engineers which have to deal with shortest-path problems in practical applications, such as in robotics (e.g., when programming an industrial robot), in routing (i.e., when selecting a path in a network), in gene technology (e.g., when studying structures of genes), or in game programming (e.g., when optimising paths for moves of players)—just to cite four of such application areas.

The authors thank (in alphabetical order) Tetsuo Asano, Donald Bailey, Chanderjit Bajaj, Partha Bhowmick, Alfred (Freddy) Bruckstein, Thomas Bülow, Xia Chen, Yewang Chen, David Coeurjolly, Eduardo Destefanis, Michael J. Dinneen, David Eppstein, Claudia Esteves Jaramillo, David Gauld, Jean-Bernard Hayet, David Kirkpatrick, Wladimir Kovalevski, Norbert Krüger, Jacques-Olivier Lachaud, Joe Mitchell, Akira Nakamura, Xiuxia Pan, Henrik G. Petersen, Nicolai Petkov, Fridrich Sloboda, Gerald Sommer, Mutsuhiro Terauchi, Ivan Reilly, the late Azriel Rosenfeld, the late Klaus Voss, Jinlong Wang, and Joviša Žunić for discussions or comments that were of relevance for this book.

The authors thank Chengle Huang (ChingLok Wong) for discussions on rubberband algorithms; he also wrote C++ programs for testing Algorithms 7 and 8. We thank Jinling Zhang and Xinbo Fu for improving C++ programs for testing Algorithm 7. The authors acknowledge computer support by Wei Chen, Wenze Chen, Yongqian Du, Wenxian Jiang, Yanmin Luo, Shujuan Peng, Huijuan Pi, Huazhen Wang, and Jian Yu.

The first author thanks dean Weibin Chen at Huaqiao University for supporting the project of writing this book. The second author thanks José L. Marroquín at CIMAT Guanajuato for an invitation to this institute, thus providing excellent conditions for working on this book project.

Parts of Chap. 4 (on relative convex hulls) are co-authored by Gisela Klette, who also contributed comments, ideas and criticisms throughout the book project.

We are grateful to Garry Tee for corrections and valuable comments, often adding important mathematical or historic details.

Huaqiao, People's Republic of China   Fajie Li
Auckland, New Zealand   Reinhard Klette

Contents

Abbreviations

Symbols

$\lvert S \rvert$	Cardinality of set S
$\parallel$	Relation sign for being parallel
∂S	Frontier of set S
$\wedge$	Logical 'and'
$\vee$	Logical 'or'
$\cap$	Intersection of sets
$\cup$	Union of sets
pq	Straight line segment with endpoints p and q
pqr	Triangle with vertices $p, q,$ and r
$\triangle$	Trapezoid or triangle (in a partitioning)

$\overline{pq}$	Straight line defined by points p and q; orientation "from p to q"
$\sphericalangle(pqr)$	Angle formed by rotating segment pq clockwise into segment qr
$\square$	End of a proof or of an example
a, b, c	Real numbers
$\mathcal{A}(\cdot)$	Area of a measurable set (as a function)
α, β	Angles
$\mathbb{C}$	Set of complex numbers $a + i \cdot b$, with $i = \sqrt{-1}$ and $a, b \in \mathbb{R}$
d_m	Minkowski metrics, for $m = 1, 2, \ldots$ or $m = \infty$
d_e	Euclidean metric; note that $d_e = d_2$
D	Determinant; see page 96
δ	Real number greater than zero
e	Edge (e.g., of a graph); real constant $e = \exp(1) \approx 2.7182818284$
E	Set of edges of a graph
ε	Real number greater than zero (the accuracy of an RBA)
ε_s	Real number greater than zero (a shift distance)
f, g, h	Functions, for example from $\mathbb{N}$ into $\mathbb{R}$
g	Simple cube-curve
$\mathbf{g}$	Tube (i.e., the union of cubes) of a simple cube-curve g
F	Face of a polyhedron
$\mathcal{F}$	Set of faces of a polyhedron
G	Graph
$\mathbb{G}$	Set $\{\ldots, -2, -1, 0, 1, 2, \ldots\}$ of integers
γ	Curve in Euclidean space (e.g., straight line, polyline, smooth curve)
H	Half plane; see page 95
i, j, k, l, m, n	Natural numbers
i	Number of iterations, e.g., in a rubberband algorithm
j	Natural number; index of points or vertices in a path
k	Natural number; total number of items
κ	Function in ε
L	Length (as a real number)
$\mathcal{L}(\cdot)$	Length of a rectifiable curve (as a function)
$l(\rho)$	Length of a path ρ
λ	Real number; e.g., between 0 and 1
n	Natural number; e.g., defining the complexity of the input
N	Neighbourhood (in the Euclidean topology, or in grids)
$\mathbb{N}$	Set $\{0, 1, 2, \ldots\}$ of natural numbers
$\mathcal{O}(\cdot)$	Asymptotic upper time bound
p, q	Points in $\mathbb{R}^2$ or $\mathbb{R}^3$, with coordinates x_p, y_p, or z_p
$p(x)$	Polynomial in x
$p.x, p_i.x$	x-coordinate of point p or point p_i
$p.y, p_i.y$	y-coordinate of point p or point p_i
$p.z, p_i.z$	z-coordinate of point p or point p_i
P	Polygon

$P^\bullet$, P°	Closure and topological interior of polygon P
π	Plane in $\mathbb{R}^3$; real constant $\pi = 4 \times \arctan(1) \approx 3.14159265358979$
Π	Polyhedron
$\Pi^\bullet$, Π°	Closure and topological interior of polyhedron Π
$\prod_{i=1}^{k} S_i$	Product of sets S_i
$\mathcal{P}$	Partitioning (of the plane or of a simple polygon)
r	Radius of a disk or sphere; point in $\mathbb{R}^2$ or $\mathbb{R}^3$
$\mathbb{R}$	Set of real numbers
ρ	Path with finite number of vertices; see page 11
s	Point in $\mathbb{R}^2$ or $\mathbb{R}^3$
S, S_i	Sets
$\mathcal{S}$	Family of sets
t	Point in $\mathbb{R}^2$ or $\mathbb{R}^3$
T	Tree; threshold (a real number)
τ	Threshold (real number)
$\mathcal{T}$	Trapezoidation or triangulation (of the plane or of a simple polygon)
v	Vertex or node; a point in $\mathbb{R}^2$ or $\mathbb{R}^3$, with coordinates x_v, y_v, or z_v
V	Set of vertices of a graph
$V(G)$, $V(T)$	Set of vertices of graph G or tree T
$V(\rho)$	Set of vertices of a path ρ
vu, $\overrightarrow{vu}$	Undirected or directed line segment between points v and u
x	Real variable
$\mathbf{x}$	Vector
y	Real variable
$\mathbf{y}$	Vector
$\mathbb{Z}$	Set of integers

Part I
Discrete or Continuous Shortest Paths

The road system of the historic city of Guanajuato is mostly underground, forming a network of tunnels. Optimising routes from one part of the city to another part is an exciting task not only for visitors of the city, but even for locals. In 2010, there was not yet a "3D route planner" available for calculating shortest connections in this historic city.

The first part of the book defines shortest paths in the geometry that we practise in our daily life, known as *Euclidean geometry*. Finding a shortest path between two given points and avoiding existing obstacles can be a challenge. We provide basic definitions and we propose the class of *rubberband algorithms* for solving various Euclidean shortest-path problems, either defined in the plane or in the 3-dimensional space.

Chapter 1
Euclidean Shortest Paths

> Ptolemy once asked Euclid whether there was any shorter way
> to a knowledge of geometry than by a study of the Elements,
> whereupon Euclid answered that there was no royal road to
> geometry.
>
> Proclus Diadochus (410...412–485)

This introductory chapter explains the difference between shortest paths in finite
graphs and shortest paths in Euclidean geometry, which is also called 'the common
geometry of our world'. The chapter demonstrates the diversity of such problems,
defined between points in a plane, on a surface, or in the 3-dimensional space.

1.1 Arithmetic Algorithms

Technology is evaluated by applying various *measures*, which are mapping components of the technology into real numbers, such as kilometres per hour, a maximum
error in millimetres, or a shape descriptor. We describe algorithms with respect to
run time, deviations of results from being optimal, or necessary constraints on input
data.

> The shortest path algorithms in this book are designed to be fast, accurate,
> and for a wide range of input data.

The time for calculating a shortest path on a computer depends on parameters
such as available memory space or execution time per applied operation. It is mean-
ingless to express the required time as an absolute value in, say, micro- or nano-
seconds, because computers differ in their parameters, they change frequently, and
your computer is certainly not identical to the one used by someone else.

F. Li, R. Klette, *Euclidean Shortest Paths*,
DOI 10.1007/978-1-4471-2256-2_1, © Springer-Verlag London Limited 2011

Measures for computing time need to be independent from the configuration of computers for expressing the quality of an algorithm. For example, we apply an abstract measure to estimate the running time of algorithms, also called *time complexity* or *computational complexity*. This is a common approach in computer science for more than 50 years. We define a set of *basic operations*, thus specifying a particular *abstract computer*, and we assign *one time unit* to each basic operation. Such a 'unit of time' is not measured in a physical scale; it is an abstract unit for each basic operation. For those abstract units, the estimation of run time of an algorithm is independent of progress in computer technology.

Definition 1.1 An *algorithm* is a finite sequence of basic operations.

We consider basic operations that are executable on 'a normal sequential computer', and not, for example, on some kind of specialised processor. Basic operations are classified into numerical operations, logical tests, and control instructions; alphanumerical operations on letters or other symbols are not a subject in this book. Logical tests are comparisons in magnitude (such as "*length* < *threshold*"?). Control instructions are of the type if-then-else, while-do, or do-until. It only remains to define a particular computer model by specifying the set of numerical operations.

Definition 1.2 An *arithmetic algorithm* is defined by having addition, subtraction, multiplication, division, and the nth root (for any integer $n > 1$) as its basic operations.

A *scientific algorithm* expands an arithmetic algorithm by the following additional basic operations: trigonometric and inverse trigonometric functions, exponential and logarithmic functions, factorials, and conversions of numbers between different representation schemes (binary, decimal, or hexadecimal).

This book applies arithmetic algorithms for solving shortest-path problems.

(There is just one case in the book where trigonometric functions are used when optimising a point location in an ellipse.)

Arithmetic algorithms are further specified by the range of numbers they are working on. For example, assuming the unbounded range of *rational numbers* a/b (for integers a and b) goes beyond the capabilities of existing computers, which can only represent rational numbers up to some finite number of bits per number. Even if divisions a/b are not performed, and assuming that there are no roots (i.e., all calculations remain in the area of integers) then there is still the problem of a limited range, having only a finite number of bits for representing all available integers.

Definition 1.3 An *arithmetic algorithm over the rational numbers* uses the basic operations as available in an arithmetic algorithm and operates on the set of all rational numbers.

1.2 Upper Time Bounds

We assume, as is common in algorithmic complexity theory, that the performance of each basic operation requires one time unit.

Consider functions f and g, defined on the set $\mathbb{N} = \{0, 1, 2, \ldots\}$ of *natural numbers* and into the set of non-negative reals; the input parameter $n \geq 0$ stands for the *problem size*, such as, for example, the number $n = |S|$ of points in an input set S.

> The book applies *asymptotic upper time bounds* $\mathcal{O}(f(n))$ for characterising the run time of an algorithm.

Definition 1.4 $\mathcal{O}(f(n))$ is the class of all functions g with

$$g(n) \leq c \cdot f(n)$$

for some *asymptotic constant* $c > 0$ and all $n \geq n_0$, for some $n_0 \geq 0$.

Example 1.1 This example is an exercise on the given upper bound. Most of the time complexities considered in this book are polynomial, but we also include higher complexities in this example. Let

$$g_1(n) = 5n + 10,$$

$$g_2(n) = 5n \log_{10} n + 1{,}000n,$$

$$g_3(n) = \frac{3}{7}n^3 + 5{,}000n^2,$$

$$g_4(n) = 3n! + 100n^{100},$$

$$g_5(n) = 100 \binom{n^2}{25} + 100n^4,$$

$$g_6(n) = \frac{1}{2}2^n + 5{,}000n^2.$$

Recall that

$$n! = n(n-1)(n-2) \cdots 1 \sim \sqrt{2\pi n}\left(\frac{n}{e}\right)^n \quad \text{(Stirling's formula)}$$

with $\pi = 4 \times \arctan(1) \approx 3.14159265$ and $e = \exp(1) \approx 2.7182818284$. Furthermore,

$$\binom{n^2}{25} = \frac{n^2(n^2 - 1) \cdots (n^2 - 24)}{25!}$$

denotes a binomial coefficient.

It follows that the given six functions are all in the class $\mathcal{O}(n^{n+1/2})$. We also say that they are *upper bounded by* $n^{n+1/2}$. Except for the function g_4, the other five functions are all in the exponential class $\mathcal{O}(2^n)$. The functions g_1, g_2, g_3, and g_5 are upper bounded by the polynomial bound n^{50}. The functions g_1, g_2, and g_3 are in $\mathcal{O}(n^3)$. Finally, the functions g_1 and g_2 are in $\mathcal{O}(n \log n)$, and the function g_1 is of *linear complexity*. □

1.3 Free Parameters in Algorithms

Finite sets of *free parameters*, such as an upper limit for the number of loops or an approximation parameter $\varepsilon > 0$, specify many of the algorithms in this book. For instance, a shortest path needs to stay at a distance of at least ε from given obstacles, such that every object with a maximum diameter of 2ε can still move on the calculated path.

Free parameters may influence the time complexity of an algorithm. We illustrate by an example before we provide general statements.

Example 1.2 We need to specify the smallest positive integer m for n positive numbers $a_1, a_2, \ldots, a_n$ and a free parameter $\varepsilon > 0$ such that at least one a_i/m is smaller than $\varepsilon > 0$; we need to return $a_1/m, a_2/m, \ldots, a_n/m$, for the identified value of m. The free parameter ε is initialised when calling a solution to this task.

Brute-force principle: Systematically check all possible candidates whether they define a solution.

A brute-force algorithm is as follows: We divide all those n numbers by $m = 1, 2, 3, 4, \ldots$ in sequence, until at least one of the values a_i/m is smaller than $\varepsilon > 0$; we stop and return $a_1/m, a_2/m, \ldots, a_n/m$.

What is the asymptotic time complexity of this brute-force algorithm? Obviously, each iteration for one m-value takes linear time, because all n values are divided and tested. The number of iterations equals m_0, where m_0 is the final value of m. It follows that $\varepsilon \approx \min\{\frac{a_1}{m_0}, \frac{a_2}{m_0}, \ldots, \frac{a_n}{m_0}\} = \min\{a_1, a_2, \ldots, a_n\}/m_0$. Thus, the upper bound for the time complexity of this algorithm may also be expressed in the form $m_0 \cdot \mathcal{O}(n)$.

Definition 1.5 We call an algorithm κ-*linear* iff[1] its time complexity is $\kappa(\varepsilon) \cdot \mathcal{O}(n)$, and the function κ does not depend on the problem size n.

The brute-force algorithm is κ-linear; the function κ is here defined by dividing the minimum of all a_i by ε.

An obvious way for speeding up would be an algorithm based on the

Throw-away principle: Remove 'quickly' all the input data from any further processing which does not have any influence on the outcome of a process.

---

[1] Read "if and only if"; abbreviation proposed by *Paul Richard Halmos* (1916–2006).

Here we calculate at first the minimum a of all n numbers in $\mathcal{O}(n)$ (and 'throw away the others'), then calculate m_0 in time $\mathcal{O}(m_0)$ by repeated divisions of a, and finally $a_1/m_0, a_2/m_0, \ldots, a_n/m_0$ again in $\mathcal{O}(n)$. The resulting upper bound equals

$$\mathcal{O}(n + m_0) = \mathcal{O}\big(\max\{n, m_0\}\big) \approx \mathcal{O}\left(\max\left\{n, \frac{\min\{a_1, a_2, \ldots, a_n\}}{\varepsilon}\right\}\right).$$

Of course, we can also have an *optimised algorithm*: determine m_0 in *constant time* $\mathcal{O}(1)$ by dividing the minimum a simply by the given ε, having thus the upper bound

$$\mathcal{O}(n) + \mathcal{O}(1) = \mathcal{O}(n)$$

in total, without any influence on this upper bound by the given ε. $\qquad\square$

We discuss a range of algorithms with one or two free parameters such as $\varepsilon > 0$ in this book.

We are not able to eliminate the influence of free parameters on time complexity in general, but it is possible sometimes as shown in the example above.

1.4 An Unsolvable Problem

A task might be a great challenge or unsolvable (at some time in human history), such as making people walk on water or providing a way that blind people can see again. It is even possible to prove that problems exist that have no solution, now and at any time in future, as in the following example.

Example 1.3 Consider a polynomial $p(x) = a_n x^n + a_{n-1} x^{n-1} + \cdots + a_1 x + a_0$, with *complex numbers* $a_k = b_k + i \cdot c_k$ as its coefficients, where b_k and c_k are reals, and $i = \sqrt{-1}$ is the imaginary unit. The task is to calculate the n *roots* $x_1, x_2, \ldots, x_n$ of the polynomial equation $p(x_k) = 0$, for $k = 1, 2, \ldots, n$.

The *Fundamental Theorem of Algebra* says that every polynomial $p(x)$ of degree n has exactly n roots in the set $\mathbb{C}$ of complex numbers, with

$$p(x) = a_n(x - x_1)(x - x_2) \cdots (x - x_n) \tag{1.1}$$

where each root is counted up to its multiplicity in Eq. (1.1). Thus, those n roots exist.[2]

But calculating those n roots with an arithmetic algorithm over some defined domain of numbers (e.g., the domain of all rational numbers) is a totally different issue. If solvable this way over the given domain, then we say that this polynomial

---

[2] *Carl Friedrich Gauss* (1777–1855) proved this theorem in his PhD thesis, published in 1799.

is *solvable by radicals* over this domain. The case $n = 1$ is simple to solve. In school mathematics, we are taught how to proceed for $n = 2$ with radicals over the domain of rational numbers; general solutions are as follows:

$$x_1 = \frac{-a_1 + \sqrt{a_1^2 - 4a_2a_0}}{2a_2} \quad \text{and} \quad x_2 = \frac{-a_1 - \sqrt{a_1^2 - 4a_2a_0}}{2a_2}.$$

Polynomials of order $n = 3$ or $n = 4$ are also all solvable by radicals over the domain of rational numbers. Indeed, for $n = 1$ to 4 those solutions by radicals apply when the polynomial coefficients are complex numbers.

However, starting with $n = 5$, this is not true anymore for all polynomials of order n. The solvability of polynomials by radicals over some domain of numbers has been studied in *Galois theory*.[3] Those studies have shown that polynomials of degree 5 are not solvable by radicals over the domain of rational numbers in general.

Besides knowing this general non-existence result, it is also of interest to identify examples of *unsolvable polynomials* (i.e., there is no arithmetic algorithm over the rational numbers for calculating its roots). For example, the 5th degree polynomial

$$p(x) = x^5 - x - 1 \tag{1.2}$$

is unsolvable.[4]   □

> We cannot obtain any (whatever time complexity) computable *exact solution* for some easy to formulate numerical problems within a given class of algorithms, because such a solution (provably) does not exist.

For some computational problems it is not yet known whether they can be solved at all by an algorithm in some class of algorithms (e.g., the class of arithmetic algorithms with polynomial time complexity). If it is known that a problem is not (exactly) solvable at all within defined constraints, or where it is not yet decided whether it is solvable or not, we still can aim at defining *approximate solutions* which might be 'practically sufficient'.

For example, the numerical value of π cannot be expressed in a current computer with arbitrary accuracy, but a finite representation such as 3.14 is typically sufficient in real-world applications for approximating the value of π.

---

[3]Named after *Èvariste Galois* (1811–1832).

[4]Shown by *Bartel Leendert Van der Waerden* (1903–1996).

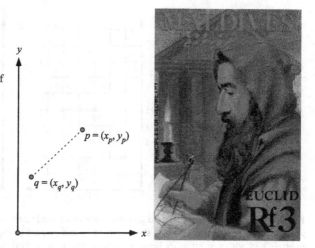

Fig. 1.1 *Right*: An artist's imagination of Euclid, on a stamp issued in the Maldives (image is in the public domain). That stamp depicts Euclid working by the light of a candle, but he would have used an oil lamp. *Left*: A rectangular Cartesian coordinate system in the plane, also showing two points p and q and the straight segment pq between both, being the shortest connection between two points

1.5 Distance, Metric, and Length of a Path

Euclidean geometry is the geometry in daily life, for example, for measuring a distance between two points, defining a triangle or a square, defining a sphere, and so forth.[5]

For measuring a distance in the plane we assume a rectangular xy *Cartesian coordinate system*,[6] as shown on the left of Fig. 1.1. Points $p = (x_p, y_p)$ and $q = (x_q, y_q)$ are at a *distance*

$$d_e(p, q) = \sqrt{(x_p - x_q)^2 + (y_p - y_q)^2} \qquad (1.3)$$

from one-another, and this is also the *length* of the straight segment, having p and q as its endpoints. The subscript e indicates 'Euclidean'.

There are other ways for measuring the distance between two points (or the length of an arc, connecting points p and q). The *Minkowski distance measure* d_m generalises the Euclidean distance,[7] with $m = 1, 2, \ldots$ or $m = \infty$:

$$d_m(p, q) = \sqrt[m]{|x_p - x_q|^m + |y_p - y_q|^m} \quad (m = 1, 2, \ldots),$$
$$d_\infty(p, q) = \max\{|x_p - x_q|, |y_p - y_q|\}.$$

Distance values decrease with increases in the value of m which is used:

$$d_{m_1}(p, q) \le d_{m_2}(p, q) \quad \text{for all } 1 \le m_2 \le m_1 \le \infty \text{ and all } p, q \in \mathbb{R}^2. \qquad (1.4)$$

---

[5]*Euclid of Alexandria* (see Fig. 1.1, right) was living around -300. He wrote either as an individual, or as the leader of a team of mathematicians a multi-volume book *Elements* that established Euclidean geometry and number theory.

[6]Introduced by *René Descartes* (in Latin: *Cartesius*; 1596–1650).

[7]Named after *Hermann Minkowski* (1864–1909).

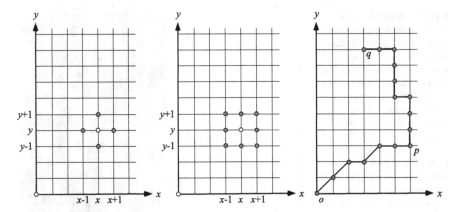

Fig. 1.2 A regular orthogonal grid and illustrations of 4-adjacent (*left*) and 8-adjacent (*middle*) grid points. The path (*right*) is of shortest length with respect to d_∞ between origin o and grid point p, and with respect to d_1 between p and grid point q; both segments of this path are not 'shortest' in the sense of Euclidean geometry

It follows that $m = 2$ defines the Euclidean distance, $d_2 = d_e$. The cases $m = 1$ and $m = \infty$ are also of frequent importance if distances are calculated in a *regular orthogonal grid* where grid points have integer coordinates only; see Fig. 1.2.

Horizontal $[(x - 1, y)$ and $(x + 1, y)]$ and vertical $[(x, y - 1)$ and $(x, y + 1)]$ neighbours of a grid point (x, y) define the set of *4-adjacent* grid points of (x, y), and if we also take the four diagonal neighbours $(x - 1, y - 1)$, $(x + 1, y - 1)$, $(x - 1, y + 1)$, and $(x + 1, y + 1)$, then we have the set of *8-adjacent* grid points of (x, y). It is easily verified that two 2-dimensional (2D) grid points p and q are 4-adjacent iff $d_1(p, q) = 1$, and 8-adjacent iff $d_\infty(p, q) = 1$.[8]

Definition 1.6 A *metric* d on a set S is defined by the following:

M1 For all $p, q \in S$ we have $d(p, q) \geq 0$, and $d(p, q) = 0$ iff $p = q$.
M2 For all $p, q \in S$ we have $d(p, q) = d(q, p)$.
M3 For all $p, q, r \in S$ we have $d(p, r) \leq d(p, q) + d(q, r)$.

Axiom M1 specifies *positive definiteness* (i.e., a metric can only take non-negative values, and value zero identifies that there was no move away from the start point) of d on S, axiom M2 describes *symmetry* (i.e., the distance remains the same, no matter whether we go 'left-to-right', or 'right-to-left'), and axiom M3 postulates *triangularity*; axiom M3 is also called *the triangle inequality*.

All the Minkowski distance measures d_m, for $1 \leq m \leq \infty$, satisfy those three axioms, and are thus metrics on $\mathbb{R}^2$.

---

[8]Adjacency relations and metrics on grid points were introduced by the US computer scientist and mathematician *Azriel Rosenfeld* (1931–2004), a pioneer of computer vision.

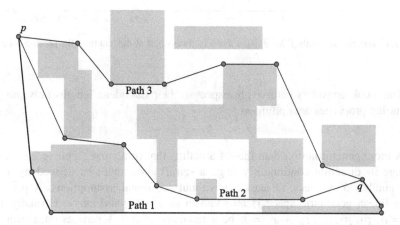

Fig. 1.3 Three paths from p to q which must not enter the *shaded obstacles*. Paths 2 and 3 are of equal and minimum length

There are many more ways for defining a distance measure (i.e., a metric). For example, let

$$d_f(p,q) = \begin{cases} |y_p| + |x_p - x_q| + |y_q|, & \text{if } x_p \neq x_q, \\ |y_p - y_q|, & \text{otherwise.} \end{cases} \qquad (1.5)$$

Read: "down the tree by $|y_p|$ at x_p, walk $|x_p - x_q|$ to the next tree at x_q, and up that tree by $|y_q|$". The three axioms of a metric are valid for this measure; let us call it the *forest distance*. The shortest connection between two points is here a move as described, and it is not a straight segment as in Euclidean geometry.

This book optimises the length of paths for the Euclidean distance measure.

Definition 1.7 In Euclidean geometry, a *path* from a point p to a point q is a finite sequence of *vertices*; it proceeds from vertex to vertex, starts at vertex p and ends at vertex q. Its *length* is the sum of the Euclidean distances between pairs of subsequent vertices on that path. A path between two vertices that has minimum length is called a *Euclidean shortest path* (ESP).

Figure 1.3 shows in bold lines an example of a path (called Path 1) from p to q which must not enter the shown shaded *obstacles*; the figure also shows two different shortest paths in thin lines (called Path 2 and Path 3; both are of identical length) from p to q. The paths have different numbers of vertices. Path 1 is not of optimum length, but has the smallest number of vertices. Path 2 goes through two 'narrow' sections between shaded obstacles. The 'quality' of a path may also be characterised by the number of vertices, the width of passages, the angles of turns at vertices, and other properties.

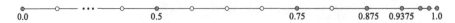

Fig. 1.4 Vertices on a path (*filled dots*): a move is always half of the length of the previous one

> This book considers paths with respect to their Euclidean lengths only; no further properties are optimised.

A more general notion than that of a path is that of a *curve*; a curve may also change its direction continuously (e.g., a spiral) and cannot be defined by a finite number of vertices, at least not without additional assumptions. A path in a Euclidean geometric space is also known as a *polygonal curve*. Formally, let $p_0 = p$, $p_1, p_2, \ldots, p_{n-1}, p_n = q$ be a *sequence* of $n + 1$ vertices on a path ρ, also denoted by $\rho = \langle p_0, p_1, \ldots, p_n \rangle$. Note that this is different from set notation $\{p_0, p_1, \ldots, p_n\}$ which does not imply any order of elements in the set. From Definition 1.7 we obtain that

Corollary 1.1 *The length of path* $\rho = \langle p_0, p_1, \ldots, p_n \rangle$ *is equal to*

$$\mathcal{L}(\rho) = \sum_{i=0}^{n-1} d_e(p_i, p_{i+1}).$$

If $p = q$ *and* $n = 0$, *we have a path of length zero.*

Example 1.4 Consider an algorithm which calculates a path on $\mathbb{R}$ from the origin $p = 0$ to point $q = 1$, starting with a first move to $p_1 = 0.5$, then to $p_2 = 0.75$, to $p_3 = 0.875$, and so forth; the following move is always half of the length of the previous move; see Fig. 1.4.

Mathematically, the algorithm never arrives at $q = 1.0$ if moves are defined on the real straight line with arbitrary accuracy; after any final number of moves, the length of the path is still less than 1. Of course, for an implementation of this algorithm we can actually assume that it 'practically' arrives at q after a finite number of steps, and this happens the earlier the fewer bits have been used on the given computer for representing a real number.   □

1.6 A Walk in Ancient Alexandria

As mentioned above, Euclid lived in Alexandria. Figure 1.5[9] represents the ancient city of Alexandria in post-Euclidean time, but still about 2,000 years ago. Assume that Euclid wanted to walk from p, the Canopic gate, to q, the entrance of the

---

[9]From http://hdl.handle.net/1911/9343, available as specified on http://creativecommons.org/licenses/by/2.5/.

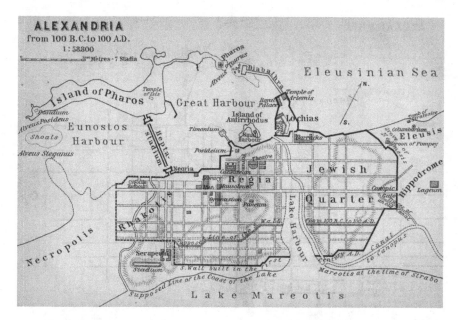

Fig. 1.5 Sieglin, W. (1908). The post-Euclidean Alexandria from −100 to +100. From *Travellers in the Middle East Archive*. For an example of a shortest-path problem, assume that the *Canopic Gate* defines the source, and the entrance to the *Serapeum* the destination. The printed scale 1:58,800 applies only to the original 1908 map

Serapeum; see the figure for both. For simplifying the geometry, we assume that Alexandria was totally flat at that time (i.e., we are walking on a plane). We are interested in finding a shortest path [with respect to the Euclidean distance measure, as in Eq. (1.3)] from p to q.

Our answer depends on the *geometric scale* or *accuracy* for defining the positions of those vertices. The vertices could identify the road crossings at a rough scale or any position somewhere on the road at a finer scale. Today's motion planning algorithms often require a high accuracy and vertices are defined at a fine scale.

Let us describe the layout of ancient Alexandria by a *weighted undirected graph*. We identify each road intersection with one *node*, and road sections between two nodes with one *edge*, which is labelled by the Euclidean distance (a *weight*) between both nodes defining this edge. We assume that there is no shortcut for walking through the city (such as crossing one of the insulae of houses).

One additional way via the Paneum is shown in Fig. 1.6. The specification of the graph could also cover distances for the non-planar case of Alexandria if distances are measured in 3-dimensional (3D) space, also representing changes in elevation.

The weight of an edge could, for example, also correspond to the expected time needed for walking from one node to an adjacent one; this weight could deliver a better model of Alexandria which is actually not on a plane. In that case, the time for walking between adjacent nodes would depend upon the direction, and

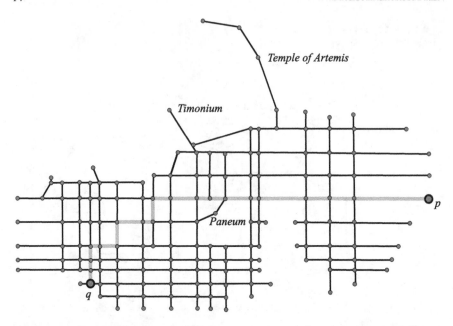

Fig. 1.6 Example of an undirected graph. Weights of edges correspond to their drawn length (i.e., the shown city is assumed to be on a plane). The figure also shows one path from p to q, which could be shortened when walking via the Paneum instead (of course, ignoring that this was on a hill in ancient Alexandria; today there is a huge roundabout at this elevated place)

hence the undirected graph would need to be replaced by a directed graph, with each undirected edge replaced by a pair of directed edges.

The following description is from an article about the Hellenistic Alexandria on ArchNet:[10]

> The city was physically divided by the intersection of two main thoroughfares: the east–west Canopic Way and the Street of the Soma (Sema). The surrounding streets of the ancient city were laid out in a Hippodamian grid. The Canopic Way connected the Canopic Gate and the Necropolis Gate of the city wall. The Street of the Soma ran between the Moon Gate and the Sun Gate of the city wall. Archaeologists estimate that both streets measured between 25 and 70 meters, and were lined with marble colonnades and paved with granite blocks.

The roads were actually very wide, and assigning a single node to a location of a road intersection is a very rough approximation. Furthermore, on the same webpage we read about the time of the Ptolemaic Dynasty (-304 to -29):

> The land use program for the city under the Ptolemaic dynasty was primarily residential. This street grid was divided into insulae (blocks), each averaging 36.5 by 182.5 meters, or 100 by 500 Ptolemaic feet. In Alexandria, a quarter accommodated six insulae intersected by two minor roads. Housing plots measured 22 by 22 meters, and each insula could hold as many as 20 houses.

---

[10]See http://www.archnet.org/.

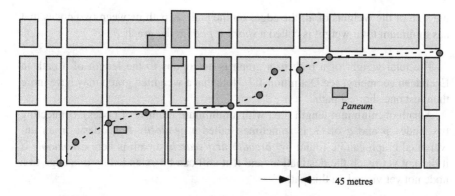

Fig. 1.7 Sketch of a shortest Euclidean path. The pass is crossing roads diagonally and is using possible shortcuts through insulae. Vertices can be located at corners or in the middle of roads or in the middle of insulae. The size '45 metres' is given for illustrating a possible scale

Figure 1.7 provides a sketch for a shortest path assuming a more detailed layout compared with the weighted graph representing the map in Fig. 1.5. It is allowed to cross roads diagonally or to walk through insulae by using access paths to those houses.

> Shortest paths may be defined as a combinatorial problem in finite weighted graphs, or as a continuous optimisation problem in Euclidean geometry.

1.7 Shortest Paths in Weighted Graphs

Definition 1.8 An (undirected) *graph* $G = [V, E]$ is defined by a finite set V of *nodes* and a set $E \subseteq \{\{p, q\} : p, q \in V \land p \neq q\}$ of *edges* between those nodes. Two nodes are *adjacent* in G if connected by an edge.

Nodes of a graph are also called *vertices* in other publications (thus symbol V here). If $G = [V, E]$ is a planar simple graph, then $|E| \leq 3|V| - 6$.

We assign a positive real *weight* $w(e) = w(p_1, p_2)$, such as the distance between positions in $\mathbb{R}^2$ represented by p_1 and p_2, to each edge $e = \{p_1, p_2\} \in E$. This map is called a *weight function* $w(e) = w(p_1, p_2)$, for $e = \{p_1, p_2\} \in E$.

If p_1 and p_2 are not adjacent (i.e., not connected by an edge), we define $w(p_1, p_2)$ to be infinite. Thus we have defined a *weighted graph* $G = [V, E, w]$. An (unweighted) graph can also be regarded as a weighted graph in which $w(p_1, p_2) = 1$ for all edges $\{p_1, p_2\} \in E$.

Definition 1.9 In a weighted graph, a *path* is a finite sequence $\rho = \langle p_0, p_1, \ldots, p_n \rangle$ of *nodes*, where p_i and p_{i+1} are adjacent, for $i = 0, \ldots, n - 1$. Its *total weight* is the sum of the weights of all the edges on the path. A path between two nodes that has minimum total weight is called a *shortest path* in the graph.

The total weight of a path in a graph is analogous to the length of a path in Euclidean geometry; see Definition 1.7. Note that a weighted graph may have more than just one shortest path.

A path of minimum length (i.e., with a minimum number of nodes), connecting two nodes p and q of G, is sometimes called a *geodesic*. A geodesic in an unweighted graph can be found by *breadth-first search*; a path is here *extendable* if it did not yet reach the destination, and if it still can be extended by one edge to a node not yet visited on this path:

1: Start at the source; this defines a path of length zero.
2: **while** destination is not yet reached **do**
3:    Extend all extendable paths by one edge to a node not yet visited on this path.
4: **end while**

We may record paths by storing just the predecessor at a node (i.e., from where this node was reached), and the minimum path can be calculated by going backward, from destination to source.

Obviously, this $\mathcal{O}(|E|)$ strategy will not deliver shortest paths in weighted graphs because the total weight of a path is not necessarily in a fixed relationship with the number of edges on it; a path with many nodes on it may have a smaller total weight than a path with fewer nodes on it.

Of course, we may try to extend the described breadth-first search thus that we calculate at first *all* potential candidates of paths, such that we can be sure that a shortest path can be identified in this set, and traced back to the source:

1: Start at the source; this defines a path of length zero.
2: **while** there is at least one extendable path **do**
3:    Extend all extendable paths by one edge to a node not yet visited on this path.
4:    Calculate the total weights of the extended paths.
5: **end while**
6: Compare the total weights of paths for selecting an optimal path.

We illustrate this straightforward extension on the breadth-first search by an example.

Example 1.5 Consider the graph in Fig. 1.8 with the shown source p and destination q. Breadth-first search starts at first with paths $\rho_1 = \langle p, s \rangle$, $\rho_2 = \langle p, r \rangle$, $\rho_3 = \langle p, t \rangle$, and weights $w_1 = 3$, $w_2 = 2$, and $w_3 = 1$. In the next iteration, we extend those three paths and obtain

$$\rho_{11} = \langle p, s, u \rangle \quad \text{with } w_{11} = 5; \qquad \rho_{12} = \langle p, s, r \rangle \quad \text{with } w_{12} = 4;$$
$$\rho_{21} = \langle p, r, s \rangle \quad \text{with } w_{21} = 3; \qquad \rho_{22} = \langle p, r, u \rangle \quad \text{with } w_{22} = 6;$$

Fig. 1.8 A weighted graph
$G = [V, E, w]$ with
$V = \{p, q, r, s, t, u, v\}$,
weights between 1 and 7 for
its edges, source p and
destination q. For example,
the path $\langle p, s, u, v, q \rangle$ does
have a total weight of 8

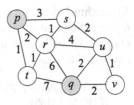

$$\rho_{23} = \langle p, r, q \rangle \quad \text{with } w_{23} = 8; \qquad \rho_{24} = \langle p, r, t \rangle \quad \text{with } w_{24} = 3;$$

$$\rho_{31} = \langle p, t, r \rangle \quad \text{with } w_{31} = 2; \qquad \rho_{32} = \langle p, t, q \rangle \quad \text{with } w_{32} = 8.$$

At this point we already know that ρ_{23} and ρ_{32} are geodesics between p and q. However, the algorithm proceeds further.

Extendable paths are ρ_{11} (to r, q, and v), ρ_{12} (to t, q, u), ρ_{21} (to u), ρ_{22} (to s, q, and v), ρ_{24} (to q), and ρ_{31} (to s, u, or q). Paths ρ_{23} and ρ_{32} have a total cost of 8.

Terminating paths at the destination q are now $\rho_{112} = \langle p, s, u, q \rangle$ with $w_{112} = 7$, $\rho_{122} = \langle p, s, r, q \rangle$ with $w_{122} = 10$, $\rho_{222} = \langle p, r, u, q \rangle$ with $w_{222} = 8$, $\rho_{241} = \langle p, r, t, q \rangle$ with $w_{241} = 10$, and $\rho_{313} = \langle p, t, r, q \rangle$ with $w_{313} = 8$.

Path $\langle p, r, u, s \rangle$ is a first example of a path which stops without reaching q. (This will also happen to path $\langle p, t, r, u, s \rangle$ in the next iteration.) Paths $\langle p, s, u, r, q \rangle$, $\langle p, s, u, v, q \rangle$, $\langle p, s, r, t, q \rangle$, $\langle p, s, r, u, q \rangle$, $\langle p, r, s, u, q \rangle$, and $\langle p, r, u, v, q \rangle$ terminate at q.

In the next iteration, we have paths $\langle p, s, u, r, t, q \rangle$, $\langle p, s, r, u, v, q \rangle$, $\langle p, r, s, u, v, q \rangle$, $\langle p, t, r, s, u, q \rangle$, and $\langle p, t, r, u, v, q \rangle$ terminating at q. Finally, path $\langle p, t, r, s, u, v, q \rangle$ also terminates at q.

We obtained the complete set of all 19 possible paths from p to q for this example of a weighted graph. We select one of those which has the minimum total cost of 7 and trace its nodes back from q to the source p. □

This extended breadth-first search algorithm contains many avoidable calculations. For example, when knowing that ρ_{22} has cost 6, but ρ_{11} has cost 5, we do not have to consider extensions of path ρ_{22}. Before discussing an optimised solution, we are generalising from a single-source–single-destination problem to a single-source–multiple-destination problem:

The (general) *shortest-path problem* of graph theory is as follows: Given a connected weighted graph $G = [V, E, w]$ and a node $p_0 \in V$, find a shortest path from p_0 to each $p \in V$.

Dijkstra's algorithm (see Fig. 1.9) solves this general shortest-path problem.[11] This algorithm does not perform redundant calculations as in the extended breadth-

---

[11] The Dutch scientist *Edger Wybe Dijkstra* (1930–2002) has made many substantial contributions to computer science.

Algorithm 1 (Dijkstra algorithm, 1959)

Input: A finite undirected weighted graph $G = [V, E, w]$ and a start node $p_0 \in V$.
Output: A labelling of all nodes such that a shortest path can be traced back from any $p \in V$ to p_0.

```
 1: Let V = {p₀, ..., pₙ}.
 2: if |V| > 1 then
 3:    Let i = 0, V₀ = {p₀}, D(p₀) = 0, and D(p) = +∞ for p ≠ p₀.
 4:    while i < |V| − 1 do
 5:       for each p ∈ V \ Vᵢ do
 6:          Update D(p) by min{D(p), D(pᵢ)+w(pᵢ, p)}. If D(p) is replaced, put
             a label [D(p), pᵢ] on p. (This allows for the tracking of shortest paths.)
             Overwrite the previous label, if there is one.
 7:       end for
 8:       Let pᵢ₊₁ be a node that minimises {D(p) : p ∈ V \ Vᵢ}.
 9:       Let Vᵢ₊₁ = Vᵢ ∪ {pᵢ₊₁}.
10:       Replace i with i + 1.
11:    end while
12: end if
```

Fig. 1.9 Dijkstra algorithm for solving the shortest-path problem of graph theory

first algorithm above. Each node in graph G, which is reachable from the source, is labelled at some stage by the minimum cost of a path from the source to this node, and never reconsidered later.

Example 1.6 We consider again the weighted graph in Fig. 1.8. At the beginning we initialise $V = \{p, q, r, s, t, u, v\}$, $i = 0$, $V_0 = \{p\}$, $D(p) = 0$, and $D(q) = \cdots = D(v) = +\infty$. We start the loop with $i = 0 < 7 - 1 = 6$.

For each node $x \in V \setminus V_0 = \{q, r, s, t, u, v\}$ we calculate $\min\{D(x), D(p) + w(x, p)\}$. This leads to updates $D(s) = 3$, $D(r) = 2$, and $D(t) = 1$. We select node t for $V_1 = \{p, t\}$ and have $i = 1 < 6$.

For each node $x \in V \setminus V_1 = \{q, r, s, u, v\}$, we calculate $\min\{D(x), D(t) + w(x, t)\}$. This leads to update $D(q) = 8$. $D(r)$ remains at 2 and is not updated. We select node r for $V_2 = \{p, t, r\}$ and have $i = 2 < 6$.

For each node $x \in V \setminus V_2 = \{q, s, u, v\}$, we calculate $\min\{D(x), D(r) + w(x, r)\}$. This leads to update $D(u) = 6$. $D(q)$ remains at 8 and is not updated. We select node s for $V_3 = \{p, t, r, s\}$ and have $i = 3 < 6$.

For each node $x \in V \setminus V_3 = \{q, u, v\}$, we calculate $\min\{D(x), D(s) + w(x, s)\}$. This leads to update $D(u) = 5$. We select node u for $V_4 = \{p, t, r, s, u\}$ and have $i = 4 < 6$.

For each node $x \in V \setminus V_3 = \{q, v\}$, we calculate $\min\{D(x), D(u) + w(x, u)\}$. This leads to updates $D(q) = 7$ and $D(v) = 6$. We select node v for $V_4 = \{p, t, r, s, u, v\}$ and have $i = 5 < 6$.

We only consider $\min\{D(q), D(v) + w(q, v)\}$. There is no update; $D(q) = 7$ is the minimum. This minimum was obtained by coming from u. The minimum

Fig. 1.10 Path planning in mobile devices is based on weighted graphs. The figure shows a recommended path for driving in Auckland, New Zealand

$D(u) = 5$ was obtained by coming from s. The minimum $D(s) = 3$ was obtained by coming from p. ☐

The Dijkstra algorithm as specified above has a computational complexity of $\mathcal{O}(|V|^2)$. Applying a heap data structure for the set $\{p \in V \setminus V_i : D(p) < +\infty\}$ of remaining nodes improves the time complexity of Dijkstra's algorithm to $\mathcal{O}(|E| + |V| \log |V|)$. (This is $\mathcal{O}(|V| \log |V|)$ if the graph G is planar.) The labels assigned in Line 6 allow the construction of a shortest path from any node $p \in V$ back to p_0 as illustrated in the example above.

The time complexity $\mathcal{O}(|E| + |V| \log |V|)$ is often critical for graphs with a very large number of nodes. A practical example is the optimisation of paths for a GPS system as available in cars or mobile phones; see Fig. 1.10.

We only mention here that a heuristic $A^\star$ *search algorithm* was designed for reducing this time complexity by taking the risk that a calculated path is only sub-optimal, and not necessarily an optimal solution.

A more in-depth discussion of graph-theoretical algorithms is beyond the scope of this book, but we will make use of the Dijkstra algorithm later in the book. We also note that treating time against optimality is already a common approach in this area, and we will discuss similar developments in the field of shortest paths in Euclidean geometry.

1.8 Points, Polygons, and Polyhedra in Euclidean Spaces

In this section, we discuss the geometric objects of Euclidean geometry, polygons in 2D space, polyhedra in 3D space, or sets of points in 2D or 3D space, that allow us to introduce various types of Euclidean shortest path (ESP) problems (in the next section, and throughout the book).

Points or vertices of polygons or polyhedra are identified by coordinates. A Cartesian coordinate system operates on a set of axes for rectangular (or oblique) coordinates. Hereafter we consider only rectangular coordinate systems.

Fig. 1.11 Right-hand 3D coordinate system (courtesy of Fay Huang, Yi-Lan, Taiwan)

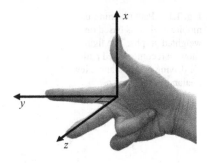

The coordinate axes intersect orthogonally at point o. It is called the *origin* of the coordinate system. Its coordinates have value 0 since point o is at distance 0 from all the axes.

The real *2D space*, also called the real *plane*, in a rectangular Cartesian coordinate system is a product

$$\mathbb{R}^2 = \mathbb{R} \times \mathbb{R}$$

of two real lines where $\mathbb{R}$ is the set of all real numbers. In $\mathbb{R}^2$, points are expressed, for example, by $p_i = (x_i, y_i)$ or $p_i = (p_i.x, p_i.y)$, whatever suits the local context. The real *3D space* is the product

$$\mathbb{R}^3 = \mathbb{R} \times \mathbb{R} \times \mathbb{R}$$

of three real lines. Points in $\mathbb{R}^3$ are denoted, for example, by p_i with $i \geq 0$ and xyz-coordinates $p_i = (x_i, y_i, z_i)$ or $p_i = (p_i.x, p_i.y, p_i.z)$.

A *right-hand coordinate system* is a rectangular Cartesian coordinate system in which the positive x-axis is identified with the thumb (pointing outward in the plane of the palm), the positive y-axis with the forefinger (pointing outward in the plane of the palm), and the positive z-axis with the middle finger of the right hand (pointing away from the plane of the palm; see Fig. 1.11).

If d is a metric on a set S then the pair $[S, d]$ defines a *metric space*. The Euclidean distance, defined in Eq. (1.3), is a metric. The Euclidean distance between two points $p = (p.x, p.y, p.z)$ and $q = (q.x, q.y, q.z)$ in the 3-dimensional space

$$d_e(p, q) = \sqrt{(p.x - q.x)^2 + (p.y - q.y)^2 + (p.z - q.z)^2} \qquad (1.6)$$

is also a metric on the set $\mathbb{R}^3$. The context specifies whether d_e is a function in 2D or 3D space.

Definition 1.10 The metric space $[\mathbb{R}^2, d_e]$ is the *Euclidean plane*, and the metric space $[\mathbb{R}^3, d_e]$ is the *Euclidean 3D space*.

A Euclidean space can be of any dimension $m \geq 1$. This book only discusses the cases $m = 2$ or $m = 3$ of our 'daily life geometry'.

Shortest-path problems in the book are defined in Euclidean 2D or 3D space.

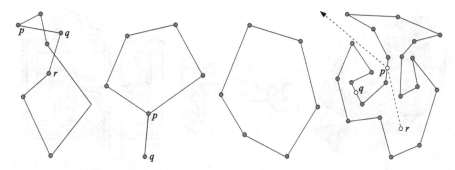

Fig. 1.12 Polygonal curves. Two of these are simple and defining simple polygons, and two of them are non-simple, defining non-simple polygons. The polygon on the *right* also illustrates visibility: point p is visible from the outside (see the *dashed ray*), but point q is not; point r, selected in the interior of the polygon, sees point p (see the *dashed line* segment), but it cannot see point q

The definition of shortest-path problems in the plane makes in general use of *simple polygons*, for defining the region of potential moves, a sequence of regions to be visited, obstacles to be avoided, and so forth.

A polygon is defined by a *loop* which is a path $\rho = \langle p_0, p_1, \ldots, p_n \rangle$ with $p_0 = p_n$. Vertices of this loop are endpoints of line segments, and the loop describes thus a *polygonal curve*.

Definition 1.11 Let ϕ be a mapping of an interval $[a, b] \subset \mathbb{R}$ into the real plane, $\phi : [a, b] \to \mathbb{R}^2$, such that $a \neq b$, $\phi(a) = \phi(b)$, and $\phi(s) \neq \phi(t)$ for all s, t with $a \leq s < t \leq b$. The set $\gamma = \{(x, y) : \phi(t) = (x, y) \wedge a \leq t \leq b\}$ is a *simple curve*.[12]

Informally speaking, a simple curve is "not intersecting itself"; any simple curve can be generated from a circle by topological deformation.

Definition 1.12 A *simple polygon P* is defined by a simple polygonal curve; this polygonal curve defines the *frontier* ∂P of P, which circumscribes the bounded *interior P°* of P.

Figure 1.12 shows two non-simple polygonal curves on the left; in the leftmost case, path $\langle p, q, r \rangle$ intersects two other line segments of the polygonal curve, and in the other example, segment pq is traversed twice in the loop. The two examples shown on the right are simple curves.

A point p on the frontier of a polygon is *visible from the outside* iff there is a ray starting at p that intersects the polygon in no other point than at p. A point p in a polygon *sees* a point q in this polygon iff the straight segment pq is contained in the polygon. See Fig. 1.12 on the right. An edge $e_i = p_i p_{i+1}$ is *visible* from another

---

[12]This was defined by *Camille Jordan* (1838–1922) in 1893, and γ is also called a *Jordan curve* today.

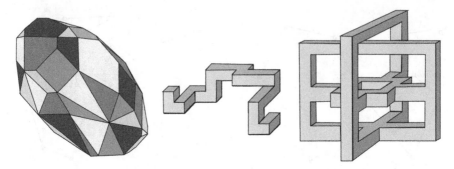

Fig. 1.13 *Left*: A simple polyhedron with randomly-rendered faces. *Middle*: A simple polyhedron defined in a regular orthogonal 3D grid. *Right*: A non-simple polyhedron (which could be, e.g., the layout for a "world" of a computer game)

edge $e_j = p_j p_{j+1}$ iff there is a point on e_i that sees a point on e_j. Those visibility concepts are useful for discussing geometric configurations.

The notion of a simple polygon can be generalised to 3D space. A polygon is a bounded area in the plane whose frontier is defined by a polygonal curve; a *polyhedron* is a bounded volume in 3D space whose surface (also called *frontier*) is the union of a set of polygons which only intersect at frontiers, but not at points in their interior (i.e., the polygons define a non-overlapping tessellation of the surface).

Definition 1.13 A *simple polyhedron* Π is defined by a finite set of polygonal *faces* which tessellates its surface completely, and which can be topologically transformed into the surface of a sphere; the union of this set of polygons defines the *frontier* $\partial \Pi$ of Π, which encloses the bounded *interior* Π° of Π.

Informally speaking, "topologically transformed into the surface of a sphere" means that the frontier may be "inflated" such that it becomes a "spherical balloon". Figure 1.13 shows two simple (left and middle) and one non-simple polyhedron. The non-simple polyhedron on the right[13] can be generated as a union of five tori. A polyhedral torus cannot be topologically transformed into a sphere.

A point p on the surface of a polyhedron is *visible from the outside* iff there is a ray starting at p that intersects the polyhedron in no other point than at p. A point p in a polyhedron *sees* a point q in this polyhedron iff the straight segment pq is contained in the polyhedron.

Objects in our real world (see Fig. 1.14) can be modelled as polyhedra by approximating curved surface patches at some selected scale by polygons. The two objects shown are of low shape complexity compared with the dimensions and variations of shapes in the whole universe.

We assume that some integer $n > 0$ characterises the complexity of input data, such as the number of points in a set, the number of vertices of a polygon, or the

---

[13]Used by *Johann Benedict Listing* (1808–1882) in 1861 when illustrating the skeletonisation of shapes in $\mathbb{R}^3$.

Fig. 1.14 *Left*: A tree (at Tzintzuntzan, Mexico) can be modelled as a simple polyhedron, assuming that nature was not producing any torus when growing this tree. *Right*: A sponge can only be modelled as a non-simple polyhedron

number of faces of a simple polyhedron. The upper limit for n is unknown because progress in technology may shift feasible values further up right now, or in a few years from now. Those values of n are so 'little' compared with the universe, or even to the infinity of the set of all integers. It would make sense to define an upper limit for n, say $n < 10^{80}$, which is the estimated number of protons in the observable universe.[14] We could conclude that all our algorithms only need to work for inputs with $n < 10^{80}$; larger inputs are out of scope.

This would not simplify considerations such as "this algorithm is correct for any $n > 0$", thus also, for example, a proof of correctness for

$$n = 10^{10^{10^{10^{10}}}}.$$

We prove results for numbers n which will never be experienced by humankind, just for the sake of mathematical simplicity.

> The complexity of input data (e.g., sets of points, polygons, or polyhedra) will be characterised by integers, without taking into account any limitation for those integers.

1.9 Euclidean Shortest Paths

Euclidean shortest paths (ESP) have been specified in Definition 1.7. In this section, we illustrate only a few examples of ESP problems, for demonstrating subjects to be considered in this book. Polygons or polyhedra are simple in what follows unless otherwise stated.

---

[14]See, e.g., www.madsci.org/posts/archives/oct98/905633072.As.r.html.

Fig. 1.15 Sketch of an ESP
problem in the plane defined
by a polygonal search
domain, source p and
destination q, an ordered set
of attractions (shown by
shaded ellipses), and a set of
obstacles, shown as *shaded
polygons* of constant shape.
The figure shows a possible
(not yet length-minimised)
path, connecting p with q via
the given sequence of
attractions, avoiding all the
obstacles, and staying in the
given search domain

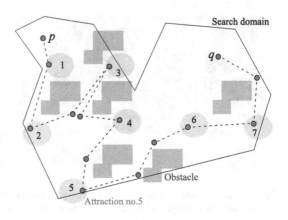

Obstacles or *attractions* are bounded or unbounded subsets in Euclidean space.
When calculating a path, it must not enter any of the obstacles, but it has to visit all
the given attractions in a specified order. A path ρ *visits* a set S iff ρ has a non-empty
intersection with S.

We also assume a *search domain* that is the space of possible moves: the source
and destination for the path are in the search domain, all the attractions need to have
at least a non-empty intersection with the search domain, and obstacles can possibly
be outside of the search domain. See Fig. 1.15 for an example.

We consider *polygonal or polyhedral search domains*, defined by a finite number
$n \geq 0$ of lines or line segments or polygonal faces, respectively. For example, a
half-plane or a 3D half-space are also possible; they are defined by one straight line
in 2D space, or one plane in 3D space (i.e., $n = 1$). Finally, all $\mathbb{R}^2$ or $\mathbb{R}^3$ are also
possible, they are defined by having *no* limiting straight line or segment or polygon
(i.e., $n = 0$).

Definition 1.14 Assume that the Euclidean space $[\mathbb{R}^2, d_e]$ or $[\mathbb{R}^3, d_e]$ contains a finite set of polygonal or polyhedral obstacles and also an ordered set of polygonal or
polyhedral attractions. We consider two points p (the *source*) and q (the final *destination*) within the given polygonal or polyhedral search domain. The *ESP problem*
is to compute a path ρ between p and q in such a way that the path ρ does not
intersect the interior of any obstacle, visits all the attractions in the specified order,
does not leave the search domain, and is of minimum Euclidean length.

An *exact solution* is path ρ that solves such an ESP problem. See Fig. 1.15. Due
to a constraint about the size of the object that is expected to move along a calculated
shortest path, we may consider expanded obstacles rather than the original obstacles;
see Fig. 1.16. Vertices of a calculated path, connecting p and q, can move freely in
the search domain, not only in the given attractions.

We provide five examples of ESP problems. Those and others are discussed in
this book:

Fig. 1.16 The same ESP problem as in Fig. 1.15 but after expanding all the obstacles in x- and y-directions, thus 'giving the moving object less space'. The figure shows a possible path from p to q

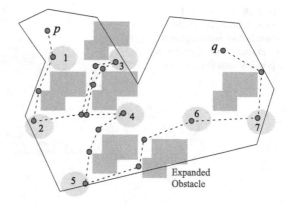

Fig. 1.17 A rectangular sheet with five embedded polygonal shapes (i.e., being the attractions) and a non-optimised path of a cutting head. At each of the five points on frontiers of the shapes, the cutting head would start (and also end) when cutting out a shape from the sheet

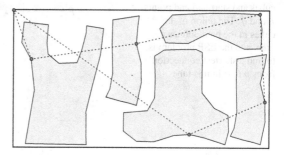

1. Find an ESP between points p and q in a simple polygon (the polygon defines the search domain); there are no obstacles or attractions. This is the well-known problem of *finding a shortest path in a simple polygon*.
2. Attractions are pairwise-disjoint polygons (see Fig. 1.17), all within a rectangular 'sheet' of material (e.g., textile, metal), $p = q$, and there are no obstacles. This is known as a *parts-cutting problem*, where a 'cutting head' needs to travel from one polygonal shape to the other for cutting them 'out' from the rectangular 'sheet'; this rectangle is the search domain.
3. The search domain is the interior or the surface of a polyhedron, and there are no obstacles or attractions. This defines either the problem of *finding a shortest path in a simple polyhedron*, or the problem of *finding a shortest path on the surface of a simple polyhedron*.
4. The whole $\mathbb{R}^3$ is our search domain, and the obstacles are a finite set of *stacked rectangles* (i.e., rectangles in parallel planes), without any attractions. We need to *find a shortest path which avoids the stacked rectangles*. See Fig. 1.18.
5. The 3D space is subdivided into a regular orthogonal grid forming cubes (e.g., voxels in 3D medical imaging, or geometric units in gene modelling). These cubes form a *simple path* iff each cube in this path is face-adjacent to exactly two other cubes in the path, except the two end cubes which are only face-adjacent to one other cube in the path. The two end cubes contain start and end vertex p and q. See Fig. 1.19.

Fig. 1.18 Five stacked
rectangles (i.e., being the
obstacles) in parallel planes
in 3D space. The figure shows
a possible path from p to q

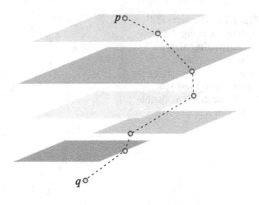

Fig. 1.19 A simple path of
cubes and start and end points
p and q. The union of the
cubes in the path defines a
tube, and the ESP problem is
to find a shortest connection
from p to q in this tube

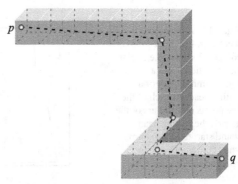

A given source p and destination q specify a *fixed ESP problem*. If there is no fixed
start or end point, then this defines a *floating ESP problem* with a higher degree of
uncertainty, thus a larger computational challenge.

1.10 Problems

Problem 1.1 Let $g(n) = n \log_{10} n$ and $f(n) = n \log_2 n$. Show that $g(n) \in \mathcal{O}(f(n))$
and $f(n) \in \mathcal{O}(g(n))$ (i.e., both functions are *asymptotically equivalent*).

Problem 1.2 Show that the Minkowski distance measure d_1 satisfies all the three
axioms M1, M2, and M3 of a metric.

Problem 1.3 A shortest d_1-path is defined by minimising the total distance

$$\mathcal{L}(\rho) = \sum_{i=0}^{n} d_1(p_i, p_{i+1}) \quad \text{with } p_0 = p \text{ and } p_{n+1} = q$$

between the source p and destination q. Figure 1.20 illustrates three (of many) options for shortest d_1-paths between p and q. Specify the area defined by the union

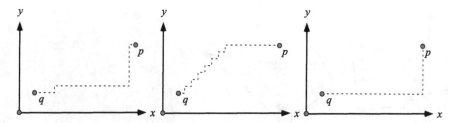

Fig. 1.20 Three shortest d_1-paths between p and q. The coordinate *axes* define the only two possible directions of moves (also called *isothetic moves*)

of all shortest d_1-paths from grid point p to grid point q. What is the analogous area for Minkowski metric d_∞?

Problem 1.4 How to define the length of a shortest path between two points p and q in $\mathbb{R}^3$ when applying the forest distance d_f rather than d_e? Generalise the given 2D distance d_f at first to a metric in $\mathbb{R}^3$.

Problem 1.5 Do some experiments with the Dijkstra algorithm. Download a source from the net, run it on weighted graphs with different values of $|E|$, and measure the actual run time (e.g., by running it on the same input 1,000 times and divide measured time by 1,000). Generate a diagram which plots values of $|E|$ together with the measured time. After a sufficient number of runs you should have a diagram showing a 'dotted curve'. Discuss the curve in relation to the upper time bounds specified above for the Dijkstra algorithm.

Problem 1.6 (Programming exercise) Implement a program which 'randomly' generates a simple polygon with $n > 0$ vertices, with only specifying the value of parameter n at the start of the program. The program should also contain the option that only the vertices are generated (i.e., a set of points) as a set of n randomly generated points.

Aim at generating a large diversity of shapes of polygons (e.g., more than just star-shaped polygons; see Fig. 1.21). For controlling the output of your program, draw the resulting polygons on screen (e.g., by drawing with OpenGL in an OpenCV window). The generated polygons will be useful for testing ESP programs later on, for programming exercises listed in subsequent chapters.

1.11 Notes

For algorithm design in general, see, for example, the books [5, 10, 19]. Finding a general solution to the general ESP problem (starting with dimension $m = 3$, without predefined areas of destination, and with having the whole $\mathbb{R}^m$ as search space) is known to be NP-hard [3]. The survey article [16] informs about ESP algorithms for

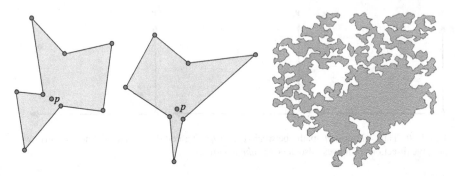

Fig. 1.21 *Left and middle*: Two star-shaped polygons (note: in a star-shaped polygon there is at least one point p 'such that a 'guard' at p sees' all the frontier of this polygon). *Right*: A randomly generated simple polygon (courtesy of Partha Bhowmick, Indian Institute of Technology Kharagpur, India)

3D space. There are arithmetic δ-approximation algorithms solving the Euclidean shortest path problem in 3D in polynomial time, see [4].

Shortest paths or path planning in 3D robotics (see, for example, [11–13], or the annual ICRA conferences in general) is dominated by decision-theoretic planning, stochastic algorithms, or heuristics. Describing a robot with its degrees of freedom (in moving) requires a high-dimensional space; the robot is a point of parameters in this space. ESP algorithms need then to be applied to those high-dimensional spaces. The problem is that there is typically no prior complete knowledge about the space; when going from p to q it might be just possible to notice that a straight path is not possible because there are some obstacles on the way (without having complete geometric knowledge about the scene).

Path planning is the subject of the book [12]. The book contains illustrations of shortest paths depending upon the chosen distance measure, and provides, in general, a very good connection between geometric problems and various robotics applications. The calculation of Euclidean shortest paths is not a subject in [12].

The pioneering paper for breadth-first search is [17]. The calculation of shortest paths is also a subject in graph theory [5] (Chaps. 24 and 25); here, a shortest path connects vertices in a given graph, where edges of the graph are labelled by weights; this situation differs from the Euclidean shortest path problem, where possible vertices are not within a predefined finite set, and thus we also do not have a finite set of predefined weights (representing distances). Dijkstra's algorithm was published in [6]. The heuristic $A^\star$-search algorithm was published in [7].

For approximation algorithms, see the books [1, 8, 9, 14, 20], or the website [18]; the so-called "absolute" or "relative approximation" schemes are basically not much different from the discussed concept of δ-approximation. Approximation algorithms for ESP calculations are specified in [15].

For random curve generation, as addressed in Problem 1.6, see [2].

References

1. Ausiello, G., Crescenzi, P., Gambosi, G., Kann, V., Marchetti-Spaccamela, A., Protasi, M.: Complexity and Approximation. Springer, New York (1999)
2. Bhowmick, P., Pal, O., Klette, R.: A linear-time algorithm for the generation of random digital curves. In: Proc. PSIVT, pp. 168–173. IEEE Comput. Soc., Los Alamitos (2010)
3. Canny, J., Reif, J.H.: New lower bound techniques for robot motion planning problems. In: Proc. IEEE Conf. Foundations Computer Science, pp. 49–60 (1987)
4. Choi, J., Sellen, J., Yap, C.-K.: Approximate Euclidean shortest path in 3-space. In: Proc. ACM Conf. Computational Geometry, pp. 41–48 (1994)
5. Cormen, T.H., Leiserson, C.E., Rivest, R.L., Stein, C.: Introduction to Algorithms, 2nd edn. MIT Press, Cambridge (2001)
6. Dijkstra, E.W.: A note on two problems in connection with graphs. Numer. Math. 1, 269–271 (1959)
7. Hart, P.E., Nilsson, N.J., Raphael, B.: A formal basis for the heuristic determination of minimum cost paths. IEEE Trans. Syst. Sci. Cybern. 4, 100–107 (1968)
8. Hochbaum, D.S. (ed.): Approximation Algorithms for NP-Hard Problems. PWS, Boston (1997)
9. Hromkovič, J.: Algorithms for Hard Problems. Springer, Berlin (2001)
10. Kleinberg, J., Tardos, E.: Algorithm Design. Pearson Education, Toronto (2005)
11. Latombe, J.-C.: Robot Motion Planning. Kluwer Academic, Boston (1991)
12. LaValle, S.M.: Planning Algorithms. Cambridge University Press, Cambridge, UK (2006)
13. Li, T.-Y., Chen, P.-F., Huang, P.-Z.: Motion for humanoid walking in a layered environment. In: Proc. Conf. Robotics Automation, vol. 3, pp. 3421–3427 (2003)
14. Mayr, E.W., Prömel, H.J., Steger, A. (eds.): Lectures on Proof Verification and Approximation Algorithms. Springer, Berlin (1998)
15. Mitchell, J.S.B.: Geometric shortest paths and network optimization. In: Handbook of Computational Geometry, pp. 633–701. Elsevier, Amsterdam (2000)
16. Mitchell, J.S.B., Sharir, M.: New results on shortest paths in three dimensions. In: Proc. SCG, pp. 124–133 (2004)
17. Moore, E.F.: The shortest path through a maze. In: Proc. Int. Symp. Switching Theory, vol. 2, pp. 285–292. Harvard University Press, Cambridge (1959)
18. Rabani, Y.: Approximation algorithms. http://www.cs.technion.ac.il/~rabani/236521.04.wi.html (2004). Accessed 28 October 2004
19. Skiena, S.S.: The Algorithm Design Manual. Springer, New York (1998)
20. Vazirani, V.V.: Approximation Algorithms. Springer, Berlin (2001)

Chapter 2
Deltas and Epsilons

I mean the word proof not in the sense of the lawyers, who set two half proofs equal to a whole one, but in the sense of a mathematician, where half proof = 0, and it is demanded for proof that every doubt becomes impossible.

Carl Friedrich Gauss (1777–1855)

The introduction ended with recalling concepts in discrete mathematics as used in this book. This second chapter adds further basic concepts in continuous mathematics that are also relevant for this book, especially in the context of approximate algorithms.

2.1 Exact and δ-Approximate Algorithms

The term *exact algorithm* seems to be clear to everybody: the algorithm computes exactly the true solution for each input. However, the true solution is not always uniquely defined. For example, algorithms for mapping a 2D or 3D binary picture into topologically equivalent *skeletal curves* (see Fig. 2.1) are often based on iterative thinning methods (i.e., object pixels in 2D, or object voxels in 3D pictures are changed into background elements in one iteration if this operation does not change the topology of the picture). Researchers have published different proposals for *thinning* strategies. Correctly implemented algorithms deliver satisfactory results with respect to topology preservation or other predefined rules. But resulting skeletal curves are different for equal inputs and different algorithms. There is no unique "true solution" for thinning algorithms.

We do have a criterion of truth for ESPs. The length of a shortest path is always uniquely defined.

Definition 2.1 An algorithm is *exact* if it delivers a true solution for each input with respect to an existing criterion that is independent from the algorithm itself.

F. Li, R. Klette, *Euclidean Shortest Paths*,
DOI 10.1007/978-1-4471-2256-2_2, © Springer-Verlag London Limited 2011

Fig. 2.1 Results of three different iterative thinning algorithms (*dark pixels*) on the same binary picture shown in *grey*. The top part of the *second* result differs significantly from that of the *first* and *third* results. (Courtesy of Gisela Klette, Auckland)

This definition requires exact correctness. Different measures specify approximate algorithms to express "how close" a solution needs to be to a true solution. This book is about minimisation problems. We say in short "a solution" for "the length of the calculated path" for the class of ESP problems.

Definition 2.2 An algorithm is a δ-*approximate algorithm* for a given minimisation problem iff, for each input instance of this problem, the algorithm delivers a solution that is at most δ times the optimum solution.

Obviously, $\delta < 1$ is impossible, and $\delta = 1$ defines an exact algorithm. In general, we may assume that $\delta = 1 + \varepsilon_0$, for some $\varepsilon_0 \geq 0$. For example, $\delta = 1.15$ defines $\varepsilon_0 = 0.15$, and thus an error limit of 15%.

Definition 2.3 An algorithm is *without guarantee* (of exactitude) iff there exists a real number $\varepsilon > 0$ and an input instance such that the difference between the output for this input and the true solution is larger than ε, for any choice of the algorithm's free parameters.

Note that a δ-approximate algorithm might even be an algorithm without guarantee, for $\varepsilon < (\delta - 1)$.

> A δ-approximate algorithm does not necessarily calculate a solution arbitrarily close to an optimal solution.

Definition 2.4 An algorithm is *within guaranteed error limits* iff for any $\varepsilon > 0$ and each input instance, there is a choice of the free parameters for this algorithm such that the difference between output for this input and true solution is smaller than or equals ε.

Corollary 2.1 *Any algorithm within guaranteed error limits is also a δ-approximation algorithm, for any $\varepsilon_0 > 0$ and $\delta = 1 + \varepsilon_0$.*

Algorithm 2 (Papadimitriou algorithm, 1985)
Input: A finite set of simple polyhedra (the obstacles) with n vertices and $\mathcal{O}(n)$ edges, a start point p that is not on the surface of any polyhedron, and an end point q, all in 3D Euclidean space. Let E_Π be the set of all edges of the polyhedral obstacles, and $\varepsilon > 0$ the accuracy constant.
Output: A path (a polyline) from p to q.

1: **for** each edge $e \in E_\Pi$ **do**
2:     Compute a point r in e such that $d_2(p, r) = \min\{d_2(p, r') : r' \in e\}$.
3:     Break e into shorter line segments with endpoints

$$(x_r + d_i, y_r + d_i, z_r + d_i) \quad \text{and} \quad (x_r - d_i, y_r - d_i, z_r - d_i)$$

    where

$$d_i = d_e(p, r) \times \frac{\varepsilon}{4\sqrt{3}n}\left(1 + \frac{\varepsilon}{4n}\right)^{i-1}$$

    for $i = 1, 2, \ldots$.
4:     Let V be the set of midpoints of such shorter line segments. [A line segment
    is called the *associated* segment of its midpoint.]
5:     Put p and q also into V.
6: **end for**
7: Construct a *visibility-weighted undirected graph* $G = [V, E, w]$ as follows: for
    any pair of points u and v in V, there is an edge $\{u, v\} \in E$ iff their associated
    segments are visible from one-another. In this case, define the weight of $\{u, v\}$
    as $d_2(u, v)$. Otherwise, let $+\infty$ be the weight of edge $\{u, v\}$.
8: Apply the Dijkstra algorithm for computing a shortest path from p to q in
    graph G

Fig. 2.2 Papadimitriou's algorithm for solving the general 3D ESP problem. The calculated shortest path in graph G is an approximate ESP for the given problem

Example 2.1 Consider a *general 3D ESP problem* where p and q are points in 3D space, and a shortest path from p to q cannot pass through a finite set of simple polyhedral obstacles having n vertices in total, and $\mathcal{O}(n)$ edges. The start point p is not on the surface of any of those polyhedra. The *Papadimitriou algorithm* is a δ-approximate algorithm for solving this problem; see Fig. 2.2 for a pseudocode of this algorithm.

    This algorithm maps the given continuous ESP problem into a discrete problem by subdividing the edges involved into a finite number of segments, where the scale of the subdivisions is defined by increments d_i, depending on the selected ε. The particular definition of increments d_i is chosen for numerical considerations in the original paper that are outside the scope for our discussion here. The visibility between edges was introduced when commenting on Fig. 1.12.

    The Papadimitriou algorithm is an algorithm within guaranteed error limits. Let $\delta = 1 + \varepsilon_0$. The algorithm produces a path guaranteed to be not longer than $(1 + \varepsilon_0)$

Algorithm 3 (Control structure of an iterative ESP algorithm)
Input: A search domain, obstacles, attractions, a start point p and an end point q, all in 2D or 3D Euclidean space.
Output: A path (a polyline) from p to q.

1: INITIALISATION: path ρ_0 from p to q, and L_0 be the length of ρ_0; $i = 1$.
2: **while** STOP CRITERION = false **do**
3:    UPDATE produces a new path ρ_i.
4:    Let L_i be the length of ρ_i.
5:    Let $i = i + 1$.
6: **end while**

Fig. 2.3 Defining control structure of an iterative ESP algorithm

times the length of a true ESP. Without proof we just state that it has the time complexity

$$\mathcal{O}\left(\frac{n^4[b + \log(n/\varepsilon_0)]^2}{\varepsilon_0^2}\right) \tag{2.1}$$

where b is the number of bits representing the coordinates of the vertices in the polyhedral search domain Π (i.e., b is the base-2 logarithm of the largest integer appearing as a coordinate for one of the vertices of Π), and n is the total number of edges of Π.   □

The Papadimitriou algorithm within guaranteed error limits is able to come arbitrarily close to an optimum solution, at the cost of an increase in time complexity of the algorithm: for small ε_0, the numerator in Eq. (2.1) is characterised by $n^4 \cdot b$ and the denominator by a very small number ε_0^2. Note that the given time complexity $f(n, \varepsilon_0)$ in Eq. (2.1) cannot be split into a product of two functions $f_1(n)$ and $f_2(\varepsilon)$, such that f_1 is independent of ε, and f_2 independent of n.

2.2 Approximate Iterative ESP Algorithms

Numerics makes frequent use of the concept of repeated iterations for defining algorithms within guaranteed error limits. An *iterative ESP algorithm* is defined by performing at first a constant number of operations before running into a loop; the loop performs a finite number of iterations, each time ending with the specification of an ESP of length L_i, where i runs from 0 to the maximum number of iterations; see Fig. 2.3.

This is a general control structure. Any iterative ESP algorithm needs to be specified by explaining the initialisation of path ρ_0, defining the stop criterion, and the update (i.e., how to obtain path ρ_{i+1} from path ρ_i and some optimisation strategy).

Fig. 2.4 A sketch of a
distribution of L_i-values, in
relation to the exact length L.
The distance between L_i and
L_{i+1} is shown to be less than
ε, but the distance δ between
L_{i+1} and L remains unknown

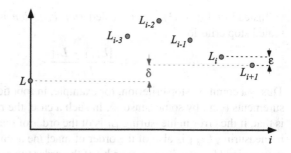

The stop criterion should ensure that the algorithm is not running into an infinite
loop. The time complexity of an iterative ESP algorithm is then simply given as
follows:

$$f(\mathbf{x}) = f_{\text{initialisation}}(\mathbf{x}) + f_{\text{stop}}(\mathbf{x}) \sum_{i=1}^{i_{\max}(\mathbf{x})} f_{\text{update}}(\mathbf{x}, i)$$

where $\mathbf{x}$ is a vector combining all the algorithm's parameters (i.e., the problem
complexity $n > 0$ and possibly some free parameters). This further simplifies if
$f_{\text{update}}(\mathbf{x}, i)$ is basically independent of the iteration number i, and $f_{\text{stop}}(\mathbf{x})$ assumed
to be a constant $c > 0$:

$$f(\mathbf{x}) = f_{\text{initialisation}}(\mathbf{x}) + c \cdot i_{\max}(\mathbf{x}) \cdot f_{\text{update}}(\mathbf{x}).$$

Because we are interested in asymptotic complexities, we can also ignore the constant c (see Definition 1.4) and obtain

$$f(\mathbf{x}) = f_{\text{initialisation}}(\mathbf{x}) + i_{\max}(\mathbf{x}) \cdot f_{\text{update}}(\mathbf{x}). \tag{2.2}$$

The update should guarantee that the new path is a better solution (i.e., of reduced
length) compared to the previous path. If implementing both ideas properly, then we
may stop the algorithm by using an *accuracy parameter* $\varepsilon > 0$:

Definition 2.5 An iterative ESP algorithm is *approximate* iff, for any $\varepsilon > 0$, there
exists a natural number i_ε such that if $i \geq i_\varepsilon$ then

$$|L_{i+1} - L_i| < \varepsilon. \tag{2.3}$$

Let L be the (exact) length of an ESP. Definition 2.5 does not compare L_i with L,
but L_i with L_{i+1}. This is illustrated in Fig. 2.4. The figure also shows a case where
L_{i-2} is larger than L_{i-3}. This should actually be avoided by a proper specification
of the algorithm.

Because we do not know L, we can only compare previously calculated L_i-
values. Definition 2.5 is thus different to that of an algorithm within guaranteed error
limits. The figure also illustrates a distance ε between L_i and L_{i+1}, and a distance
δ between L_{i+1} and L, indicating that δ could possibly still be larger than ε.

Instead of Eq. (2.3), the unscaled *stop criterion* in Definition 2.5, consider a scaled stop criterion

$$\frac{|L_{i+1} - L_i|}{L_{i+1}} < \varepsilon. \tag{2.4}$$

This is a common stop criterion, for example, in robotics, where values L_i are measurements (e.g., by some sensors). In such a case, the reason for applying Eq. (2.4) is that, if the error in measuring L_i is of the order of machine accuracy and the error in measuring L_{i+1} is also of the order of machine accuracy, then the resulting error in unscaled $|L_{i+1} - L_i|$ can even be of the order twice of that of machine accuracy. However, we stay with Eq. (2.3) for quantities L_i calculated in our algorithms.

The following section shows that any approximate ESP algorithm is also an algorithm within guaranteed error limits, and thus also an $(1 + \varepsilon_0)$-approximate algorithm, for any $\varepsilon_0 > 0$.

2.3 Convergence Criteria

Let f be a function that maps values from some subset $S \subseteq \mathbb{R}$ into $\mathbb{R}$, and let $c \in \mathbb{R} \cup \{-\infty, +\infty\}$ be a constant. The notation $x \to c$ is short for 'value x goes arbitrarily close to c': (i) for $c \in \mathbb{R}$ and any $\varepsilon > 0$ there is a value $x \in S$ with $|x - c| < \varepsilon$; (ii) for $c = -\infty$ and any $T < 0$ there is a value $x \in S$ with $x < T$; (iii) for $c = +\infty$ and any $T > 0$ there is a value $x \in S$ with $x > T$. Let

$$\lim_{x \to c} f(x)$$

be defined and equal to a real number a iff $f(x)$ can go arbitrarily close to a as $x \to c$, which means that for any $\delta > 0$ there is

(case $c \in \mathbb{R}$) an $\varepsilon > 0$ such that for any $x \in S$ with $|x - c| < \varepsilon$ it follows that $|f(x) - a| < \delta$;
(case $c = -\infty$) a $T_0 < 0$ such that for any $x \in S$ with $x < T_0$ it follows that $|f(x) - a| < \delta$;
(case $c = +\infty$) a $T_0 > 0$ such that for any $x \in S$ with $x > T_0$ it follows that $|f(x) - a| < \delta$.

Definition 2.6 Function f *converges* (or *is convergent*) as $x \to c$ iff $\lim_{x \to c} f(x)$ is defined. If it is defined and equals a then a is called the *limit* of f as $x \to c$.

For example, f can be defined on the set of natural numbers $\mathbb{N} = \{0, 1, 2, \ldots\}$, defining a sequence $a_i = f(i)$ of real numbers, for $i \geq 0$. This sequence $\{a_0, a_1, a_2, \ldots\}$ may, for example, 'oscillate somehow' around a value $a \in \mathbb{R}$ by coming closer and closer, say, but for index $i + 1$ a bit more than for index i:

$$|a - a_{i+1}| < |a - a_i|$$

for all $i = 0, 1, 2, \ldots$; see Fig. 2.5. The infinite sequence $\{a_0, a_1, a_2, \ldots\}$ is converging to the limit a. In this case, we have that constant c equals $+\infty$.

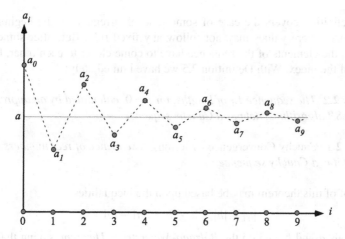

Fig. 2.5 A sequence of reals with constantly decreasing distances from the real number a

Example 2.2 Let b be any real; the sequence $a_{i+1} = (a_i + b/a_i)/2$ of rational numbers, for any given initial rational number a_0, is converging[1] to $a = \sqrt{b}$. Let $b = 2$ and $a_0 = 1$, for example, then the first elements are

$$a_0 = 1,$$
$$a_1 = 1.5,$$
$$a_2 = 1.41666666666\ldots,$$
$$a_3 = 1.41421568627\ldots,$$
$$a_4 = 1.41421356237\ldots,$$

and the sequence converges to the limit $\sqrt{2}$. □

Now let us 'remove' the real a from the definition of convergence, and we just claim for a given sequence $\{a_0, a_1, a_2, \ldots\}$ of real numbers and for all $i = 0, 1, 2, \ldots$ that

$$|a_{i+1} - a_{i+2}| < |a_i - a_{i+1}|.$$

This is a first step towards a *Cauchy sequence*.[2] (In general, a Cauchy sequence can also be defined in terms of a function f defined on $S \subseteq \mathbb{R}$; however, we only need this special case of $S = \mathbb{N}$ in this book.)

Definition 2.7 A *Cauchy sequence* is an infinite sequence $\{a_0, a_1, a_2, \ldots\}$ of real numbers, such that for any $\varepsilon > 0$ there is an index i_0 with $|a_i - a_j| < \varepsilon$, for any $i, j \geq i_0$.

---

[1] *Heron of Alexandria* (ca. 10–70) described this approximation method, which is also known as *Babylonian method*.

[2] Named after *Baron Augustin-Louis Cauchy* (1789–1857), who was central for establishing the infinitesimal calculus (e.g., of convergence of real numbers).

This definition covers the case of some kind of 'irregular' finite beginning of the sequence where values may not follow any fixed rule. But, after some finite beginning, the elements of the sequence have to come closer to each other, for any increase of the index. With Definition 2.5 we have immediately:

Corollary 2.2 *The sequence L_i of lengths, for $i \geq 0$, calculated by an approximate iterative ESP algorithms, defines a Cauchy sequence.*

Theorem 2.1 (Cauchy Convergence Criterion) *A sequence of real numbers is convergent iff it is a Cauchy sequence.*

A proof of this theorem may be based upon the inequalities

$$|a| - |b| \leq |a - b| \leq |a| + |b|$$

for any reals a and b, and on the *Bolzano–Weierstrass Theorem*, saying that every bounded sequence has a convergent subsequence.[3] A sequence $\{a_0, a_1, a_2, \ldots\}$ of real numbers is *bounded* iff there are real numbers a and b such that $a \leq a_i \leq b$, for all $i \geq 0$.

Let $a_1, a_2, a_3, \ldots$ be a convergent sequence. Thus, for any $\varepsilon > 0$ there is a k_ε such that $|a_k - a| < \varepsilon$, for all $k \geq k_\varepsilon$. Now consider $\varepsilon/2$. Let $i, j > k_{\varepsilon/2}$, with $|a_i - a| < \varepsilon/2$ and $|a_j - a| < \varepsilon/2$. Applying the above second inequality, we obtain that

$$|a_i - a_j| = \left| (a_i - a) - (a_j - a) \right| \leq |a_i - a| + |a_j - a| \leq \varepsilon/2 + \varepsilon/2 = \varepsilon.$$

This shows that the sequence is Cauchy.

The other direction of the proof is not so short, and we just mention that every Cauchy sequence is bounded, and possesses thus a convergent subsequence (see the Bolzano–Weierstrass Theorem above). The proof can then be completed by showing that, if a Cauchy sequence has a subsequence that is convergent to a, then the whole sequence is also convergent to a.

We state the Cauchy Convergence Criterion above, and the following basic results of infinitesimal calculus without (complete) proof, but for possible reference later in the book. Theorem 2.1 and Corollary 2.2 show that any approximate iterative ESP algorithm is also an algorithm with guaranteed error limits.

Definition 2.8 A sequence $\{a_0, a_1, a_2, \ldots\}$ of real numbers is *monotonically decreasing* (*monotonically increasing*) iff $a_i \geq a_{i+1}$ ($a_i \leq a_{i+1}$), for all $i \geq 0$. Such a sequence is called *monotone* iff it is either monotonically de- or increasing.

Theorem 2.2 (Monotone Convergence Criterion) *A monotone sequence of real numbers is convergent iff it is bounded.*

Corollary 2.3 *A monotonically decreasing sequence of real numbers is convergent iff it is lower bounded.*

---

[3] Named after *Bernard Placidus Johann Nepomuk Bolzano* (1781–1848) and *Karl Theodor Wilhelm Weierstrass* (1815–1897).

Fig. 2.6 *Left*: Graph of a non-convex function. *Right*: Graph of a convex function. The *points* shown are local minima

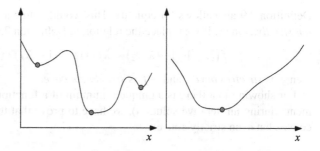

In this book, convergence proofs for iterative ESP algorithms are always done by showing that a sequence is monotonically decreasing and lower bounded.

Let L_i, for $i \geq 0$, be the sequence of lengths calculated by an iterative ESP algorithm. In particular, according to Corollary 2.3, we only have to show the existence of a lower bound if a sequence is monotonically decreasing; then the sequence is convergent, also a Cauchy sequence, and the algorithm is approximate, thus also an algorithm with guaranteed error limits.

There is always a uniquely specified minimum length L for any input instance of the considered ESP problem (i.e., a lower bound). However, Corollary 2.3 does not say that the sequence is converging towards this lower bound; it might be another real number larger than the lower bound.

2.4 Convex Functions

Measurements during an iterative solution define discrete samples L_i, for iterations $i \geq 0$. We may assume that a continuous function $f(x)$, defined on the interval $\mathbb{R}^+$ of all non-negative real numbers, is fitting these samples accurately: $f(i) = L_i$, for all $i \geq 0$.

Definition 2.9 A function f, defined on an interval of real numbers, is *convex* if

$$f\left(\lambda x_1 + (1 - \lambda)x_2\right) \leq \lambda f(x_1) + (1 - \lambda)f(x_2)$$

for any two reals x_1, x_2 in this interval, and $0 \leq \lambda \leq 1$.

(In general, a convex function is defined over a convex subset of $\mathbb{R}^n$.)

A convex function possesses the important property that every local minimum is also a global minimum. This property is used in this book for showing whether solutions obtained by iterative ESP algorithms are always the global solutions. If f is a convex function then $-f$ is called a *concave function*. For example, the graph of $y = x^3$ is convex for $x \geq 0$ but concave for $x < 0$.

See Fig. 2.6 for examples of two unary functions $f(x)$; the graph of a convex function may also run parallel to the x-axis in one segment, because the relation in

Definition 2.9 also allows for equality. This would define a convex function that is *not strictly convex*. If we replace the relation in Definition 2.9 by

$$f\big(\lambda x_1 + (1 - \lambda)x_2\big) < \lambda f(x_1) + (1 - \lambda)f(x_2) \tag{2.5}$$

then f is *strictly convex*, and $-f$ is *strictly concave*.

For showing that there is a unique minimum of a function (defined by measurements during an iterative solution), we have to prove that this function is not only convex but even strictly convex.

2.5 Topology in Euclidean Spaces

For showing that there is a unique minimum of a function, we have to prove that this function is not only convex but even strictly convex; this is sometimes possible by applying results from topology in Euclidean spaces. We also need topology for defining "topological equivalence", or for describing accurately the "interior" of a set and its "frontier". (We already used those two notions in the context of polygons and polyhedra.)

We start our considerations in the 1-dimensional set $\mathbb{R}$ of reals. Let $a < b$ be two reals. The interval $[a, b] = \{x : a \le x \le b\}$ is *closed* (it also contains both its endpoints a and b), and the interval $(a, b) = \{x : a < x < b\}$ is *open*. We can also define "half-open" (or "half-closed") intervals such as $[a, b) = \{x : a \le x < b\}$. Points a and b define the *frontier* of the open (i.e., (a, b)) or closed (i.e., $[a, b]$) interval; the open interval (a, b) is also called the *interior* of $[a, b]$.

For a real c and $\varepsilon > 0$ we define with $N_\varepsilon(c) = \{x : |x - c| < \varepsilon\}$ the ε-*neighbourhood* of c, which is an open interval.

Definition 2.10 A set $S \subseteq \mathbb{R}$ is *open* if, for any point $c \in S$, there is an $\varepsilon > 0$ such that $N_\varepsilon(c) \subseteq S$.

Note that this is not true for the points a and b in the frontier of $[a, b]$. The property is true for any real in the open interval (a, b), and also for any real in an arbitrary (finite or infinite) union

$$(a_0, b_0) \cup (a_1, b_1) \cup (a_2, b_2) \cup \cdots$$

of open intervals, which do not have to be pairwise disjoint, or a finite intersection

$$(a_0, b_0) \cap (a_1, b_1) \cap (a_2, b_2) \cap \cdots \cap (a_m, b_m)$$

of open intervals.

If S is an open set, then $\mathbb{R} \setminus S$ is called a *closed set*. Note that closed intervals, as defined above, are also closed sets according to this definition. The set $\mathbb{R}$ is open, and the empty set $\emptyset$ is also open; thus, these two sets are also closed. These are the only two examples where a set is both closed and open. We summarise:

- Any ε-neighbourhood of a point $c \in \mathbb{R}$ is open.

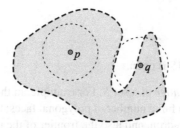

Fig. 2.7 The *dashed line* shows the frontier of an open set S. The shown neighbourhood of p is completely contained in S, but the shown neighbourhood for point q is not. For point q, a smaller radius of the neighbourhood needs to be chosen for having also a neighbourhood completely in S

- The union of any finite or infinite number of open sets is again open.
- The intersection of any finite number of open sets is again open.
- $\mathbb{R}$ and $\emptyset$ are both open and closed.

Now we generalise these notions of open or closed sets to the Euclidean plane or 3D space using the Euclidean metric $d_e = d_2$ for defining ε-neighbourhoods

$$N_\varepsilon(p) = \{q : d_e(p, q) < \varepsilon\}$$

of points $p \in \mathbb{R}^m$, for $m = 2$ or $m = 3$. These neighbourhoods form open disks in the plane (see Fig. 2.7) or open spheres in 3D space.

Definition 2.11 Let $m = 2$ or $m = 3$. A set $S \subseteq \mathbb{R}^m$ is *open* iff, for any point $p \in S$, there is an $\varepsilon > 0$ such that $N_\varepsilon(p) \subseteq S$.

Again, if $S \subseteq \mathbb{R}^m$ is an open set, then $\mathbb{R}^m \setminus S$ is called a *closed set*. It follows that

- Any ε-neighbourhood of a point $p \in \mathbb{R}^m$ is open.
- The union of any finite or infinite number of open sets is again open.
- The intersection of any finite number of open sets is again open.
- $\mathbb{R}^m$ and $\emptyset$ are both open and closed.

This defines a consistent approach for dealing with open or closed sets in $\mathbb{R}^m$, for $m = 1$ and also for $m = 2$ or $m = 3$, and this can actually be extended to any $m \geq 1$ for defining topologies in Euclidean spaces $[\mathbb{R}^m, d_e]$.

A set $S \subseteq \mathbb{R}^m$ is *bounded* if there is some point $p \in \mathbb{R}^m$ and a radius $r > 0$ such that S is completely contained in the r-neighbourhood of p: $S \subseteq N_r(p)$. A set is *compact* if it is bounded and closed. For example, simple polygons or simple polyhedra are all bounded sets.

We conclude this section with introducing some commonly used notation. Let $S \subseteq \mathbb{R}^m$. We define:

$$\begin{aligned}
(\textit{interior}) \quad & S^\circ = \{p \in S : \exists \varepsilon (\varepsilon > 0 \wedge N_\varepsilon(p) \subseteq S)\}, \\
(\textit{frontier}) \quad & \partial S = \{p \in \mathbb{R}^m \setminus S^\circ : \forall \varepsilon (\varepsilon > 0 \rightarrow N_\varepsilon(p) \cap S \neq \emptyset)\}, \\
(\textit{closure}) \quad & S^\bullet = S^\circ \cup \partial S.
\end{aligned}$$

For any $S \subseteq \mathbb{R}^m$ it follows that

- S is open iff $S = S°$.
- S is closed iff $S = S^{\bullet}$.
- $S° \cap \partial S = \emptyset$.
- If S is closed then $S \setminus \partial S$ is open.

The set ∂S defines the *frontier* of a set S. For example, in the case $m = 3$, a simple polyhedron is defined by a finite number of polygonal faces; the union of those faces is the surface of the polyhedron, and also the frontier of the polyhedron. In the case $m = 2$, a simple polygon is defined by a polygonal loop; this loop is the frontier of the polygon. In the case $m = 1$, the frontier of an interval (a, b), $(a, b]$, $[a, b)$, or $[a, b]$ is the set $\{a, b\}$.

Definition 2.12 S is called (*topologically*) *connected* iff it is not the union of two disjoint nonempty open subsets (or, equivalently, closed subsets) of S.

Maximum connected subsets of S are called *components* of S.

A path *visits* a set S iff this path and S do have a nonempty intersection. Because a path does have an orientation (i.e., from start to end, or by a defined loop), it also visits S at some point *for the first time*, if there is a nonempty intersection, and if S is topologically closed. If the path does not start in S, then the first visit of a closed set S will be in ∂S.

Definition 2.13 A family $\mathcal{F}$ of subsets of a set $S_0 \subseteq \mathbb{R}^m$ defines a *topological space* or a *topology* iff it satisfies the following axioms:

T1 $\{\emptyset, S_0\} \subseteq \mathcal{F}$.
T2 The union of any finite or infinite number of sets in $\mathcal{F}$ is again in $\mathcal{F}$.
T3 The intersection of any finite number of sets in $\mathcal{F}$ is again in $\mathcal{F}$.

A set $S \in \mathcal{F}$ is *open*, and $S_0 \setminus S$ is *closed*.

According to axiom T1, the base set S_0 is open in its own topology. However, S_0 does not need to be open in another topology. For example, a compact set S_0 in $\mathbb{R}^m$ is not open in the topology of $\mathbb{R}^m$, but may define its own topology (i.e., all the open subsets of S_0 only), and S_0 is then open in its own topology.

Definition 2.14 A topological space or topology $\mathcal{F}$, defined by all open subsets of a set $S_0 \subseteq \mathbb{R}^m$, is called the *topology on S_0*.

The important concept of "topological equivalence" will be specified in the following section, after defining continuous mappings between topological spaces.

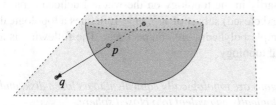 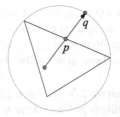

Fig. 2.8 *Left*: A perspective projection from the centre of the open (i.e., not containing its frontier) *half-sphere* onto a *plane tangent* to the *half-sphere* (which is parallel to the base of the hemisphere) defines a homeomorphism; point *p* maps onto point *q*. *Right*: Projection of a *triangle* onto a *circle* [R. Klette and A. Rosenfeld, 2004]

2.6 Continuous and Differentiable Functions; Length of a Curve

Consider a function f from a domain $D \subseteq \mathbb{R}$ into $\mathbb{R}$. Informally speaking, f is continuous on D iff, for $a_1, a_2 \in D$, values $f(a_1)$ and $f(a_2)$ are "close" to each other if a_1 and a_2 are close to each other. More formally:

for any $\varepsilon > 0$ there is some $\delta > 0$ such that $|f(a_1) - f(a_2)| < \varepsilon$, for any $a_2 \in D$ with $|a_1 - a_2| < \delta$.

This says that the open δ-neighbourhood of a_1 is mapped into an open ε-neighbourhood of $f(a_1)$. An exact general definition of continuous functions is as follows:

Definition 2.15 A function f from a topology on S_0 into a topology on S_1 is called *continuous* iff, for any open set $S \subseteq S_1$, the set $f^{-1}(S) = \{a \in S_0 : f(a) \in S\}$ is also open in S_0.

In the previous section, we defined topologies on Euclidean spaces $\mathbb{R}^m$, for $1 \leq m \leq 3$. A continuous function maps one topology (or topological space) into another.

Definition 2.16 A function f from a topology on S_0 into a topology on S_1 is called a *homeomorphism* iff it is one-to-one, onto S_1, continuous, and f^{-1} is also continuous.

A homeomorphism allows us to map sets one-to-one "smoothly" from S_0 into S_1 and also "backward", from all S_1 into S_0.

Definition 2.17 Two sets S_0 and S_1 are *topologically equivalent* iff they can be mapped by a homeomorphism from S_0 onto S_1.

For example, the Euclidean plane $\mathbb{R}^2$ is topologically equivalent to an open half-sphere, and a triangle (i.e., just the polyline) is topologically equivalent to a circle; see Fig. 2.8. A circle with one point removed is topologically equivalent to $\mathbb{R}^1$. The base sets of these examples of topological spaces are all open with respect to their topology, but not necessarily in the topology on the whole Euclidean space. Any compact (i.e., bounded and closed) subset of $\mathbb{R}^m$, $m \geq 1$, defines a topological space. Simple polygons or simple polyhedra are compact sets. The following is a basic theorem in combinatorial topology:

Theorem 2.3 *All simple polygons are topologically equivalent to a closed disk, and all simple polyhedra are topologically equivalent to a closed sphere.*

Derivatives   Let f be a function from $\mathbb{R}$ into $\mathbb{R}$. The *derivative* $f'(x)$ is the slope of the tangent to the graph of a function f at $(x, f(x))$, defined by the limit

$$f'(x) = \frac{\mathrm{d}f(x)}{\mathrm{d}x} = \lim_{\varepsilon \to 0} \frac{f(x+\varepsilon) - f(x)}{\varepsilon}$$

of *Newton's difference quotient*[4] of function f at x. If this limit is defined for all x and $x + \varepsilon$ in $[a, b]$, then f is not only continuous but also *differentiable* at $x \in [a, b]$.

Now assume that f is a function in variables $x_1, x_2, \ldots$, and x_m; the partial derivative of f with respect to variable x_i is denoted and defined by

$$\frac{\partial f(x_1, x_2, \ldots, x_m)}{\partial x_i} = \lim_{\varepsilon \to 0} \frac{f(x_1, \ldots, x_i + \varepsilon, \ldots, x_m) - f(x_1, \ldots, x_i, \ldots, x_m)}{\varepsilon}.$$

For example, the polynomial $p(\lambda_1, \lambda_2, \lambda_3) = 4\lambda_1^2\lambda_2 + \lambda_1\lambda_3 + 5\lambda_2^3\lambda_3$ has the partial derivative $8\lambda_1\lambda_2 + \lambda_3$ with respect to λ_1, $4\lambda_1^2 + 15\lambda_2^2\lambda_3$ with respect to λ_2, and $\lambda_1 + 5\lambda_2^3$ with respect to λ_3.

Length of a curve   In 2D with rectangular Cartesian coordinates, consider a parameterised curve $\gamma(\lambda) = (x(\lambda), y(\lambda))$, for $a \leq \lambda \leq b$. Denote the derivatives of $x(\lambda)$ and $y(\lambda)$ by $\dot{x}$ and $\dot{y}$.

Definition 2.18 The *length* of the curve $\gamma(\lambda)$ in 2D is defined as

$$\mathcal{L}(\gamma) = \int_{x(a)}^{x(b)} \sqrt{1 + \left(\frac{\mathrm{d}y}{\mathrm{d}x}\right)^2}\, \mathrm{d}x.$$

The integral over x gives signed length of any part of the curve for which y is a single-valued function of x. The following integral over λ gives positive length over a general parameterised curve:

$$\mathcal{L}(\gamma) = \int_a^b \sqrt{\dot{x}^2 + \dot{y}^2}\, \mathrm{d}\lambda.$$

In 3D, consider a parameterised curve $\gamma(\lambda) = (x(\lambda), y(\lambda), z(\lambda))$, for $a \leq \lambda \leq b$. Denote the derivative of $z(\lambda)$ by $\dot{z}$.

---

[4]Named after *Isaac Newton* (1642–1727 in the Julian calendar, which was then used in England).

Definition 2.19 The *length* of the curve $\gamma(\lambda)$ is defined by

$$\mathcal{L}(\gamma) = \int_a^b \sqrt{\dot{x}^2 + \dot{y}^2 + \dot{z}^2}\,d\lambda.$$

In 2D or 3D, the integral over λ gives the same length of a curve for all rectangular Cartesian coordinate axes.

Example 2.3 Consider a line segment between points $p = (p_x, p_y, p_z)$ and $q = (q_x, q_y, q_z)$. A parameterised form of the segment is given by $\gamma(\lambda) = p + \lambda(q - p)$, for $0 \le \lambda \le 1$. We obtain that $x(\lambda) = p_x + \lambda(q_x - p_x)$, and $\dot{x} = q_x - p_x$, with similar coordinate differences for $\dot{y}$ and $\dot{z}$. It follows that the length $\mathcal{L}(\gamma)$ equals (as expected) the Euclidean distance between p and q. ☐

> The example shows that the definition of the length of a path (see Definition 1.7) is consistent with the general definition of the length of a curve.

2.7 Calculating a Zero of a Continuous Function

Continuous functions are important in a numerical context. For example, sometimes we want to calculate a zero x of a continuous function f (i.e., $f(x) = 0$), defined on a domain $D \subset \mathbb{R}$, without having a general formula for calculating those zeros (compare Sect. 1.4; such a general formula may even not exist).

Assume a continuous function f that is defined in the interval $[a, b]$, with $a < b$, and that satisfies $f(a)f(b) < 0$. *Bolzano's Theorem* proves that f has at least one zero in $[a, b]$.

n-section method A straightforward method is to subdivide $[a, b]$ uniformly into $n > 0$ sections of equal length and to test f at the resulting endpoints of these sections; see Fig. 2.9.

> *Binary search* is a general strategy for reducing time in a search routine: divide the search space recursively into halves and apply the search criterion at borders of the resulting subspaces.

Binary-search method We replace the subdivision into n equal sections by an iterated division into halves, testing the sign of the product of two f-values once (for the left endpoint of the current segment and its midpoint). We know that there is at least one zero of f somewhere in the current segment, and this product tells us whether we have to continue the search in the left half of the segment or in the right half. The search stops when the value at the midpoint is sufficiently close to zero.

Algorithm 4 (n-Section Method)
Input: Reals a and b, integer n, an accuracy constant $\varepsilon > 0$; we also have a way to
calculate $f(x)$, for any $x \in [a, b]$.
Output: Value $c \in [a, b]$ such that $|f(c)| < \varepsilon$.

1: Set *flag* = *false* and $i = 0$.
2: **while** $i < n$ **do**
3:      **if** $|f(a + i \cdot (b - a)/n)| < \varepsilon$ **then**
4:          Let $c = a + i \cdot (b - a)/n$, $i = n$, and *flag* = *true*.
5:      **end if**
6: **end while**
7: **if** *flag* = *false* **then**
8:      "Value of n was to small."
9: **end if**

Fig. 2.9 n-section method for finding the zeros of a continuous function f satisfying
$f(a)f(b) < 0$

Algorithm 5 (Binary Search Method)
Input and *Output* as for Algorithm 4.

1: Set $l = a$ and $r = b$.
2: **while** $|f(l + (r - l)/2)| \geq \varepsilon$ **do**
3:      **if** $f(l) \cdot f(l + (r - l)/2) < 0$ **then**
4:          $r = l + (r - l)/2$
5:      **else**
6:          $l = l + (r - l)/2$
7:      **end if**
8: **end while**

Fig. 2.10 Binary-search method for finding zeros of a function f satisfying $f(a)f(b) < 0$

That binary-search method is much more time-efficient than the 'crude' n-section
method. See Fig. 2.10 for pseudocode. It finds always a zero within the predefined
accuracy limit.

Newton–Raphson method   Regarding the calculation of zeros of f, let us suppose that we also have access to a calculation of derivatives $f'(x)$, for $x \in [a, b]$.
The algorithm is shown in Fig. 2.11.[5] Values of the derivative may be approximated
by difference quotients.

---

[5]It is named after *Isaac Newton* (see footnote on page 44) and *Joseph Raphson* (about 1648–about
1715).

Algorithm 6 (Newton–Raphson Method)

Input: Reals a and b; we also have a way to calculate $f(x)$ and $f'(x)$, for any $x \in [a, b]$.

Output: Value $c \in [a, b]$ as an approximate zero of f.

1: Let $c \in [a, b]$ be an initial guess for a zero.
2: **while** STOP CRITERION = false **do**
3:     Replace c by $c - \frac{f(c)}{f'(c)}$
4: **end while**

Fig. 2.11 Newton–Raphson method for finding one zero of a smooth function f satisfying $f(a)f(b) < 0$, and having a derivative of constant sign in $[a, b]$

> If f is a smooth function then the Newton–Raphson method is more time-efficient than the binary search method.

The initial value of c can be specified by a small number of binary-search steps for reducing run-time. A small $\varepsilon > 0$ is used for specifying the stop criterion in the Newton–Raphson method (i.e., "$|f(c)| > \varepsilon$?").

The caption of Fig. 2.11 states a condition which ensures that f has a single zero z in (a, b)—but the Newton–Raphson method will converge only if c is 'sufficiently close' to z. There is no practical way of deciding beforehand that c will give convergence to z, unless f satisfies additional conditions (which are not easy to test in general). For instance, apply Newton–Raphson to a quite simple function, and the values of c for which the algorithm does converge give the Mandelbrot set. If f satisfies the additional condition that $f''(x)$ has constant sign in $[a, b]$, then, if $f(b)$ has the same sign as $f''(x)$, the startpoint $c = b$ gives convergence to z, but otherwise the startpoint $c = a$ gives convergence to z.[6]

2.8 Cauchy's Mean-Value Theorem

The line segment $(a, f(a))(b, f(b))$ is a *chord* of the graph of f, and $\frac{f(b)-f(a)}{b-a}$ is the *slope* of this chord; see Fig. 2.12.

For $a < b$, assume a continuous function f that maps the closed interval $[a, b]$ into $\mathbb{R}$. Let f be differentiable on the open interval (a, b). *Cauchy's Mean-Value Theorem* says that there exists a real c, $a < c < b$, such that

$$f'(c) = \frac{f(b) - f(a)}{b - a}.$$

---

[6]This paragraph was provided by *Garry Tee*, who also pointed out that a clear account of such convergence conditions (with illustrations) is given in Sim Borisovich Norkin's textbook *The Elements of Computational Mathematics*, Pergamon Press, Oxford, 1965.

Fig. 2.12 Graph of a
function f with a chord
$pq = (a, f(a))(b, f(b))$. The
dashed line shows the tangent
at $(c, f(c))$

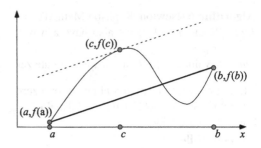

This theorem is not difficult to show;[7] basically, it says that there is a c such that the
tangent at $(c, f(c))$ is parallel to the chord $(a, f(a))(b, f(b))$. In Fig. 2.12, there
are two different points $(c, f(c))$ possible and only one of those is shown.

For the Newton–Raphson method it follows (from the mean-value theorem) that,
if $f(a)$ and $f(b)$ have different signs and $f'(x)$ has a constant sign on $[a, b]$, then
f has exactly one zero in (a, b).

2.9 Problems

Problem 2.1 Assume that we want to measure the length of straight line segments
in the plane in a regular orthogonal grid after digitising them as follows (the so-
called *grid intersection digitisation*): go on the given straight line segment from one
end to the other; for every intersection of a grid line with the straight line, take the
closest grid point as the next grid point (if there are two at equal distance, decide for
the one closer to the origin); this maps a straight line segment into a polyline with a
finite number of vertices (at grid points); see Fig. 2.13 for three examples.

Assume that the regular orthogonal grid has unit distance between grid lines. The
resulting polylines have segments either of length 1 or of length $\sqrt{2}$. Use the total
length of the polyline (i.e., sums of 1s and of $\sqrt{2}$s) as an estimator for the length of
the original straight line segments. Does this define a δ-approximation? For what δ?
Is this an algorithm with guaranteed error limits?

Problem 2.2 Show that any 2nd order polynomial $f(x) = a_0 + a_1x + a_2x^2$ is
convex over the entire real line if $a_2 > 0$, and that any 3rd order polynomial
$f(x) = a_0 + a_1x + a_2x^2 + a_3x^3$ is not convex over the entire real line if $a_3 > 0$.

Problem 2.3 (Programming exercise) Your program compares two approaches for
measuring the length of a curve.

(1) Consider a parabola $y = c_0 + c_1x + c_2x^2$, allowing that parameters c_0, c_1 and
c_2 be selected by a user of the program.

---

[7]*Vatasseri Paramesvara* (ca. 1380–1460) studied already mean-value formulas for the sine function. The theorem is due to *A.-L. Cauchy* (see footnote on page 37).

Fig. 2.13 *Top*: Three *straight line* segments with assigned *grid points* when using grid-intersection digitisation. *Bottom*: Resulting polylines

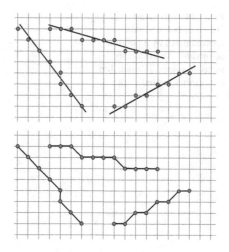

(2) Arclength of a parabola can readily be expressed in terms of elementary functions; see Definition 2.18. Use this analytic approach for calculating the arclength for $a = 0$ and $b = 100$.

(3) Now, in the main part of your program, provide an alternative way for measuring this length based on approximating the graph of the given polynomial on the interval $[0, 100]$ by a polyline at uniformly distributed x-values; take the sum of the lengths of all straight segments of the polyline as your approximate length estimate.

For (3), use different numbers $n_i > 0$ of uniformly distributed x-values, thus producing different length estimates L_i. Discuss the *speed of convergence* of those estimates towards the length obtained in Step (ii) in dependence of an increase in values n_i, using a sufficiently large number of n_i-values.

For an additional challenge, replace (i) by the following: For generating the input (i.e., the curve), now specify a 3rd order polynomial $p(x)$ by selecting parameters c_0, c_1, c_2, and c_3.

Calculate the length $\mathcal{L}(\gamma)$ of the graph $\gamma(\lambda) = (x(\lambda), p(x(\lambda)))$ of this polynomial, with $0 \leq \lambda \leq 1$, $x(0) = 0$, and $x(1) = 100$, now by using a numerical integration procedure for the integral in Definition 2.18. See Fig. 2.14 for one 3rd order polynomial and two examples with $n_1 = 5$ and $n_2 = 10$, as of relevance for Step (iii).

2.10 Notes

There are many different proposals for iterative thinning procedures; see, e.g., [9].

For materials on approximation algorithms, see the books [3, 7, 8, 12, 17] and the website [14]. For Definition 2.2, see, e.g., [7]. [6] describes a so-called "2-approximation linear algorithm" for calculating a shortest path on the surface of

Fig. 2.14 *Top*: Graph of a
3rd-order polynomial
approximated by a polyline
defined by five uniformly
sampled values. *Bottom*: Ten
uniformly sampled values

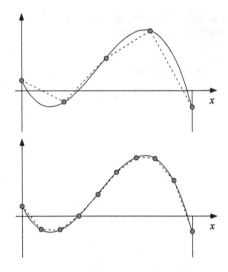

a convex polyhedron, which is an example of an algorithm without guarantee. Example 2.1 (the Papadimitriou algorithm) is due to [13]; published in 1985, this is the first δ-approximation algorithm for solving a general 3D ESP problem.

Definition 2.5 differs from those in [3, 8, 12]; δ-approximation algorithms are considered in [7]. Some books, such as [3, 7, 8, 12, 14], also consider so-called "absolute" or "relative approximation", and so forth. These are schemes that are basically not much different from the concept of δ-approximation; for this reason we will not recall these concepts, and use δ-approximation as *the* concept of approximation algorithms in general in this book.

For basic definitions and theorems of mathematical analysis (Cauchy sequences, convergence, and so forth), see, e.g., [4]. Topology was introduced in [11]; for textbooks on topology, see, e.g., [1, 2]. For topology and the length of curves, see also [10]; this book also discusses topology and the *grid intersection digitisation* as used in Problem 2.1. For a discussion of convex functions, see [5, 15, 16].

References

1. Aleksandrov, P.S.: Combinatorial Topology, vol. 1. Graylock Press, Rochester (1956)
2. Aleksandrov, P.S.: Combinatorial Topology, vol. 2. Graylock Press, Rochester (1957)
3. Ausiello, G., Crescenzi, P., Gambosi, G., Kann, V., Marchetti-Spaccamela, A., Protasi, M.: Complexity and Approximation. Springer, New York (1999)
4. Bartle, R.G., Sherbert, D.: Introduction to Real Analysis, 2nd edn. Wiley, New York (2000)
5. Boyd, S., Vandenberghe, L.: Convex Optimization. Cambridge University Press, Cambridge, UK (2004)
6. Hershberger, J., Suri, S.: Practical methods for approximating shortest paths on a convex polytope in $\mathbb{R}^3$. In: Proc. ACM-SIAM Sympos. Discrete Algorithms, pp. 447–456 (1995)
7. Hochbaum, D.S. (ed.): Approximation Algorithms for NP-Hard Problems. PWS, Boston (1997)
8. Hromkovič, J.: Algorithms for Hard Problems. Springer, Berlin (2001)

9. Klette, G.: Skeletal Curves in Digital Image Analysis. VDM, Saarbrücken (2010)
10. Klette, R., Rosenfeld, A.: Digital Geometry. Morgan Kaufmann, San Francisco (2004)
11. Listing, J.B.: Vorstudien zur Topologie. Göttinger Studien, 1. Abteilung math. und naturw. Abh., pp. 811–875. Several missing proofs were later published by Tait, P.G.: On knots. Proc. R. Soc. Edinb. **9**, 306–317 (1875–1878). A more recent review: Tripodi, A.: L'introduzione alla topologia di Johann Benedict Listing. Mem. Accad. Naz. Sci. Lett. Arti Modena **13**, 3–14 (1971)
12. Mayr, E.W., Prömel, H.J., Steger, A. (eds.): Lectures on Proof Verification and Approximation Algorithms. Springer, Berlin (1998)
13. Papadimitriou, C.H.: An algorithm for shortest path motion in three dimensions. Inf. Process. Lett. **20**, 259–263 (1985)
14. Rabani, Y.: Approximation algorithms. http://www.cs.technion.ac.il/~rabani/236521.04.wi.html (2006). Accessed July 2011
15. Roberts, A.W., Varberg, V.D.: Convex Functions. Academic Press, New York (1973)
16. Rockafellar, R.T.: Convex Analysis. Princeton University Press, Princeton (1970)
17. Vazirani, V.V.: Approximation Algorithms. Springer, Berlin (2001)

Chapter 3
Rubberband Algorithms

> *There are two ways of constructing a software design: One way is to make it so simple that there are obviously no deficiencies, and the other way is to make it so complicated that there are no obvious deficiencies. The first method is far more difficult.*
> *Sir Charles Antony Richard Hoare (born 1934)*

This chapter introduces a class of algorithms, called *rubberband algorithms* (RBAs). They will be used frequently in the remainder of this book.

3.1 Pursuit Paths

Pursuit paths have attracted quite some attention in mathematics, for optimising tactics to catch, for example, a pursued animal.

Definition 3.1 A *pursuee* and a *pursuer* move in a plane. The pursuer describes a *pursuit path* by being always directed toward the pursuee.

A pursuit path is uniquely defined by the movement of the pursuee, its distance to the pursuer at the time when the pursuit starts, and the *chasing tactics* (e.g., uniform velocity of pursuee and pursuer). To be precise, 'being always directed toward the pursuee' means that the tangent of the pursuit path at time t points to the position of the pursuee at time t.

Example 3.1 A hare starts at the origin $o = (0, 0)$ of the Euclidean plane and runs along the y-axis. A pursuing dog starts chasing the hare at the same time at position $p = (k, 0)$. The chasing tactic is to maintain constant distance from the hare. The pursuit path of the dog is the *tractrix*

$$y = \pm k \log_e \left(\frac{k + \sqrt{k^2 - x^2}}{x} \right) - \sqrt{k^2 - x^2}.$$

F. Li, R. Klette, *Euclidean Shortest Paths*,
DOI 10.1007/978-1-4471-2256-2_3, © Springer-Verlag London Limited 2011

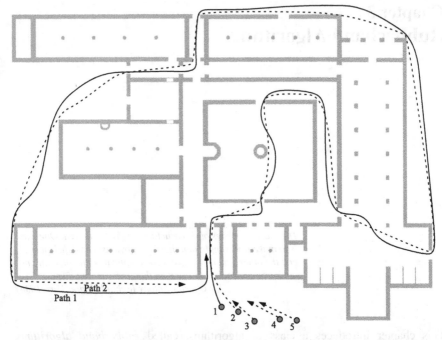

Fig. 3.1 The pursuit path of the second runner shortens the path of the followed first runner. Runners *3*, *4*, and *5* shorten their paths even more (architectural plan of the monastery in Bebenhausen near Tübingen)

That was the first occasion in which a differential equation was solved in terms of known functions.[1]   □

In extension of this classical problem, let a first runner be pursued by several other runners, where each runner chases the previous runner; see Fig. 3.1 for an example. The runners have to cross line segments that define *steps* (i.e., particular attractions in the sense of an ESP problem) identified by the path of the first runner. Figure 3.2 shows 18 steps, where some are degenerated into single points (i.e., corners or tangential points).

We know that the shortest path between two points is a straight segment. Thus, an ESP connecting those steps in sequence will be a polyline; see Fig. 3.2. Interestingly it can be shown that pursuit paths converge toward a polyline, assuming that all runners move with the same constant speed.

We assume that the first runner starts at time $\tau_1 = 0$ and selects a path γ_1 from p in one step S_j to point q in the next step S_{j+1} (e.g., between steps S_{16} and S_{17} in Fig. 3.2). The second runner starts at p at time $\tau_2 > \tau_1 = 0$ and selects a path γ_2

---

[1]*Florimond de Beaune* (1601–1652) mapped physical problems, such as this pursuit problem, into a mathematical description. His derivation of the tractrix was done about 1639.

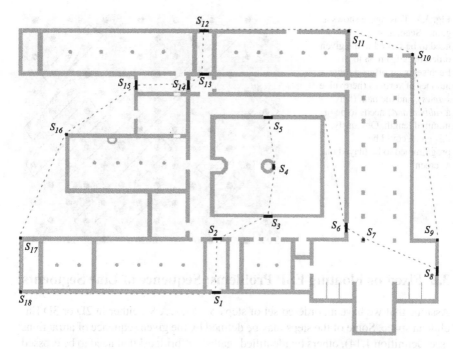

Fig. 3.2 A sequence of 18 steps $S_1, S_2, \ldots, S_{18}$ constrains a path. The shown path is a *polyline*, consisting of straight segments between subsequent steps. This path is not yet of minimum length; vertices of the polyline may still slide (a little) within steps for further optimisation

while chasing the first runner. The third runner starts at p at time $\tau_3 > \tau_2$, and so forth. For $i \geq 1$,

$$\gamma_i = \left\{ (x_i(\lambda), y_i(\lambda)) : \tau_i \leq \lambda \leq T_i \right\}$$

for some $T_i > 0$, with $(x_i(\tau_i), y_i(\tau_i)) = p$ and $(x_i(T_i), y_i(T_i)) = q$. It follows that the path γ_{i+1} of the $(i+1)$st runner is defined by the solution of the differential equation

$$\frac{d\gamma_{i+1}(\lambda)}{d\lambda} = \frac{\gamma_i(\lambda) - \gamma_{i+1}(\lambda)}{d_e(\gamma_i(\lambda), \gamma_{i+1}(\lambda))}$$

for $\lambda \geq \tau_{i+1}$ and $\gamma_i(\lambda) \neq \gamma_{i+1}(\lambda)$, with the initial condition that $\gamma_{i+1}(\tau_{i+1}) = p$.

If the chaser catches up with the chased runner then both paths are assumed to be identical until reaching point q. We provide the following theorem[2] without proof:

Theorem 3.1 *For $i = 1, 2, \ldots$, the sequence of pursuit paths γ_i, defined by an initial path γ_1 from p to q and start times τ_i, converges to the straight segment pq.*

Convergence is defined by considering the maximal Euclidean distance between points $\gamma_i(\lambda)$, for $\tau_{i+1} \leq \lambda \leq T_i$, and the straight segment pq.

---

[2]Published by *Alfred M. Bruckstein* in 1993; see [3].

Fig. 3.3 This figure shows a game: Segments S_1 to S_9 need to be visited in the given order, back from S_9 to S_1. Each segment offers a finite number of vertices (here: nine *screws*), and the path (here: a *rubberband*) needs to be of minimal length. Obviously, this game could be programmed to be played on a screen

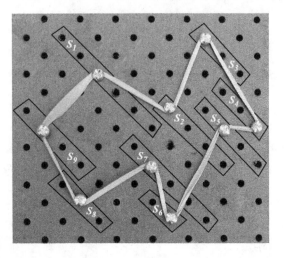

3.2 Fixed or Floating ESP Problems; Sequence of Line Segments

Assume that we have an ordered set of steps $S_1, S_2, \ldots, S_k$, either in 2D or 3D Euclidean space. Some of the steps may be defined by the given sequence of attractions (see Definition 1.14), others by identified 'gates' or 'bridges' that need to be crossed by any shortest, or approximately-shortest path.

Definition 3.2 We have two cases for defining an ESP that visits those steps in the given order:

Fixed ESP Problem: The ESP needs to run from p to S_1, then S_2, and so forth, ending at S_k, and finally point q, for given points p and q.

Floating ESP Problem: The ESP has to form a loop. It visits S_1, then S_2, and so forth, finally S_k, and it returns to the already defined vertex in S_1, without having given points p and q.

Figure 3.3 illustrates a floating ESP problem. For identifying an ESP, we may start at any step, select a vertex there, continue with selecting a vertex on the next step, and so forth. The figure illustrates a case where there is only a finite number of options for selecting a vertex on each step. This can be modelled by a weighted undirected graph, and the Dijkstra algorithm solves the problem of finding an ESP.

Let $S_1, S_2, \ldots,$ and S_k be k straight line segments, $k \geq 1$, in the Euclidean 3D space, defining an ordered set of *steps*. Any line segment is closed (i.e., it contains both of its endpoints) if not otherwise stated. We consider a fixed ESP problem. Let $S_0 = \{p\}$ and $S_{k+1} = \{q\}$ be two more steps; defined this way only for technical reasons.

Let L be the length of a shortest path, starting at p, then visiting segments $S_1, \ldots,$ and S_k in order, and finally ending at q. The task is to compute such a path of length L, or (at least) a path whose length is "very close" to L in the sense of Sect. 2.1.

Fig. 3.4 Points p and q, an ordered set of steps $[S_1, \ldots, S_6]$, and an initial path $\langle p, p_1, p_2, \ldots, p_6, q \rangle$, where points p_j have been selected as being the midpoints of those segments S_j, for $j = 1, 2, \ldots, 6$

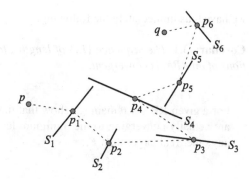

Theorem 3.2 *There is exactly one shortest path for the fixed ESP problem of k pairwise disjoint line segments.*

This uniqueness theorem[3] is cited without proof here, but will be also verified later in this chapter by our discussion of an approximate algorithm for calculating such a shortest path.

Figure 3.4 illustrates a possible initialisation (in 2D only) for solving this fixed ESP problem. A first set of points p_j is here initialised by selecting, for example, the midpoints of segments S_j, for $j = 1, \ldots, 6$. This can be then further processed by an iterative ESP algorithm; see the definition of the general control structure in Fig. 2.3, with the intention that the resulting algorithm is also approximate; see Definition 2.5.

3.3 Rubberband Algorithms

Assume that the dashed polyline in Fig. 3.4 is a rubberband that can freely move on the given steps. For example, point p_4 would slide into the endpoint on the right of S_4 if p_5 and p_3 are considered to be fixed. Actually, point p_3 will move to the left on S_3, thus limiting the possible space where p_4 can slide in on S_4.

A *rubberband algorithm* attempts to emulate the behaviour of such a flexible band on a given set of steps.

Definition 3.3 A *rubberband algorithm* (RBA) is an iterative ESP algorithm where a path through all steps (in the given order) is selected for initialisation; in each iteration $i \geq 1$, this path is transformed into a new path such that the length L_i of the path is reduced from length L_{i-1}.

Thus, the lengths L_i of the sequence of calculated paths are monotonously decreasing; they are also lower bounded by the minimum length L. With Corollary 2.3

---

[3]First published by *Micha Sharir* and *Amir Schorr* in 1986; see [15].

we have thus immediately the following:

Corollary 3.1 *The sequence* $\{L_i\}$ *of lengths, for* $i = 1, 2, \ldots$, *calculated in iterations of an RBA, is convergent.*

> For a given RBA it remains to show that the lengths calculated by this RBA are actually converging to the minimum length L, for any input, and not to any other real larger than L.

There are, obviously, many options for *initialisation*. For example, if steps are line segments, then we may always choose one of the end points of the segments, always the midpoint, or always a random point on each line segment. For a given RBA, the convergence behaviour needs to be studied with respect to the selected initialisation method for specifying the first path.

> The convergence behaviour of an RBA must not depend on the selected initialisation method.

In every iteration $i \geq 1$, the update of the calculated path is typically done by subsequent local optimisations, considering three steps S_{j-1}, S_j, and S_{j+1}, for $j = 1, \ldots, k$, and optimising locally the position of the vertex of the path on step S_j. An iteration $i \geq 1$ is finished when arriving at $j = k$.

A *local optimisation approach* using three steps is *only one option*: we could also consider any fixed number $2l + 1 \geq 3$ of subsequent line segments, $S_{j-l}, \ldots, S_j, \ldots, S_{j+l}$, and optimise positions of vertices on $2l - 1$ segments $S_{j-l+1}, \ldots, S_j, \ldots, S_{j+l-1}$.

However, the use of $l > 1$ will increase the complexity of local optimisation, not only with respect to implementation, but also (and this is actually more important for our discussion of RBAs) with respect to convergence proofs. Thus we decided for $l = 1$ (i.e., only optimising a vertex on one step S_j), and we may have potentially a few more iterations compared with $l > 1$.

> In this book, local optimisation of RBAs is for three subsequent steps.

In each iteration i, we calculate a path of length L_i. In the stop criterion, we may compare this length with the length of the path calculated in the previous iteration, and the standard way is by using an accuracy constant ε. If the difference between previous length and current length is smaller than or equal to ε then we stop and the result is a path of length $L_{\text{final}} = L_i$.

We illustrate the concept of rubberband algorithms by providing in the next section a first example of an RBA; there will be many more RBAs in this book, also

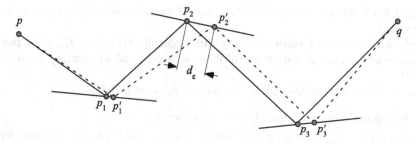

Fig. 3.5 Local upper error bound d_ε between a path (*full line*) and an optimal path (*dashed line*)

for steps which are not straight line segments. The actual challenge is not only to ensure the convergence of the L_is toward the minimum length L (say, for any kind of initialisation):

> We need to ensure that the resulting time complexity is "feasible" which means that the algorithm is of relevance for real-world applications.

For example, a κ-linear solution (see Sect. 1.3) is feasible in dependency of the practical behaviour of function $\kappa(\varepsilon)$ and the size of the asymptotic constant $c > 0$ of the linear component in $\mathcal{O}(n)$; a theoretical run-time analysis of an algorithm needs to be accompanied by experiments illustrating the actual run-time behaviour.

It is practically important to understand the relation between the chosen ε and the actual accuracy of the calculated path ρ_{final} with length L_{final}. See Fig. 2.4; the difference $L_{\text{final}} - L$ from the true minimum value L might be larger than ε.

Let $d_\varepsilon \geq 0$ be the upper error bound for distances between any vertex p_i in the calculated path and its corresponding optimal vertex p_i' in step S_i which means that $d_e(p_i, p_i') \leq d_\varepsilon$, for all $i = 0, 1, \ldots, k - 1$; see Fig. 3.5.

Definition 3.4 The upper bound d_ε characterises the *local accuracy* of an RBA, and the difference $L_{\text{final}} - L$ between the length L_{final} of a calculated path and the optimum length L of the ESP characterises the *global accuracy*.

3.4   A Rubberband Algorithm for Line Segments in 3D Space

We continue the discussion of the fixed line-segment ESP problem in 3D space. The input is defined by a sequence of $k > 0$ pairwise disjoint line segments $S_1, S_2, \ldots, S_k$ in 3D space, and two points p and q that are not on any of the k segments. Let $S_0 = \{p\}$ and $S_{k+1} = \{q\}$. We also use an accuracy constant $\varepsilon > 0$.

For a pseudocode of the algorithm, see Fig. 3.6. We specify the δ-value of the output further below. The given RBA is an example of an iterative ESP algorithm.

Figure 3.7 illustrates all steps for the first iteration. The initial configuration is shown at the top, left. The first iteration starts at the top, right and it ends at the

Algorithm 7 (RBA for the fixed ESP problem of pairwise disjoint line segments in 3D space)

Input: A sequence of k pairwise disjoint line segments $S_1, S_2, \ldots, S_k$ in 3D; two points p and q that are not on any of those segments, and an accuracy constant $\varepsilon > 0$.

Output: A δ-approximation path $\langle p, p_1, p_2, \ldots, p_k, q \rangle$ of an ESP.

1: For each $j \in \{0, 1, \ldots, k+1\}$, let p_j be a point S_j.

2: $L_{\text{current}} \leftarrow \sum_{j=0}^{k} d_e(p_j, p_{j+1})$, where $p_0 = p$ and $p_{k+1} = q$; and let $L_{\text{previous}} \leftarrow \infty$.

3: **while** $L_{\text{previous}} - L_{\text{current}} \geq \varepsilon$ **do**

4:   **for** each $j \in \{1, 2, \ldots, k\}$ **do**

5:     Compute a point $q_j \in S_j$ such that $d_e(p_{j-1}, q_j) + d_e(q_j, p_{j+1}) = \min\{d_e(p_{j-1}, p) + d_e(p, p_{j+1}) : p \in S_j\}$

6:     Update the path $\langle p, p_1, p_2, \ldots, p_k, q \rangle$ by replacing p_j by q_j.

7:   **end for**

8:   Let $L_{\text{previous}} \leftarrow L_{\text{current}}$ and $L_{\text{current}} \leftarrow \sum_{j=0}^{k} d_e(p_j, p_{j+1})$.

9: **end while**

10: Return $\{p, p_1, p_2, \ldots, p_k, q\}$.

Fig. 3.6 RBA for solving the fixed ESP problem for a sequence of pairwise disjoint line segments in 3D space. Note that the initialisation remains unspecified; we show below that the result is "not much" influenced by the chosen initial path

bottom, left. The first step of the second iteration (bottom, right of Fig. 3.7) defines already the shortest path (i.e., not only an approximate, but an accurate solution). Points $p_2, \ldots, p_6$ cannot move into any 'better' position anymore. At the end of the second iteration, we would have that $L_{\text{current}} = L_{\text{previous}}$, saying that this is already the optimum. This means that we could replace a positive ε, such as, for example, $\varepsilon = 10^{-15}$ even (theoretically; ignoring numerical inaccuracies of a given computer) by $\varepsilon = 0$ for this input example. Value $\varepsilon = 0$ or a case of $L_{\text{current}} = L_{\text{previous}}$ defines an *accurate result* rather then just an *approximate result*.

However, such a (quick) termination with an accurate result is an exception in the general set of possible input configurations. In Sect. 3.9, we even provide an example for an ESP problem in 3D space where no RBA can ever stop (after any finite number of iterations) with an exact solution.

3.5 Asymptotic and Experimental Time Complexity

We provide a theoretical analysis of the time complexity of Algorithm 7. This was prepared at a general level by Eq. (2.2), for any iterative ESP algorithm, listing total time $f(\mathbf{x})$, time $f_{\text{initialisation}}(\mathbf{x})$ for initialisation, the number $i_{\max}(\mathbf{x})$ of iterations, and the time $f_{\text{update}}(\mathbf{x})$ for updates. We have $\mathbf{x} = (k, \varepsilon)$, combining the input complexity k (i.e., here the number of line segments) and the free parameter ε.

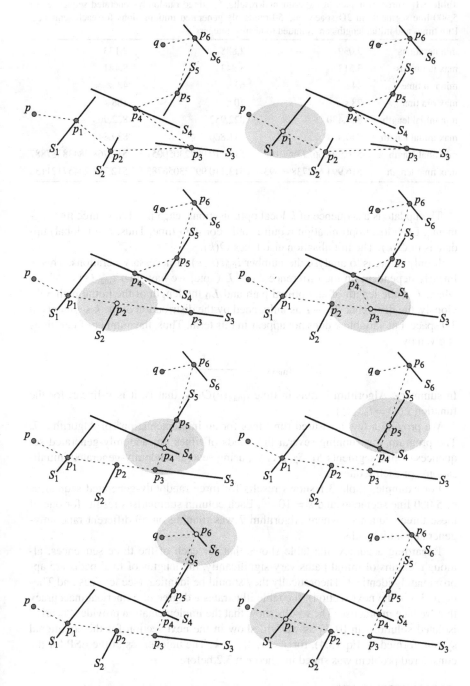

Fig. 3.7 2D illustration of the provided RBA for line segments in 3D space

Table 3.1 Three examples of experimental results, for three randomly-generated sequences of 5,000 line segments in 2D space, and 50 randomly-generated initialisations for each sequence. Run times and initial lengths are rounded to nearest integers

min iterations	2,039	2,888	2,133
max iterations	3,513	3,243	8,441
min run time	44 s	63 s	48 s
max run time	78 s	70 s	188 s
min initial length	827,430	822,952	822,905
max initial length	846,928	841,860	839,848
min final length	516,994.66273890162	513,110.99723050051	512,768.28438387887
max final length	516,994.66273896693	513,110.99723056785	512,768.28457121132

The update is a sequence of k local optimisations, each involving three line segments. Each local optimisation requires only constant time. Thus, one (global) update is in $\mathcal{O}(k)$. The initialisation also takes $\mathcal{O}(k)$.

It only remains to analyse the number $i_{\max}(k, \varepsilon)$ of necessary iterations. This is linearly dependent on the difference $L_0 - L$ (note: we have that $i_{\max}(\varepsilon) = \frac{L_0 - L}{\varepsilon}$) where L is the length of an optimal path and L_0 the length of the initial path. Obviously, the difference $L_0 - L$ is influenced by the distribution of the k segments in 3D space, but variable k does not appear in this term. Thus, the number of iterations is given by

$$i_{\max}(\varepsilon) = \frac{L_0 - L}{\varepsilon}.$$

In summary, Algorithm 7 runs in time $i_{\max}(\varepsilon)\mathcal{O}(k)$, that is, it is κ-linear, for the function $\kappa(\varepsilon) = i_{\max}(\varepsilon)$.

We provide a few measured run times for an implementation[4] of Algorithm 7. The program was running several thousands of times on randomly-generated sequences of line segments $S_1, S_2, \ldots, S_k$, using several randomly-generated initialisations for each sequence.

For example, Table 3.1 shows results for three randomly-generated sequences of 5,000 line segments and $\varepsilon = 10^{-15}$. Each column summarises results for one of these three sequences where Algorithm 7 was running on 50 different randomly-generated initial paths.

Regarding accuracy, the table shows that for each of the three sequences, although lengths of initial paths vary significantly, the lengths of final paths are approximately identical. Theoretically they should be identical (see lemmas and Theorem 3.3 in the next section); the table illustrates a degree of numerical inaccuracy that we face in practise. The results show that the implementation provides "nearly" isolated solutions in $[0, 1]$ as we will show in the next section for the equational system formed by Eq. (3.4), for $j = 1, 2, \ldots, k$. The uniqueness of the ESP for the considered problem was stated in Theorem 3.2 before.

---

[4]The source code can be downloaded at www.mi.auckland.ac.nz; follow the link at the 2009 MItech Report 51.

Fig. 3.8 Three subsequent
line segments of the given
sequence of steps

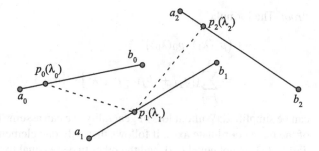

Regarding time complexity, the table illustrates that we can expect about $k/2$ iterations for k line segments and $\varepsilon = 10^{-15}$. On 2010 computer technology, this was about one minute for 5,000 line segments, where the implementation used was straightforward, not aiming at any run-time optimisation.

3.6 Proof of Correctness

We have to show that the repeated local optimisation of Algorithm 7 ensures that the calculated path converges as $\varepsilon \to 0$ to a shortest path, independent of the chosen initialisation.

We provide a few auxiliary results first. Let S_0, S_1, and S_2 be three pairwise disjoint line segments of local optimisation, the two endpoints of S_i being $a_i = (a_{1i}, a_{2i}, a_{3i})$ and $b_i = (b_{1i}, b_{2i}, b_{3i})$. Points $p_i \in S_i$, for $i = 0, 1, 2$, are defined by individual values λ_i as follows:

$$p_i(\lambda_i) = a_i + (b_i - a_i)\lambda_i$$
$$= \big(a_{1i} + (b_{1i} - a_{1i})\lambda_i, a_{2i} + (b_{2i} - a_{2i})\lambda_i, a_{3i} + (b_{3i} - a_{3i})\lambda_i\big)$$

with $0 \le \lambda_i \le 1$. Let

$$d(\lambda_0, \lambda_1, \lambda_2) = d_e\big(p_1(\lambda_1), p_0(\lambda_0)\big) + d_e\big(p_1(\lambda_1), p_2(\lambda_2)\big);$$

see Fig. 3.8. The equation

$$\frac{\partial d(\lambda_0, \lambda_1, \lambda_2)}{\partial \lambda_1} = 0 \tag{3.1}$$

identifies a minimum of $d(\lambda_0, \lambda_1, \lambda_2)$, and there is actually only one minimum. More about this later; at first we consider consequences of Eq. (3.1).

Definition 3.5 If an expression is derived from a finite number of polynomials in x by only applying operations "+", "−", "×", "÷", or "$\sqrt{\ }$" finitely often then we say that this expression is a *simple compound of polynomials* in x.

Lemma 3.1 *Equation* (3.1) *implies that λ_2 is a simple compound of polynomials of λ_0 and λ_1.*

Proof The formula

$$d_e\big(p_1(\lambda_1),\, p_0(\lambda_0)\big)^2$$

$$= \sum_{j=1}^{3}\big([a_{j1} + (b_{j1} - a_{j1})\lambda_1] - [a_{j0} + (b_{j0} - a_{j0})\lambda_0]\big)^2 \qquad (3.2)$$

can be simplified: Without loss of generality, we can assume that S_1 is parallel to one of the three coordinate axes. It follows that only one element of the set $\{b_{j1} - a_{j1} : j = 1, 2, 3\}$ is not equal to 0, and the other two are equal to 0. Thus, we can assume that the expression on the right of Eq. (3.2) can be written in the form

$$\big\{[a_{11} + (b_{11} - a_{11})\lambda_1] - [a_{10} + (b_{10} - a_{10})\lambda_0]\big\}^2$$
$$+ \big\{a_{21} - [a_{20} + (b_{20} - a_{20})\lambda_0]\big\}^2$$
$$+ \big\{a_{31} - [a_{30} + (b_{30} - a_{30})\lambda_0]\big\}^2.$$

Thus, we have that

$$d_e(p_1, p_0) = |A_1|\sqrt{(\lambda_1 + B_0\lambda_0 + C_0)^2 + D_0\lambda_0^2 + E_0\lambda_0 + F_0}$$

where A_1 is a function of a_{j1} and b_{j1}; B_0, C_0, D_0, E_0 and F_0 are functions of a_{j0}, b_{j0}, a_{j1} and b_{j1}, all for $j = 0, 1, 2$. Analogously, we have that

$$d_e(p_1, p_2) = |A_1|\sqrt{(\lambda_1 + B_2\lambda_2 + C_2)^2 + D_2\lambda_2^2 + E_2\lambda_2 + F_2}$$

with the same A_1, and B_2, C_2, D_2, E_2, and F_2 are now functions of a_{j1}, b_{j1}, a_{j2} and b_{j2} for $j = 0, 1, 2$. Because of the assumed Eq. (3.1), or equivalently,

$$\frac{\partial(d_e(p_1, p_0) + d_e(p_1, p_2))}{\partial\lambda_1} = 0,$$

we have that

$$\frac{\lambda_1 + B_0\lambda_0 + C_0}{\sqrt{(\lambda_1 + B_0\lambda_0 + C_0)^2 + D_0\lambda_0^2 + E_0\lambda_0 + F_0}}$$
$$+ \frac{\lambda_1 + B_2\lambda_2 + C_2}{\sqrt{(\lambda_1 + B_2\lambda_2 + C_2)^2 + D_2\lambda_2^2 + E_2\lambda_2 + F_2}} = 0.$$

This equation can be written in the form

$$A\lambda_2^2 + B\lambda_2 + C = 0$$

where A, B, and C are polynomials in λ_0, λ_1, and in $a_{j0}, b_{j0}, a_{j1}, b_{j1}, a_{j2}$ and b_{j2} for $j = 0, 1, 2$.   □

All the λ_0, λ_1 and λ_2 are in $[0, 1]$. For keeping λ_2 inside of the interval $[0, 1]$ (see Fig. 3.9), let $\lambda_2 = 0$ if $\lambda_2 < 0$ when solving $A\lambda_2^2 + B\lambda_2 + C = 0$, and $\lambda_2 = 1$ if $\lambda_2 > 1$; this defines the endpoints of segment S_2.

Fig. 3.9 The solution for λ defines a point (here point p_2') on the *straight line* that is incident with this step, but this point may be outside of the step. Then we take the nearest endpoint of the step

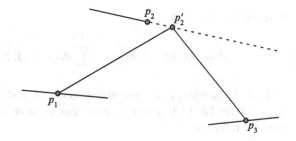

Lemma 3.2 *Equation* (3.1) *implies that* λ_1 *is a continuous function of* λ_0 *and* λ_2.

Proof We may translate two points $p_0(\lambda_0)$ and $p_2(\lambda_2)$, and line segment S_1 such that the endpoint a_1 of S_1 is identical with the origin. Then rotate $p_0(\lambda_0)$, $p_2(\lambda_2)$, and S_1 so that the other endpoint b_1 of S_1 is (also) on the x-axis. Let $p_0(\lambda_0) = (p_{10}, p_{20}, p_{30})$, $p_2(\lambda_2) = (p_{12}, p_{22}, p_{32})$. After translation and rotation, we have that $a_1 = (0, 0, 0)$ and $b_1 = (b_{11}, 0, 0)$. Thus, $p_1(\lambda_1) = (b_{11}\lambda_1, 0, 0)$, and

$$d_e(p_1, p_0) = \sqrt{(b_{11}\lambda_1 - p_{10})^2 + p_{20}^2 + p_{30}^2},$$

$$d_e(p_1, p_2) = \sqrt{(b_{11}\lambda_1 - p_{12})^2 + p_{22}^2 + p_{32}^2}.$$

Equation (3.1) is short for

$$\frac{\partial(d_e(p_1, p_0) + d_e(p_1, p_2))}{\partial \lambda_1} = 0.$$

From this we obtain that

$$\frac{b_{11}\lambda_1 - p_{10}}{\sqrt{(b_{11}\lambda_1 - p_{10})^2 + p_{20}^2 + p_{30}^2}} + \frac{b_{11}\lambda_1 - p_{12}}{\sqrt{(b_{11}\lambda_1 - p_{12})^2 + p_{22}^2 + p_{32}^2}} = 0.$$

This equation has a unique solution

$$\lambda_1 = \frac{p_{10}\sqrt{p_{22}^2 + p_{32}^2} + p_{12}\sqrt{p_{20}^2 + p_{30}^2}}{b_{11}(\sqrt{p_{22}^2 + p_{32}^2} + \sqrt{p_{20}^2 + p_{30}^2})}$$

where p_0 and p_2 are defined by λ_0 and λ_2, respectively. $\square$

Again, to keep λ_1 inside of [0, 1], let $\lambda_1 = 0$ if we have to satisfy $\lambda_1 < 0$; and let $\lambda_1 = 1$ if we have to satisfy $\lambda_1 > 1$.

We consider the sequence $S_0, S_1, \ldots,$ and S_{k+1} of the given $k + 2$ line segments. All segments contain their endpoints $a_j = (a_{1j}, a_{2j}, a_{3j})$ and $b_j = (b_{1j}, b_{2j}, b_{3j})$, for $j = 0, 1, \ldots, k+1$. Points $p_j \in S_j$, for $j = 0, 1, 2, \ldots, k+1$, can be written as follows:

$$p_j(\lambda_j) = a_j + (b_j - a_j)\lambda_j$$
$$= (a_{1j} + (b_{1j} - a_{1j})\lambda_j, a_{2j} + (b_{2j} - a_{2j})\lambda_j, a_{3j} + (b_{3j} - a_{3j})\lambda_j)$$

where $0 \leq \lambda_j \leq 1$. Let

$$d(\lambda_0, \lambda_1, \lambda_2, \ldots, \lambda_{k+1}) = \sum_{j=0}^{k} d_e\big(p_j(\lambda_j), p_{j+1}(\lambda_{j+1})\big). \tag{3.3}$$

Both S_0 and S_{k+1} are just single points p and q, respectively, with $\lambda_0 = \lambda_{k+1} = 0$. In Eq. (3.3), every λ_j only appears in two subsequent terms. Thus we have the following

Lemma 3.3 *For each* $j \in \{1, 2, \ldots, k\}$,

$$\frac{\partial d(\lambda_0, \lambda_1, \lambda_2, \ldots, \lambda_k, \lambda_{k+1})}{\partial \lambda_j} = 0 \tag{3.4}$$

is equivalent to

$$\frac{\partial d(\lambda_{j-1}, \lambda_j, \lambda_{j+1})}{\partial \lambda_j} = 0 \tag{3.5}$$

where $\lambda_1, \lambda_2, \ldots, \lambda_k$ *are in* $[0, 1]$.

Equation (3.4) is now related to a global minimum property of the Euclidean path $\langle p, p_1, p_2, \ldots, p_k, q \rangle$ while Eq. (3.5) is related to a local minimum property of the same path. Therefore, Lemma 3.3 describes a relationship between global and local minimum properties of the same path.

We introduce the notions of "isolated" or "interval solutions" for the next lemma. Let $f(x_0) = 0$, and let $f(x) \neq 0$ for a sufficiently small number $\varepsilon > 0$ and all $x \neq x_0$ in the ε-neighbourhood $N_\varepsilon(x_0)$; then we say that x_0 is an *isolated root* or *isolated solution* of $f(x)$. If $I \subset \mathbb{R}$ is a bounded interval, and for all x in I, $f(x) = 0$, then we say that I is an *interval solution* of f.

We generalise those two definitions for the multivariate case: Let f be a function from $\mathbb{R}^m$ into $\mathbb{R}$, for $m \geq 1$. Assume that $f(p_0) = 0$, and for a sufficiently small number $\varepsilon > 0$ and all $p \neq p_0$ in the ε-neighbourhood $N_\varepsilon(p_0)$, let $f(p) \neq 0$; then we say that p_0 is an *isolated root* or *isolated solution* of f.

Let $I_k \subset \mathbb{R}$ be a bounded interval of non-zero length; for $i = 1, 2, \ldots, k - 1$, $k + 1, \ldots, m$, the bounded interval $I_i \subset \mathbb{R}$ is possibly just a single point. $(I_1, I_2, \ldots, I_m)$ is an *interval solution* of f if for each $x_k \in I_k$ there exist values $x_i \in I_i$, for $i = 1, 2, \ldots, k - 1, k + 1, \ldots, m$, such that $f(x_1, x_2, \ldots, x_m) = 0$. Note that the important point is that there is at least one interval involved of non-zero length.

Consider m functions $f_1, \ldots, f_m$, for $m \geq 1$. We say that p_0 is an *isolated solution for those m functions* if p_0 is an isolated solution for any of those m functions, and $(I_1, I_2, \ldots, I_m)$ is an *interval solution* to those m functions if $(I_1, I_2, \ldots, I_m)$ is an interval solution for any of those m functions.

Lemma 3.4 *The equational system* (3.4), *with* $j = 1, 2, \ldots, k$, *implies a unary equation* $f(\lambda_1) = 0$ *which has only a finite number of isolated or interval solutions in* $[0, 1]$.

Proof By Lemmas 3.3 and 3.1, λ_{j+1} is a simple compound of polynomials in λ_{j-1} and λ_j, denoted by $\lambda_{j+1} = f_j(\lambda_{j-1}, \lambda_j)$. Thus, the system formed by Eq. (3.4) (where $j = 1, 2, \ldots, k$) implies an equational system formed by $\lambda_2 = f_2(\lambda_0, \lambda_1)$, $\lambda_3 = f_3(\lambda_1, \lambda_2)$, $\lambda_4 = f_4(\lambda_2, \lambda_3), \ldots, \lambda_k = f_k(\lambda_{k-2}, \lambda_{k-1})$, and $\lambda_{k+1} = f_{k+1}(\lambda_{k-1}, \lambda_k)$.

Note that $\lambda_0 = \lambda_{k+1} = 0$. Therefore, $f(\lambda_1)$ is a simple compound of polynomials in λ_1. Function $f(\lambda_1)$ has only a finite number of monotonous intervals in $[0, 1]$, and $f(\lambda_1)$ is differentiable in each of those monotonous intervals.[5] Thus, $f(\lambda_1)$ can be approximately expressed as a linear function in a finite number of monotonous subintervals in $[0, 1]$.

Therefore, function $f(\lambda_1)$ has only a finite number of isolated or interval solutions in $[0, 1]$. $\square$

Let $S_1, S_2, \ldots,$ and S_k be k non-empty subsets of $\mathbb{R}^3$; the term

$$\prod_{j=1}^{k} S_j = S_1 \times S_2 \times \cdots \times S_k$$

denotes the *cross product*. This cross product is a subset of $\mathbb{R}^{3k}$.

Algorithm 7 runs for a limited time, influenced by the chosen accuracy parameter ε. It maps an initial path $\langle p, p_1, p_2, \ldots, p_k, q \rangle$ into a final path with a length L_{final}, assuming i_{max} iterations until termination. In this sense, Algorithm 7 defines a unique mapping from $\prod_{j=1}^{k} S_j$ into $\mathbb{R}$.

The initial path can also be characterised by λ-values, defining the positions of those points p_j on segments S_j. Let

$$\prod_{j=1}^{k} [0, 1] = [0, 1] \times [0, 1] \times \cdots \times [0, 1]$$

be the k-dimensional unit cube $[0, 1]^k$. Algorithm 7 defines thus also a unique mapping from $[0, 1]^k$ into $\mathbb{R}$.

Lemma 3.5 *Function* $f_{\text{RBA}}(\lambda_1, \lambda_2, \ldots, \lambda_k)$ *is continuous.*

Proof This follows from Lemma 3.2, and because Algorithm 7 terminates after a finite number of steps. $\square$

Lemma 3.6 *Function* $f_{\text{RBA}}(\lambda_1, \lambda_2, \ldots, \lambda_k)$ *maps all* $[0, 1]^k$ *into a finite set.*

Proof It is sufficient to prove that for each interval solution J of the equational system formed by Eq. (3.4), with $j = 1, 2, \ldots, k$, the function

$$f_{\text{RBA}}(\lambda_1, \lambda_2, \ldots, \lambda_k) : J \to \mathbb{R}$$

---

[5]Let f be a function, mapping $\mathbb{R}$ into $\mathbb{R}$. Suppose interval $J \subseteq I$ is a subinterval of interval I, and f is monotonous in J; then we say that J is a *monotonous interval* of f in the larger interval I.

has only a finite number of values.

Suppose that $f(\lambda_1) = 0$, for all λ_1 in an interval $I \subseteq [0, 1]$, and $f(\lambda_1)$ is defined as in Lemma 3.4. By Lemma 3.4, $d(\lambda_0, \lambda_1, \lambda_2, \ldots, \lambda_{k+1})$ implies a unary length function $L(\lambda_1)$, where λ_1 is where in an interval $I' \subseteq I$, $d(\lambda_0, \lambda_1, \lambda_2, \ldots, \lambda_{k+1})$ is defined as in Eq. (3.3), and

$$\frac{d[L(\lambda_1)]}{d\lambda_1} = 0$$

for all $\lambda_1 \in I' \subseteq I$. This implies that the length function $L(\lambda_1)$ is constant, for all $\lambda_1 \in I' \subseteq I$. Thus, the function $f_{\text{RBA}}(\lambda_1, \lambda_2, \ldots, \lambda_k) : J \to \mathbb{R}$ has only a finite number of values.   □

Theorem 3.3 *If the chosen accuracy constant ε is sufficiently small, then, for any initial path, Algorithm 7 outputs a unique $[1 + 2(k + 1) \cdot d_\varepsilon / L]$-approximation path, where k is the number of steps, L the optimum length, and d_ε the local accuracy bound.*

Proof Following Lemma 3.5, Algorithm 7 defines a function $f_{\text{RBA}}(p_1, p_2, \ldots, p_k)$ on $\prod_{j=1}^{k} S_j$ which is continuous and, following Lemma 3.6, maps into a finite number of positive real numbers (i.e., the lengths of calculated paths), for any initial sequence of points $p_1, p_2, \ldots, p_k$ sampled in $\prod_{j=1}^{k} S_j$. Therefore, the range of $f_{\text{RBA}}(p_1, p_2, \ldots, p_k)$ must be a singleton.

For each $j \in \{0, 1, \ldots, k\}$, the error of the difference between $d_e(p_j, p_{j+1})$ values of calculated vertices and $d_e(p'_j, p'_{j+1})$ values of optimal vertices is at most $2 \cdot d_\varepsilon$ because of $d_e(p_i, p'_i) \le d_\varepsilon$. We know that $p = p_0 = p'_0$ and $q = p_{k+1} = p'_{k+1}$. We obtain that

$$L \le \sum_{j=0}^{k} d_e(p_j, p_{j+1}) \le \sum_{j=0}^{k} \left[d_e(p'_j, p'_{j+1}) + 2d_\varepsilon \right] = L + 2(k+1)d_\varepsilon. \qquad (3.6)$$

Thus, the output path is an $[1 + 2(k + 1) \cdot d_\varepsilon / L]$-approximation path.   □

The local upper accuracy bound d_ε might be specified by applying calculations as summarised in the proofs of the lemmata prior to this theorem. It is sufficient to state that

$$\lim_{\varepsilon \to 0} d_\varepsilon = 0 \qquad (3.7)$$

at a theoretical level.

Algorithm 7 can calculate paths of any local or global accuracy, only depending on the chosen accuracy constant $\varepsilon > 0$.

This is the goal for any RBA that is controlled by an accuracy constant $\varepsilon > 0$.

Fig. 3.10 Illustration of two intersecting line segments

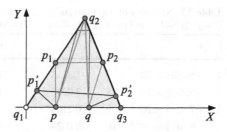

Table 3.2 Number i of iterations and resulting δs, for Example 3.2, illustrated by Fig. 3.10, with $p_1 = (1, 2)$ and $p_2 = (2.5, 2)$ as initialisation points

i	δ
1	-0.8900
2	-0.1752
3	-0.0019
4	-1.293×10^{-5}
5	-8.443×10^{-8}
6	-5.493×10^{-10}
7	-3.574×10^{-12}

3.7 Processing Non-disjoint Line Segments as Inputs

So far we assumed that input line segments are pairwise disjoint. This ensures that two subsequent vertices of the generated path are always different to each other. In this section, we discuss the consequences for non-disjoint segments. In this case, two subsequent vertices might be identical (at the intersection of both line segments). It is difficult to exclude intersections in practise. In this section, we modify Algorithm 7 for being able to handle non-disjoint line segments as well.

Before detailing the modifications of this algorithm compared to Algorithm 7, we discuss three examples of paths on non-disjoint line segments, and how they would be processed by Algorithm 7 as given above.

Example 3.2 Let the input for Algorithm 7 be as follows (see also Fig. 3.10 for a sketch of this input):

$$S_1 = q_1 q_2, \qquad S_2 = q_2 q_3, \qquad q_1 = (0, 0), \qquad q_2 = (2, 4),$$
$$q_3 = (3, 0), \qquad p = (1, 0), \quad \text{and} \quad q = (2, 0).$$

To initialise, let p_1 and p_2 be the centres of S_1 and S_2, respectively [i.e., $p_1 = (1, 2)$, and $p_2 = (2.5, 2)$]. We obtain that the length of the initialised polyline $\rho = \langle p, p_1, p_2, q \rangle$ is equal to 5.5616. Algorithm 7 finds the shortest path $\rho = \langle p, p_1', p_2', q \rangle$ where $p_1' = (0.3646, 0.7291)$, $p_2' = (2.8636, 0.5455)$ and the length of it is equal to 4.4944; see Table 3.2, which lists resulting δs (here $\delta = L_0 - L_1$ in Step 3 of Algorithm 8) for the number i of iterations. □

Table 3.3 Number i of iterations and resulting δs, for the example shown in Fig. 3.10, with $p_1 = (2-\delta', 2(2-\delta'))$ and $p_2 = (2+\delta', -4((2+\delta')-3))$ as initialisation points and $\varepsilon_s = 2.221 \times 10^{-16}$

i	δ	i	δ	i	δ	i	δ
1	-5.4831×10^{-7}	7	-1.2313	13	-7.0319×10^{-10}	19	8.8818×10^{-16}
2	-6.2779×10^{-6}	8	-2.0286	14	-4.5732×10^{-12}	20	8.8818×10^{-16}
3	-7.7817×10^{-5}	9	-0.2104	15	-3.0198×10^{-14}	21	-8.8818×10^{-16}
4	-9.6471×10^{-4}	10	-0.0024	16	-8.8818×10^{-16}	22	8.8818×10^{-16}
5	-0.0119	11	-1.6550×10^{-5}	17	8.8818×10^{-16}	23	-8.8818×10^{-16}
6	-0.1430	12	-1.0809×10^{-7}	18	-8.8818×10^{-16}	24	0

Example 3.3 The output of Algorithm 7 is false if we modify the initialisation in Example 3.2 so that $p_1 = p_2 = q_2$. In this case, the calculated path equals $\rho = \langle p, p'_1, p'_2, q \rangle$, where $p'_1 = q_2$ and $p'_2 = q_2$, and its length equals 8.1231. □

Definition 3.6 An updated polygonal path with at least two subsequent identical vertices is called *degenerate*, or a *degenerate case*.

Algorithm 7 fails in general for degenerate cases.

Our solution for a degenerate case is as follows: we do not allow that a case $p_j = p_{j+1}$ happens by removing sufficiently small line segments from the intersecting ends of segments S_j and S_{j+1}. The following example demonstrates this strategy. When moving vertices apart by some small *slide distance* $\varepsilon_s > 0$, we need to acknowledge some numerical dependency between ε_m and the chosen accuracy constant ε, which will depend in general on the chosen implementation.

Example 3.4 We modify the initialisation step of Example 3.3. Assume that we selected a very high accuracy, such as

$$\varepsilon = 10^{-100}.$$

Now let

$$\delta' = 2.221 \times 10^{-16},$$
$$x_1 = 2 - \delta' \quad \text{and} \quad y_1 = 2 \times x_1,$$
$$x_2 = 2 + \delta' \quad \text{and} \quad y_2 = -4 \times (x_2 - 3),$$
$$p_1 = (x_1, y_1) \quad \text{and} \quad p_2 = (x_2, y_2).$$

The length of the initialised polyline $\rho = \langle p, p_1, p_2, q \rangle$ is equal to 8.1231. Algorithm 7 calculates the shortest path $\rho = \langle p, p'_1, p'_2, q \rangle$, where $p'_1 = (0.3646, 0.7291)$ and $p'_2 = (2.8636, 0.5455)$, and its length equals 4.4944, as in Example 3.2. See Table 3.3 for resulting δs in dependency of the number i of iterations.

Of course, if we select a larger accuracy constant, for example, $\varepsilon = 10^{-10}$, then the algorithm will stop sooner, after fewer iterations.

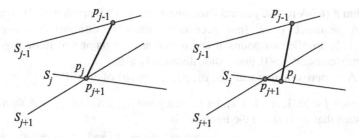

Fig. 3.11 Handling a degenerate case of identical vertices $p_j = p_{j+1}$, as shown on the left. Point p_j slides on S_j 'a little' away from its former position

We discuss a particular implementation on a Pentium 4 PC using Matlab 7.04. When we changed the value of δ' from 2.221×10^{-16} into a slightly smaller slide distance

$$\varepsilon_s = 2.22 \times 10^{-16}$$

then we obtained the same false result as that of Example 3.3. This is because this particular implementation was not able to recognise a difference between x_1 and $x_1 \mp 2.22 \times 10^{-16}$. The value

$$\delta' = 2.221 \times 10^{-16}$$

is large enough for *this* particular implementation. □

We are now ready to formally specify our method for handling degenerate cases. Let S_{j-1}, S_j and S_{j+1} be three subsequent segments in the input sequence of steps such that $S_j \cap S_{j+1} \neq \emptyset$. Assume that p_{j-1}, p_j and p_{j+1} are three subsequent vertices of an updated polygonal path (in some iteration i) such that p_j and p_{j+1} are identical; see left of Fig. 3.11.

Let ε_s be a sufficiently small positive number for moves on steps. There are at most two possible points p in S_j such that $d_e(p, p_{j+1}) = \varepsilon_s$. Select that point p for which $d_e(p, p_{j-1}) + d_e(p, p_{j+1})$ is the smaller value, and update the polygonal path by letting $p_j = p$; see right of Fig. 3.11.

We say that p_j is ε_s-*transformed* into $p \in S_j$. Analogously to the explanation of Eq. (3.6), the total error of this ε_s-transform equals $2m \cdot \varepsilon_s$ when handling $m \leq k$ degenerate cases.

We provide a pseudocode of the modified Algorithm 7 (see Algorithm 8) in Fig. 3.12. By taking the additional error into account, possibly added by a maximum of k ε_s-transforms, and following the proof of Theorem 3.3, we obtain:

Corollary 3.2 *The output of Algorithm 8 is a*

$$\left(1 + 2\big[(k+1)d_\varepsilon + k\varepsilon_s\big]/L\right)\text{-approximation path}$$

$\langle p, p_1, p_2, \ldots, p_k, q \rangle$ *which starts at p, then visits segments S_j at p_j in the given order, and finally ends at q, where L and d_ε are defined as for Algorithm 7, and ε_s is a chosen slide distance.*

Algorithm 8 (RBA for the general fixed line-segment ESP problem in 3D space)

Input: A sequence of k (not necessarily pairwise disjoint) line segments $S_1, S_2, \ldots, S_k$ in 3D; two points p and q that are not on any of those segments, an accuracy constant $\varepsilon > 0$, and a slide distance $\varepsilon_s > 0$.

Output: A δ-approximation path $\langle p, p_1, p_2, \ldots, p_k, q \rangle$ of an ESP.

1: For each $j \in \{0, 1, \ldots, k+1\}$, let p_j be a point S_j; if $S_j \cap S_{j \mp 1} \neq \emptyset$, then select p_j such that p_j is not in the intersection.

2: $L_{\text{current}} \leftarrow \sum_{j=0}^{k} d_e(p_j, p_{j+1})$, where $p_0 = p$ and $p_{k+1} = q$; and let $L_{\text{previous}} \leftarrow \infty$.

3: **while** $L_{\text{previous}} - L_{\text{current}} \geq \varepsilon$ **do**

4:      **for** each $j \in \{1, 2, \ldots, k\}$ **do**

5:          Compute a point $q_j \in S_j$ such that $d_e(p_{j-1}, q_j) + d_e(q_j, p_{j+1}) = \min\{d_e(p_{j-1}, p) + d_e(p, p_{j+1}) : p \in S_j\}$

6:          **if** $S_j \cap S_{j \mp 1} \neq \emptyset$ and q_j is in the intersection **then**

7:              ε_s-transform q_j to be another point (still denoted by q_j) in S_j.

8:          **end if**

9:          Update the path $\langle p, p_1, p_2, \ldots, p_k, q \rangle$ by replacing p_j by q_j.

10:     **end for**

11:     Let $L_{\text{previous}} \leftarrow L_{\text{current}}$ and $L_{\text{current}} \leftarrow \sum_{j=0}^{k} d_e(p_j, p_{j+1})$.

12: **end while**

13: Return $\{p, p_1, p_2, \ldots, p_k, q\}$.

Fig. 3.12 RBA for solving the general fixed line-segment ESP problem in 3D space. The initialisation includes now also that subsequent vertices need to be different from each other, and lines 6 and 7 deal with possible degenerate cases

For guaranteeing convergence toward the length of an ESP, the constant ε_s needs to be chosen sufficiently small. Analogously to the theoretical consideration in Sect. 3.5, Algorithm 8 runs in $\kappa(\varepsilon, \varepsilon_s) \cdot \mathcal{O}(k)$ time, where $\kappa(\varepsilon, \varepsilon_s)$ depends mainly upon ε, to a minor degree also on ε_s, but not on the number k of line segments.

3.8 More Experimental Studies

This section complements the run-times reported in Sect. 3.5, using an implementation of Algorithm 7. The experimental studies in this section are about accuracy, and they allow us to conclude on run-time. We vary the number k of line segments between 2 and 20,000. We explain below our approaches for obtaining the experimental results provided in Table 3.4.

For any of the listed values of k, a number n_e of randomly-selected *configurations* was tested; each configuration is defined by a sequence of k line segments and a randomly-selected initialisation of path ρ_0.

Table 3.4 Experimental results of Algorithm 7. k is the number of line segments, n_e is the number of experiments for k segments, i_{max} is the maximal number of iterations needed for any of the n_e experiments, $\delta_{200} = \max_{l=1,\ldots,n_e}\{L_{199}^l - L_{200}^l\}$, $L_{\delta_{200}} = L_{200}^{l_0}$, where l_0 satisfies $\delta_{200} = L_{199}^{l_0} - L_{200}^{l_0}$, and $L_{min} = \min_{l=1,\ldots,n_e} L^l$, where L^l is the output length of lth experiment. The last column shows just a reference number for the row (there have been more experiments)

k	n_e	i_{max}	δ_{200}	$L_{\delta_{200}}$	L_{min}	$\frac{\delta_{200}}{L_{\delta_{200}}}$	Row
2	100,000	48,062	0.027009	511.463	491.35	5.28073×10^{-5}	1
3	210,000	112,765	0.014378	498.412	491.35	2.88482×10^{-5}	2
4	10,100,000	122,403	0.253529	610.916	491.35	4.14998×10^{-4}	3
6	563,422	388,571	0.107719	1,091.16	496.02	9.87197×10^{-5}	5
8	762,066	462,346	0.281012	1,237.31	501.97	2.27115×10^{-4}	7
10	697,858	195,220	0.191796	1,207.78	581.96	1.58800×10^{-4}	8
11	135,571	115,019	0.194815	1,480.20	660.83	1.31614×10^{-4}	9
40	146,490	318,223	0.465686	3,583.48	2,326.34	1.29954×10^{-4}	15
60	100,000	159,249	0.328890	7,796.82	4,316.65	4.21826×10^{-5}	16
80	106,495	183,808	0.432228	10,005.20	5,926.99	4.32003×10^{-5}	18
90	102,820	186,644	0.542972	9,976.34	6,730.69	5.44260×10^{-5}	19
100	81,475	123,031	2.025440	10,107.10	7,642.06	2.00398×10^{-4}	20
400	18,706	197,676	1.514930	45,664.1	35,527.6	3.31755×10^{-5}	25
600	14,007	4,862,710	0.162764	63,368.0	56,021.7	2.56855×10^{-6}	26
700	8,782	89,821	0.490649	76,541.5	65,682.9	6.41023×10^{-6}	27
800	6,509	159,955	0.213358	87,739.3	75,034.3	2.43173×10^{-6}	28
900	4,370	101,469	0.167307	93,941.6	86,420.9	1.78097×10^{-6}	29
1,000	10,282	162,547	0.464297	107,397.0	95,780.1	4.32318×10^{-6}	30
2,000	4,942	93,893	0.886554	211,389.0	198,226.0	4.19395×10^{-6}	31
4,000	4,693	139,894	0.486883	414,910.0	401,991.0	1.17347×10^{-6}	32
8,000	2,218	270,534	0.438270	827,244.0	813,093.0	5.29795×10^{-7}	34
10,000	1,448	55,770	0.885872	1.05849×10^6	1.02096×10^6	8.36921×10^{-7}	35
20,000	304	31,194	0.286782	2.08472×10^6	2.05278×10^6	1.37564×10^{-7}	36

Example 3.5 The first study is about the differences in path length after only 200 iterations. For this let

$$\delta_{200}(k) = \max_{l=1,\ldots,n_e}\left\{L_{199}^l(k) - L_{200}^l(k)\right\}$$

where $L_{199}^l(k)$ and $L_{200}^l(k)$ are the lengths of the updated paths in the 199th and 200th iteration of the lth configuration for k line segments. The maximal value of δ_{200} is 2.025440 in Row 20. In other words, in an extensive set of 14,403,815 experiments in total, we always have $\kappa(\varepsilon) \leq 200$, for any accuracy constant $\varepsilon \geq 2.025440$. The fifth column shows the minimum length of paths in iteration 200 which define this maximum difference δ_{200}, for illustrating the scale of the length of the generated paths. The length in the fifth column can also be compared with the minimum length in the sixth column.

If using more than i iterations (i.e., $\kappa(\varepsilon) > i$) then the difference L_{δ_i} reduces to

$$L_{\delta_i} \times \left(1 - \frac{\delta_i}{L_{\delta_i}}\right) = 2 \times L_{\delta_i} - L_{\delta_{i-1}}.$$

We consider again $i = 200$ and the ratios $\delta_{200} : L_{\delta_{200}}$. For all the examples of k segments studied, the maximum of those ratios equals 0.000414998 in Row 3. This means that for any configuration and more than 200 iterations (i.e., $\kappa(\varepsilon) > 200$), then there are $\kappa(\varepsilon) - 200$ iterations required for reducing the length of the calculated path from $L_{\delta_{200}}$ to $L_{\delta_{200}} \times (1 - 0.000414998)$. For example, Row 26 shows that there was a configuration where the algorithm spent $\kappa(\varepsilon) - 200 = 4,862,510$ iterations for reducing the length of the calculated path from $L_{\delta_{200}} = 66,368.0$ to $L_{\delta_{200}} \times (1 - 2.56855 \times 10^{-6})$. $\qquad\qquad\square$

Example 3.6 For a second study we use the accuracy constant $\varepsilon = 10^{-15}$. As specified in the table, we have, for example, that

$$\kappa(\varepsilon) \geq 112,765 \quad \text{when } k = 3 \text{ (see Row 2)},$$
$$\kappa(\varepsilon) \geq 388,571 \quad \text{when } k = 6 \text{ (see Row 5)},$$
$$\kappa(\varepsilon) \geq 462,346 \quad \text{when } k = 8 \text{ (see Row 7)},$$
$$\kappa(\varepsilon) \geq 4,862,710 \quad \text{when } k = 600 \text{ (see Row 26)},$$

and so forth, where $\kappa(\varepsilon)$ equals the maximum number of iterations i_{max} for any of the randomly-generated configurations with k segments. The κ-values show that there is no formal dependency on k; the values depend on the geometric complexity of the generated configuration and on the chosen accuracy ε. The smaller the ε, the greater the κ-value. $\qquad\qquad\square$

3.9 An Interesting Input Example of Segments in 3D Space

Consider the following input for Algorithm 7: let $p = (1, 4, 7)$, $q = (4, 7, 4)$, $k = 2$, where the two endpoints of S_1 are $(2, 4, 5)$ and $(2, 5, 5)$, and the two endpoints of S_2 are $(4, 5, 4)$ and $(4, 5, 5)$.

It can be shown that this fixed ESP problem is equivalent (see Lemma 3.4) to finding the roots of the polynomial

$$p(x) = 84x^6 - 228x^5 + 361x^4 + 20x^3 + 210x^2 - 200x + 25$$

which is not solvable by radicals over the field of rational numbers. The proof of this unsolvability is very complicated, and we do not provide it here (but later in Sect. 9.9). As discussed in Sect. 1.4, this means the following:

Any arithmetic algorithm for solving the fixed 3D straight segment ESP problem can only be approximate and not exact.

Fig. 3.13 Illustration for
level-2 cells in 2D or 2.5D,
where $m = 7$

3.10 A Generic Rubberband Algorithm

Following Sect. 1.9, a *generic ESP from p to q* is an ESP which is in a plane (2D ESP), in the surface of a connected polyhedron (2.5D ESP), or in 3D free space (3D ESP). The *generic ESP free space* is the search space in 2D, 2.5D, or 3D space.

In the 2D case, the generic ESP free space is a polygon which may have some holes (i.e., obstacles which are simple polygons). In the 2.5D case, the generic ESP free space is the surface of a connected polyhedron (i.e., the union of some simple polygons). In the 3D case, the generic ESP free space equals $\mathbb{R}^3 \setminus \bigcup_{i=1}^{m} \Pi_i^\circ$, where Π_i (i.e., an obstacle) is a connected polyhedron, for $i = 1, 2, \ldots, m$, and $\Pi_j \cap \Pi_k = \emptyset$, where $j \neq k$, for $j, k = 1, 2, \ldots, m$.

Let Π be the generic ESP free space. In the 2D case, Π can be decomposed into some triangles (see Chap. 5) aiming at a minimum number of additional "new" vertices. In the 2.5D case, each simple surface polygon of Π can be decomposed into some triangles without adding any "new" vertex. In the 3D case, Π can be decomposed into some tetrahedrons, again with the goal to minimise the number of "new" vertices. We call each triangle (in both the 2D and the 2.5D case) or tetrahedron (in the 3D case) a *level-1 cell (with respect to Π)*. Each side of a triangle (level-1 2D or 2.5D cell), or each triangular face of a tetrahedron (level-1 3D cell) is called a *side* of a level-1 cell.

Definition 3.7 The *shortest (longest) edge* of all level-1 cells, denoted by l_C (L_C), is the shortest (longest) side of all level-1 cells if they are triangles, or the shortest (longest) side of all the triangular faces of all level-1 cells if they are tetrahedra.

We introduce an integer parameter $h > 0$ which defines *decomposition resolution*.[6] For each level-1 cell C, we have the following two cases:

Case 1 (generic ESP free space in 2D or in 2.5D). If C is a triangle, then we decompose three sides (line segments) of C into smaller line segments as follows: for each side S_C, let u and v be the two endpoints of S_C. We add cut points $\{w_i : i = 1, 2, \ldots, m - 1\}$ on the line segment uv (i.e., S_C) such that $d_e(w_{i-1}, w_i) = l_C/h$ and $d_e(w_{m-1}, w_m) \leq l_C/h$, where $w_0 = u$ and $w_m = v$, for $i = 1, 2, \ldots, m - 1$; see Fig. 3.13.

Case 2 (generic ESP free space in 3D). If C is a tetrahedron, then we decompose the four sides (i.e. the triangles) of C into smaller line segments as follows: for each side S_C, let u, v and w be the three vertices of S_C. Without loss of generality, assume that the edge uv is the longest side of the triangle $\triangle uvw$ (i.e., S_C). Let w' be the point on uv such that $w'w \perp uv$. We add cut points on both $w'w$ and uv

---

[6]This is motivated by the definition of grid resolution and the study of multigrid approaches in [8].

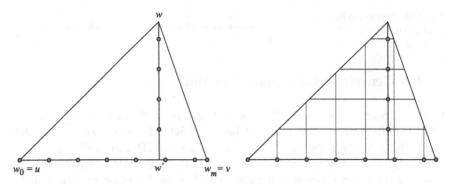

Fig. 3.14 Illustration for level-2 cells in 3D, where $m = 7$

exactly the same way as in Case 1. Then we add cut edges through each cut point such that each cut edge is parallel to $w'w$ or uv. These cut edges cut triangle $\triangle uvw$ (i.e., S_C) into smaller triangles, squares, trapezoids, or even irregular 5-gons or 6-gons; see Fig. 3.14.

We call each smaller line segment in Case 1, or smaller triangle, square, trapezoid, or irregular 5-gon or 6-gon in Case 2, a *level-2 cell (with respect to the level-1 cell C)*.

In Fig. 3.14, left, six points were added on the side uv; four points were added on the line segment $w'w$. Right, 21 level-2 cells are created with respect to the level-1 cell (i.e., the triangle $\triangle uvw$) including eight triangles, seven squares, one trapezoid, four irregular 5-gons, and one 6-gon.

Definition 3.8 For each level-1 cell C, we construct the *cell visibility (undirected) weighted graph* $G_C = [V, E, w]$ as follows: each node (also called vertex) v in V corresponds to a level-2 cell C_v^2 with respect to C. For any two nodes v and v' in V, the unordered pair $\{v, v'\}$ is an edge in E iff the corresponding level-2 cells C_v^2 and $C_{v'}^2$ of v and v' are not on the same side of the level-1 cell C. For each edge $\{v, v'\}$ in E, its weight is defined by

$$w(\{v, v'\}) = \min\{d_e(q, q') : q \in C_v^{2\bullet} \wedge q' \in C_{v'}^{2\bullet}\}.$$

C_v^2 is also called the *level-2 cell corresponding to* the node v of the graph G_C.

For example, in Fig. 3.15, left, three sides of the level-1 cell C (i.e., the triangle) are decomposed into four level-2 cells $C_{v_1}^2$, $C_{v_2}^2$, $C_{v_3}^2$, $C_{v_4}^2$, three level-2 cells $C_{v_5}^2$, $C_{v_6}^2$, $C_{v_7}^2$, and four level-2 cells $C_{v_8}^2$, $C_{v_9}^2$, $C_{v_{10}}^2$, $C_{v_{11}}^2$, respectively. Right, the cell visibility (undirected) weighted graph $G_C = [V, E, w]$, where $V = \{v_1, v_2, \ldots, v_{11}\}$. Each node v_i in V corresponds to the level-2 cell $C_{v_i}^2$ shown on the left, where $i = 1, 2, \ldots, 11$.

Definition 3.9 For a level-1 cell C which contains a point p_1 in the generic ESP free space, we can construct the *cell visibility (undirected) weighted graph for point p_1*,

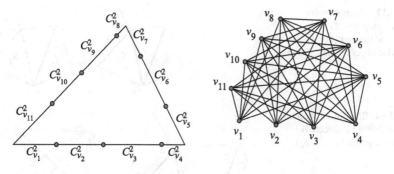

Fig. 3.15 Illustration for a cell visibility (undirected) weighted graph

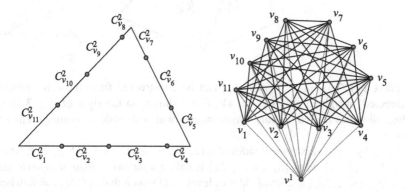

Fig. 3.16 Illustration for a cell visibility (undirected) weighted graph for point p_1

denoted by $G_C(p_1) = [V_1, E_1, w_1]$, as follows: $V_1 = V \cup \{v_1\}$, where V is defined as in the cell visibility (undirected) weighted graph G_C above, node v^1 corresponds to point p_1. For any two nodes v and v' in V_1, if $v \neq v^1$ and $v' \neq v^1$, then the unordered pair $\{v, v'\}$ is an edge in E iff the corresponding level-2 cells C_v^2 and $C_{v'}^2$ of v and v' are not on the same side of level-1 cell C, otherwise, $\{v, v'\}$ is an edge in E. For each edge $\{v, v'\}$ in E, its weight is defined by

$$w(\{v, v'\}) = \min\{d_e(q, q') : q \in C_v^{2\bullet} \wedge q' \in C_{v'}^{2\bullet}\}$$

if $v \neq v^1$ and $v' \neq v^1$, and

$$w(\{v^1, v'\}) = \min\{d_e(p_1, q') : q' \in C_{v'}^{2\bullet}\}$$

or

$$w(\{v, v^1\}) = \min\{d_e(q, p_1) : q \in C_v^{2\bullet}\}$$

otherwise.

For example, in Fig. 3.16, left, point p_1 is located in the interior of the level-1 cell C (i.e., the triangle). Right, the cell visibility (undirected) weighted graph (with

Fig. 3.17 Illustration for generic step sets. *Left*: a generic step set. *Right*: pairwise disjoint generic step set

Fig. 3.18 Illustration for a cell visibility (undirected) weighted graph $G_C[S_1, S_2]$

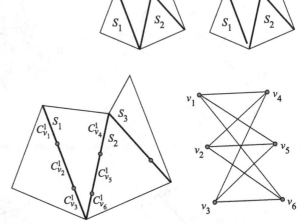

the point p_1) $G_C(p_1) = [V_1, E_1, w_1]$ can be constructed from the cell visibility (undirected) weighted graph $G_C = [V, E, w]$ shown on the right of Fig. 3.15 by adding all edges (in red colour) which are incident with node v^1 corresponding to point p_1.

If $\{S_1, S_2, \ldots, S_k\}$ is a set of sides of level-1 cells such that $S_i \cap S_{i+1} = \emptyset$, where $i = 1, 2, \ldots, k - 1$, then $\{S_1, S_2, \ldots, S_k\}$ is called a *pairwise disjoint generic step set*. If $\{S_1, S_2, \ldots, S_k\}$ is a set of sides of level-1 cells such that $S_i \cap S_{i+1} \neq \emptyset$, where $i = 1, 2, \ldots, k - 1$, then $\{S_1, S_2, \ldots, S_k\}$ is called *generic step set*.

Analogously to L_C and l_C, L_S and l_S are the longest (shortest) steps in the generic step set.

For example, in Fig. 3.17, left, a generic step set $\{S_1, S_2, S_3\}$ is shown, and on the right a pairwise disjoint generic step set $\{S_1, S_2, S_3\}$.

For each step S_i in a generic step set $\{S_1, S_2, \ldots, S_k\}$, S_i can be thought to be a side of a level-1 cell C_i. Each level-2 cell with respect to the level-1 cell C_i is also called a *level-1 cell with respect to the side (step) S_i*, where $i = 1, 2, \ldots, k$.

For two consecutive steps S_{i-1} and S_i in a generic step set $\{S_1, S_2, \ldots, S_k\}$, where $i = 2, 3, \ldots, k$, we can construct the cell visibility (undirected) weighted graph $G_C[S_{i-1}, S_i] = [V, E, w]$ as follows: Each node (also called vertex) v in V corresponds to a level-1 cell C_v^1 with respect to S_{i-1} or S_i. For any two nodes v and v' in V, the unordered pair $\{v, v'\}$ is an edge in E iff the corresponding level-1 cells C_v^1 and $C_{v'}^1$ of v and v' are not on the same step S_{i-1} or S_i. For each edge $\{v, v'\}$ in E, its weight is defined as

$$w(\{v, v'\}) = \min\{d_e(q, q') : q \in C_v^{1^\bullet} \land q' \in C_{v'}^{1^\bullet}\}.$$

For example, in Fig. 3.18, left, three steps S_1, S_2 and S_3 are decomposed into three level-1 cells $C_{v_1}^1$, $C_{v_2}^1$, $C_{v_3}^1$, three level-1 cells $C_{v_4}^1$, $C_{v_5}^1$, $C_{v_6}^1$, and two level-1 cells $C_{v_7}^1$, $C_{v_8}^1$, respectively. The same figure shows on the right the cell visibility (undirected) weighted graph $G_C[S_1, S_2] = [V, E, w]$, where $V = \{v_1, v_2, \ldots, v_6\}$.

Fig. 3.19 Illustration for a
cell visibility (undirected)
weighted graph $G_C[p_1, S_1]$

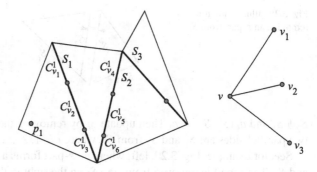

Each node v_i in V corresponds to the level-1 cell $C_{v_i}^1$ shown on the left, where
$i = 1, 2, \ldots, 6$.

If S_{i-1} is degenerated into a single point p_1, then we can construct the cell visibility (undirected) weighted graph $G_C[p_1, S_i] = [V, E, w]$ as follows: Each node v
in V corresponds to a level-1 cell C_v^1 with respect to S_i, or it corresponds to the
point p_1. For any two nodes v and v' in V, the unordered pair $\{v, v'\}$ is an edge in E
iff v corresponds to point p_1 and v' corresponds to a level-1 cell C_v^1 with respect to
S_i, or v' corresponds to point p_1 and v corresponds to a level-1 cell C_v^1 with respect
to S_i. For each edge $\{v, v'\}$ in E, its weight is defined by

$$w(\{v, v'\}) = \min\{d_e(p_1, q') : q' \in C_{v'}^{1^\bullet}\}$$

if v corresponds to the point p_1 and v' corresponds to a level-1 cell C_v^1 with respect
to S_i, or

$$w(\{v, v'\}) = \min\{d_e(q, p_1) : q \in C_v^{1^\bullet}\}$$

if v' corresponds to the point p_1 and v corresponds to a level-1 cell C_v^1 with respect
to S_i.

For example, see Fig. 3.19, left, with point p_1. Three steps S_1, S_2, and S_3 are
decomposed into three level-1 cells $C_{v_1}^1$, $C_{v_2}^1$, $C_{v_3}^1$, three level-1 cells $C_{v_4}^1$, $C_{v_5}^1$, $C_{v_6}^1$,
and two level-1 cells $C_{v_7}^1$, $C_{v_8}^1$, respectively, exactly the same as those shown in
Fig. 3.18. In the same figure on the right, the cell visibility (undirected) weighted
graph $G_C[p_1, S_i] = [V, E, w]$ is shown, where $V = \{v, v_1, v_2, v_3\}$. Node v corresponds to point p_1, and each node v_i in $V \setminus \{v\}$ corresponds to the level-1 cell $C_{v_i}^1$
shown on the left, where $i = 1, 2, 3$.

Analogously, if S_i is degenerated into a single point p_1, then we can also construct the cell visibility (undirected) weighted graph $G_C[S_{i-1}, p_1] = [V, E, w]$:

Remove an ε-part from S_i defined as follows: Consider two consecutive steps
S_{i-1} and S_i in a generic step set $\{S_1, S_2, \ldots, S_k\}$, where $i = 2, 3, \ldots, k$. In the 2D
case, $S_i \cap S_{i+1}$ is a single point, denoted by p_1, which is an endpoint of both line
segments S_i and S_{i+1}. Let p_2 be in S_i such that $d_e(p_2, p_1) = \varepsilon$. Then update S_i by
removing the small line segment $p_2 p_1$ from S_i, where $i = 1, 2, \ldots, k - 1$. In the
3D case, $S_i \cap S_{i+1}$ is a line segment, denoted by S_1', which is a common side of
both triangles S_i and S_{i+1}. Let line segment S_2' be completely inside of S_i such that

Fig. 3.20 Illustration for removing an ε-part from S_i

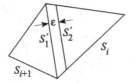 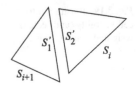

$S_2' \parallel S_1'$ and $d_e(S_2', S_1') = \varepsilon$. Then update S_i by removing the small trapezoid whose two parallel sides are S_1' and S_2' from S_i, where $i = 1, 2, \ldots, k - 1$.

See, for example, Fig. 3.20, left, where an ε-part forms a trapezoid defined by S_1' and S_2'. The ε-part is removed from step S_i on the right in this figure.

The main idea behind the generic RBA   Algorithms 7 and 8 compute a solution whose length L is an upper bound to the optimal (i.e., true) shortest path. It still remains a challenge to estimate the approximation factor of those algorithms. The following generic RBA applies the Dijkstra algorithm for computing a step set and then a lower bound L_h of the length of the optimal path. Then we can easily obtain an upper bound L/L_h of the approximation factor of our algorithms. The *key idea* is to construct a "good" visibility weighted graph by keeping all edges inside of their cells of uniform size. In this way, we can greatly reduce the time complexity of the applied Dijkstra algorithm.

> In this way, the generic RBA defines a universal path optimisation method which applies the discrete shortest path method (i.e., the Dijkstra algorithm) for solving continuous shortest path problems (i.e., the Euclidean shortest path problems).

We call the graph constructed in Algorithm 10 *indirect* visibility graph.

Algorithms 9 and 10, given in Figs. 3.21 and 3.22, follow the general concept of RBAs as outlined above, and do not require more explanations due to their simplicity. See Fig. 3.23 for an example. Line 6 in Algorithm 10 stays short for taking the union of those two graphs.

The main computation of Algorithm 9 occurs in the while-loop. The for-loop takes $\mathcal{O}(k)$ time. The number of iterations of the while-loop equals $\kappa(\varepsilon) = \frac{L_0 - L}{\varepsilon}$, where L_0 is the length of an initial path and L the length of the optimum path. Thus, we have:

Theorem 3.4 *Algorithm 9 can be computed in time $\kappa(\varepsilon) \times \mathcal{O}(k)$ for k steps.*

Theorem 3.5 *In the 2D or 2.5D case, Algorithm 10 can be computed in time $m_\delta \times \mathcal{O}(m^2 k + (mk) \log(mk) + \kappa(\varepsilon) \times k)$, where m_δ is the number of iterations taken by the while-loop, $m = (L_S/l_S) \times h$, and L_S (l_S) is the longest (shortest) step in the step set.*

Algorithm 9 (Simple generic RBA for computing the generic ESP)
Input: Source point p and target point q, a pairwise disjoint generic step set $\{S_1, S_2, \ldots, S_k\}$ such that p and q are not on S_i, for any $i = 1, 2, \ldots, k$, and an accuracy constant $\varepsilon > 0$.
Output: A path approximating a generic ESP from p to q.

1: For each $j \in \{0, 1, \ldots, k+1\}$, let p_j be a point in S_j.
2: $L_{current} \leftarrow \sum_{j=0}^{k} d_e(p_j, p_{j+1})$, where $p_0 = p$ and $p_{k+1} = q$, and $L_{previous} \leftarrow \infty$.
3: **while** $L_{previous} - L_{current} \geq \varepsilon$ **do**
4:     **for** each $j \in \{1, 2, \ldots, k\}$ **do**
5:         Compute a point $q_j \in S_j$ such that $d_e(p_{j-1}, q_j) + d_e(q_j, p_{j+1}) = \min\{d_e(p_{j-1}, p) + d_e(p, p_{j+1}) : p \in S_j\}$
6:         Update the path $\langle p, p_1, p_2, \ldots, p_k, q \rangle$ by replacing p_j by q_j.
7:     **end for**
8:     Let $L_{previous} \leftarrow L_{current}$ and $L_{current} \leftarrow \sum_{j=0}^{k} d_e(p_j, p_{j+1})$.
9: **end while**
10: Return $\{p, p_1, p_2, \ldots, p_k, q\}$.

Fig. 3.21 A generic RBA for computing a generic ESP, basically coinciding with Algorithm 7

Proof Line 1 takes constant time. In the 2D or 2.5D case, each step S_i is a line segment. Line 4 takes $\mathcal{O}(m)$ time because there are at most m points that can be added to step S_i. Lines 5 and 6 take $\mathcal{O}(m^2)$ time because there are at most m^2 edges in the cell visibility weighted graph $G_C[S_{i-1}, S_i]$. Line 7 only takes constant time. Thus, the for-loop can be computed in $\mathcal{O}(m^2k)$ time.

Line 9 can be computed in $\mathcal{O}(m^2k + (mk)\log(mk))$ time (see the time complexity of the Dijkstra algorithm in Sect. 1.7) because the indirect visibility graph G_V has $\mathcal{O}(m^2k)$ edges and $\mathcal{O}(mk)$ nodes. By Theorem 3.4, Line 10 can be computed in $\kappa(\varepsilon) \times \mathcal{O}(k)$ time. Line 11 only takes constant time. Thus, the main computation of the while-loop is done in $\mathcal{O}(m^2k + (mk)\log(mk) + \kappa(\varepsilon) \times k)$ time. □

Theorem 3.6 *In the 3D case, Algorithm 10 can be computed in time $m_\delta \times \mathcal{O}(m^4k + (m^2k)\log(m^2k) + \kappa(\varepsilon) \times k)$, where m_δ is the number of iterations taken by the while-loop, $m = (L_S / l_S) \times h$, and L_S (l_S) is the longest (shortest) side of the steps in the step set.*

Proof In the 3D case, each step S_i is a triangle. Line 4 now takes $\mathcal{O}(m^2)$ time because there are $\mathcal{O}(m)$ points that can be added on the longest side of S_i. In conclusion, Lines 5 and 6 require $\mathcal{O}(m^4)$ time because there are $\mathcal{O}(m^4)$ edges in the cell visibility weighted graph $G_C[S_{i-1}, S_i]$. Thus, the for-loop can be computed in $\mathcal{O}(m^4k)$ time.

Line 9 can be computed in $\mathcal{O}(m^4k + (m^2k)\log(m^2k))$ time due to the time complexity of the Dijkstra algorithm: the indirect visibility graph G_V has $\mathcal{O}(m^4k)$ edges and $\mathcal{O}(m^2k)$ nodes. Altogether, the main computation of the while-loop is accomplished in $\mathcal{O}(m^4k + (m^2k)\log(m^2k) + \kappa(\varepsilon) \times k)$ time. □

Algorithm 10 (Another simple generic RBA for computing a generic ESP)
Input: Source point p and target point q, generic step set $\{S_1, S_2, \ldots, S_k\}$, where p and q are not on any S_i, for $i = 1, 2, \ldots, k$, an integer parameter $h > 0$ defining the decomposition resolution, an accuracy constant $\varepsilon > 0$, and an approximation parameter $\delta > 1$.
Output: A δ-approximation path for the generic ESP from p to q.

1: Start with an empty visibility weighted graph G_V, $L = +\infty$, the given integer h, and let L_h be a sufficiently small positive real number.
2: **while** $L/L_h \geq \delta$ **do**
3:     **for** each $i \in \{1, 2, \ldots, k\}$ **do**
4:         Decompose S_i into a set $S^1(S_i, h)$ of level-1 cells with respect to S_i.
5:         Construct the cell visibility weighted graph $G_C[S_{i-1}, S_i]$.
6:         Let $G_V = G_V + G_C[S_{i-1}, S_i]$.
7:         Remove an ε-part from step S_i, and denote the resulting step by S_i'.
8:     **end for**
9:     Let G_V, p and q be the input for the Dijkstra algorithm for computing a shortest path $\rho_V(p, q, h) = (p, v_1, v_2, \ldots, v_k, q)$ from p to q in G_V. Let L_h be the length of $\rho_V(p, q, m)$. Let C_i^1 be the level-1 cell corresponding to v_i, and let u_i be the centre of C_i^1, for $i = 1, 2, \ldots, k$.
10:    Let ε, p, q, step set $\{S_1', S_2', \ldots, S_k'\}$, and initial path $(p, u_1, u_2, \ldots, u_k, q)$ be the input for Algorithm 9 for computing an approximate ESP $\rho(p, q)$. Let L be the length of $\rho(p, q)$.
11:    Let $h = 2 \times h$.
12: **end while**
13: Return the approximate ESP $\rho(p, q)$.

Fig. 3.22 A second generic RBA for computing a generic ESP which makes more explicit use of the 'key idea' presented above, also using Dijkstra's algorithm

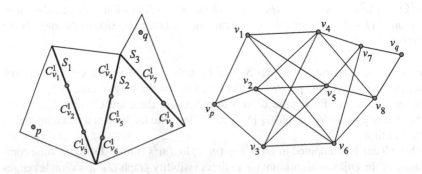

Fig. 3.23 An example of an indirect visibility graph

Algorithm 11 (see Fig. 3.24) is finally our proposal for a general time-optimised design of a generic RBA.

Again, the pseudocode contains the necessary specifications, and should be easy to follow after having discussed RBAs before. The graph constructed in Algorithm 11 is again an indirect visibility graph.

Theorem 3.7 *In the 2.5D case (i.e., the generic ESP free space Π is the surface of a connected polyhedron), Algorithm 11 can be computed in time $m_\delta \times \mathcal{O}(m^2 n + (mn) \log(mn) + \kappa(\varepsilon) \times n)$, where m_δ is the number of iterations taken by the while loop, $m = (L_S/l_S) \times h$, L_S (l_S) is the longest (shortest) edge of Π, and $n = |V(\Pi)|$ (i.e., the number of vertices of Π).*

Proof Line 1 takes constant time. In the 2.5D case, Π is the surface of a connected polyhedron. Thus, Line 2 can be computed in $\mathcal{O}(n)$ time because there are $\mathcal{O}(n)$ faces and each face F_i can be decomposed into triangles in $\mathcal{O}(|V(F_i)|)$ time where $n = |V(\Pi)|$ is the number of vertices of Π.[7]

Line 6 takes $\mathcal{O}(m)$ time because there are $\mathcal{O}(m)$ points can be added on the side S_C. Thus, Lines 5–7 take $\mathcal{O}(m)$ time because there are three sides for each cell C. Line 9 takes $\mathcal{O}(m^2)$ time because there are at most m^2 edges in the cell visibility weighted graph $G_C(p)$. Analogously, Lines 12, 14, and 17 take $\mathcal{O}(m^2)$ time. Thus, the outer for-loop (Lines 4–18) can be computed in $\mathcal{O}(m^2 n)$ time.

Line 19 can be computed in $\mathcal{O}(m^2 n + (mn) \log(mn))$ time due to the time complexity of the Dijkstra algorithm, because the indirect visibility graph G_V has $\mathcal{O}(m^2 n)$ edges and $\mathcal{O}(mn)$ nodes. By Theorem 3.5, Line 20 can be computed in $m_\delta \times \mathcal{O}(m^2 k + (mk) \log(mk) + \kappa(\varepsilon) \times k)$ time, where m_δ is the number of iterations taken by the while-loop in Algorithm 10. Line 21 takes constant time. Line 23 takes in $\mathcal{O}(n)$ time because there are $\mathcal{O}(n)$ vertices in the output path. Thus, the main computation inside the while-loop occurs in Lines 19 and 20.   □

For the 2D and 3D case, we have the following theorems, and proofs are analogous to that of Theorem 3.7. The following theorem differs from Theorem 3.7 just by the different meaning of Π in the 2D case:

Theorem 3.8 *In the 2D case (i.e., the generic ESP free space Π is a polygon which may have some holes), Algorithm 11 can be computed in $m_\delta \times \mathcal{O}(m^2 n + (mn) \log(mn) + \kappa(\varepsilon) \times n)$ time.*

Note that is not difficult to decompose Π into triangles by adding at most n "new" vertices.

---

[7]For linear-time triangulation, see the Chazelle method in Chap. 5, but also the critical discussion pointing to a missing implementation so far.

Algorithm 11 (Generic RBA for computing the generic ESP)

Input: Source point p and target point q in a generic ESP free space Π, an integer parameter $h > 0$ for decomposition resolution, an accuracy constant $\varepsilon > 0$, and an approximation parameter $\delta > 1$.

Output: A δ-approximation path for a generic ESP from p to q.

1: Initialise a visibility weighted graph as empty graph G_V, let $L = +\infty$, read the provided parameter h, and let L_h be a sufficiently small positive real number.

2: Decompose Π into a set S^1 of level-1 cells.

3: **while** $L/L_h \geq \delta$ **do**

4:    **for** each cell $C \in S^1$ **do**

5:       **for** each side S_C of C **do**

6:          Decompose S_C into a set $S^2(S_C, h)$ of level-2 cells.

7:       **end for**

8:       **if** C contains p **then**

9:          Construct the cell visibility weighted graph G_C for p, denoted by $G_C(p)$.

10:      **else**

11:         **if** C contains q **then**

12:            Construct the cell visibility weighted graph G_C for q, denoted by $G_C(q)$.

13:         **else**

14:            Construct the cell visibility weighted graph G_C.

15:         **end if**

16:      **end if**

17:      Let $G_V = G_V + G_C$.

18:    **end for**

19:    Let G_V, p and q be the input for the Dijkstra algorithm for computing a shortest path $\rho_{G_V}(p, q, h) = (p, v_1, v_2, \ldots, v_k, q)$ from p to q in G_V. Let L_h be the length of $\rho_{G_V}(p, q, h)$. Let C_i^2 be the level-2 cell corresponding to node v_i of graph G_V, and $S_{C_i^1}$ be the side of the level-1 cell corresponding to C_i^2, where $i = 1, 2, \ldots, k$.

20:    Let ε, p, q, and step set

$$\{S_{C_1^1}, S_{C_2^1}, \ldots, S_{C_k^1}\}$$

be the input for Algorithm 10 for computing an approximate ESP $\rho(p, q)$. Let L be the length of $\rho(p, q)$.

21:    Let $h = 2 \times h$.

22: **end while**

23: Return the approximate ESP $\rho(p, q)$.

Fig. 3.24 Generic RBA for computing an approximate generic ESP

Theorem 3.9 *In the 3D case [i.e., the generic ESP free space $\Pi = \mathbb{R}^3 \setminus \bigcup_{i=1}^{m'} \Pi_i^\circ$, where obstacle Π_i is a connected polyhedron, for $i = 1, 2, \ldots, m'$, and $\Pi_j \cap \Pi_k = \emptyset$, where $j \neq k$, for $j, k = 1, 2, \ldots, m'$], Algorithm 11 can be computed in $m_\delta \times \mathcal{O}(m^4 n + (m^2 n) \log(m^2 n) + \kappa(\varepsilon) \times n + (n_1 + r^2) \log r)$ time, where $n = n_1 + n_2$, $n_1 = |V(\Pi)|$.*

In the theorem, n_1 is the number of vertices of Π, n_2 is the number of "new" vertices added for decomposing Π into tetrahedrons, and r is the number of "notches" (i.e., features which cause nonconvexity in polyhedra).

Proofs of the correctness of Algorithms 9, 10, and 11 can follow exactly the same 'line-by-line' pattern as followed in the proof of Theorem 3.3.

By Theorem 3.9, Algorithm 11 may be expected to outperform other algorithms, for example, for the general 3D ESP problem which is NP-hard, and for which there does not exist any exact solution algorithm; see, e.g., Example 9.13.

We provide two final remarks for this chapter:

About $\kappa(\varepsilon)$. For Algorithm 9, we may apply the Dijkstra algorithm with a small value of the decomposition resolution parameter h to obtain a lower bound l for the length of an optimal path. Then we choose the accuracy constant ε equals $\alpha \times l$, instead of α, where α may be between 10^{-2} and 10^{-5} (instead of 10^{-15}) depending on the given practical application. Then we have that

$$\kappa(\varepsilon) = \frac{L_0 - L}{\alpha \times l}.$$

This may lead to a "good" upper bound for $\kappa(\varepsilon)$ in practical applications.

About the indirect visibility graph. In contrast to the visibility graphs introduced in computational geometry,[8] our indirect visibility graphs keep each edge inside of a single level-1 cell. This greatly reduces the total number of edges of our visibility graphs and speeds up the application of the Dijkstra algorithm. Thus, in Theorems 3.5, 3.6, 3.7, 3.8, and 3.9, the number of edges of the indirect visibility graph equals $m^2 k$, $m^4 k$, $m^2 n$, $m^2 n$, and $m^4 n$, respectively, and can be further reduced when the lengths of sides of level-1 cells are increasing differently (i.e., more nonuniformly).

3.11 Problems

Problem 3.1 In Algorithm 8, what may happen if there is no ε_s-transform q_j performed in Line 7?

---

[8]For example, see references [1, 6, 7, 9, 14, 16].

Problem 3.2 Discuss the advantages and disadvantages between the unscaled stop criterion (2.3) and the scaled stop criterion (2.4) for Algorithm 7.

Problem 3.3 State three key issues when applying Algorithm 7.

Problem 3.4 Prove that Algorithm 10 defines a sequence of intervals $\{[l_h, L_h]\}$, where l_h is the length obtained by the Dijkstra algorithm, and L_h is the length of an approximation path, such that $\lim_{h \to +\infty} l_h = \lim_{h \to +\infty} L_h$ which is the length of the optimal path.

Problem 3.5 For Algorithm 11, if the value L_C / l_C (L_C and l_C are the shortest and longest edge of all level-1 cells) is very large, then how to improve the algorithm by selecting a good decomposition resolution integer parameter $h > 0$ such that the algorithm can already terminate in fewer iterations by reducing the approximation parameter?

Problem 3.6 The value of decomposition resolution parameter $h > 0$ in Line 11 of Algorithm 10 increases exponentially. State a simple way to slow down the speed of growth of h.

Problem 3.7 Which different indirect visibility graphs have been introduced in Sect. 3.10? What are the differences between them?

Problem 3.8 Discuss the difference between indirect visibility graphs introduced in this book and visibility graphs as commonly used in computational geometry.

Problem 3.9 Can Algorithm 11 be improved by (i) modifying cells from triangles (in the 2D and 2.5D case) or tetrahedrons (in the 3D case) to more general convex subpolygons (in the 2D and 2.5D case) or to convex polyhedrons (in the 3D case), and (ii) defining a corresponding indirect visibility graph?

Problem 3.10 In Lemma 3.1, Eq. (3.1) can be written as

$$A\lambda_1^2 + B\lambda_1 + C = 0$$

where A, B, and C are polynomials in λ_0, λ_2, and also in $a_{j0}, b_{j0}, a_{j1}, b_{j1}, a_{j2}$ and b_{j2}, for $j = 0, 1, 2$. Under these conditions, is there a possibility that $\lambda_1 \notin [0, 1]$ for some λ_0, λ_2, S_0, S_1, and S_2?

Problem 3.11 Discuss other options for terminating Algorithm 7.

Problem 3.12 What would be the minimal value of $\varepsilon > 0$ for Algorithm 7 on your computer and your programming environment when implementing the algorithm?

Problem 3.13 The number of iterations of Algorithm 7 depends on the value of ε, denoted by $i_{max}(\varepsilon)$. Can you specify a non-trivial upper bound for $i_{max}(\varepsilon)$?

Problem 3.14 (Programming exercise)

(a) Implement Algorithm 7 for the 2D space. A user of the program may at first interactively specify a finite number of segments S_j, and then a start p and a destination q. This specification may be done by means of graphical input or numeric input; however, the resulting configuration has to be visualised on screen in a graphical format.
(b) Select midpoints of steps for defining the initial path.
(c) A user of the program should now be able to select either a mode defined by a fixed number $i_0 > 0$ of iterations, or a mode defined by an accuracy constant $\varepsilon > 0$.
(d) In the first case, the program runs through iterations $i = 0, \ldots, i_0$ and outputs the sequence of values $L_i - L_{i+1}$ in some appropriate way (e.g., as a table, or by means of a diagram using a logarithmic scale for those values). In this case, also visualise the calculated paths on screen.
(e) In the second case, the program runs through iterations $i = 0, 1, \ldots$ until $L_i - L_{i+1} < \varepsilon$; in this case, measure the run time until the program stops (note: a more accurate measurement may result from having this process running, say, 100 times, and then divide the time by 100). Output the resulting ε-shortest path and the number of iterations used to arrive at this result.

Problem 3.15 (Programming exercise) Implement the game as suggested by Fig. 3.3.

Problem 3.16 After having Algorithm 7 implemented for the 2D space, replace the unscaled stop criterion (2.3) by the scaled stop criterion (2.4). How does this effect the performance with respect to the number of iterations until stop is reached?

Problem 3.17 (Research problem) Specify (i.e., find a formula) a local upper accuracy bound d_ε in Theorem 3.3?

Problem 3.18 (Research problem) Provide a short proof for Theorem 3.3 based on convex analysis (e.g., to prove that Algorithm 7 defines a convex function and outputs a local minimum).

Problem 3.19 (Research problem) Consider the proof of Lemma 3.6. Answer the question whether interval I' is a proper subset of interval I or not. If a proper subset then how does this affect the proof of this lemma?

Problem 3.20 (Research problem) Let Π be a simple polygon. Insert a finite number of holes (i.e., which are simple polygons again) into Π such that frontiers are pairwise disjoint. Let n be the total number of given vertices in those polygons. Provide an algorithm which adds at most n new vertices to decompose this polygon Π with holes into triangles. Can this be done in $\mathcal{O}(n)$ time?

3.12 Notes

The rubberband was patented in England on March 17, 1845 by *Stephen Perry*. He invented the rubberband for holding papers or envelopes together.

The paper [4] proposed a rubberband algorithm for the particular case of calculating a shortest path in a 'cuboidal world', where the 3D space is subdivided into uniformly sized cubes. The paper [12] analysed this algorithm further in detail, which led to a correction of the rubberband algorithm originally published. The Ph.D. Thesis [10] showed that the 'rubberband concept' used in this algorithm actually allows us to establish a whole class of rubberband algorithms, suitable for solving various Euclidean shortest paths, defined in the plane or in 3D space; see also [11] or [13].

The paper [3] discussed sequences of pursuit paths (motivated by ant movements) and also contains a proof of Theorem 3.1.

Note that there are other techniques of proofs for the uniqueness of Euclidean shortest paths (other than the one given in this chapter), such as in [5, 15, 17], but those cannot be straightforwardly generalised, for example, from the input case of a sequence of line segments to that of a sequence of polygons. For Theorem 3.2, see [15]. For the derivation of the polynomial

$$p(x) = 84x^6 - 228x^5 + 361x^4 + 20x^3 + 210x^2 - 200x + 25$$

see [11], Sect. 7.4.1, pp. 102–106. For decomposition of polyhedrons into tetrahedrons, see [2].

References

1. Asano, T., Guibas, L., Hershberger, J., Imai, H.: Visibility of disjoint polygons. Algorithmica **1**, 49–63 (1986)
2. Bajaj, C.L., Dey, T.K.: Convex decomposition of polyhedra and robustness. SIAM J. Comput. **21**(2), 339–364 (1992)
3. Bruckstein, A.M.: Why the ant trails look so straight and nice? Math. Intell. **15**(2), 59–62 (1993)
4. Bülow, T., Klette, R.: Digital curves in 3D space and a linear-time length estimation algorithm. IEEE Trans. Pattern Anal. Mach. Intell. **24**, 962–970 (2002)
5. Choi, J., Sellen, J., Yap, C.-K.: Precision-sensitive Euclidean shortest path in 3-space. In: Proc. Ann. ACM Symp. Computational Geometry, pp. 350–359 (1995)
6. Ghosh, S.K., Mount, D.M.: An output sensitive algorithm for computing visibility graphs. SIAM J. Comput. **20**, 888–910 (1991)
7. Kapoor, S., Maheshwari, S.N.: Efficient algorithms for Euclidean shortest path and visibility problems with polygonal. In: Proc. Ann. ACM Sympos. Computational Geometry, pp. 172–182 (1988)
8. Klette, R., Rosenfeld, A.: Digital Geometry. Morgan Kaufmann, San Francisco (2004)
9. Lee, D.T.: Proximity and reachability in the plane. Ph.D. thesis, University of Illinois at Urbana–Champaign, Urbana (1978)
10. Li, F.: Exact and approximate algorithms for the calculation of shortest paths. Ph.D. thesis, Computer Science Department, The University of Auckland (2007)

11. Li, F., Klette, R.: Exact and approximate algorithms for the calculation of shortest paths. Technical report 2141, Institute for Mathematics and Its Applications, University of Minnesota (2006)
12. Li, F., Klette, R.: Analysis of the rubberband algorithm. Image Vis. Comput. **25**, 1588–1598 (2007)
13. Li, F., Klette, R.: Euclidean shortest paths in a simple polygon (invited talk). In: Bhattacharya, B.B., et al. (eds.) Algorithms, Architectures, and Information Systems Security, Indian Statistical Institute Platinum Jubilee Conf., pp. 3–24. World Scientific, Delhi (2008)
14. Overmars, M.H., Welzl, E.: New methods for constructing visibility graphs. In: Proc. Ann. ACM Sympos. Computational Geometry, pp. 164–171 (1988)
15. Sharir, M., Schorr, A.: On shortest paths in polyhedral spaces. SIAM J. Comput. **15**, 193–215 (1986)
16. Welzl, E.: Constructing the visibility graph for n line segments in $\mathcal{O}(n^2)$ time. Inf. Process. Lett. **20**, 167–171 (1985)
17. Yap, C.-K.: Towards exact geometric computation. Comput. Geom. **7**, 3–23 (1997)

Part II
Paths in the Plane

The image above shows an ant street in Argentina's Sierra Grande, near La Cumbrecita. Many ant species are able to find shortest paths (with respect to some 'obstacles') based on odorous chemical substances (called *pheromones*) deposited by ants. Ant-colony optimisation may be cited as an example where local optimisation is performed for converging towards a global optimum. However, ants will not always succeed, as controlled experiments have illustrated.

The second part of the book discusses algorithms for calculating exact or approximate ESPs in the plane. There are exact solutions for convex hulls, relative convex hulls, and more ESP problems in the plane, for example, based on a linear-time triangulation of a simple polygon, and there are also 'easy to program' δ-approximate solutions following the general RBA design.

Chapter 4
Convex Hulls in the Plane

Laughter is the shortest distance between two people.
Yakov Smirnoff (born 1951)

Convex hulls in the plane are examples for shortest paths around sets of points, or around simple polygons. Possibly these paths may be constrained by available polygonal regions. This chapter explains a few exact algorithms in this area which run typically in linear or $(n \log n)$-time with respect to a given input parameter n. However, the problems could also be solved approximately by rubberband algorithms.

4.1 Convex Sets

Assume that we have a finite set of points or polygonal sets, such as in Fig. 4.1, and the task is to find a shortest path "around" those, starting at p and ending at q, where obstacles such as the shown road need to be taken into account. Such a shortest path will follow the *convex hull* wherever possible. We define convex sets first, then convex hulls in the following section, and show then the correctness of this statement.

Definition 4.1 A subset $S \subset \mathbb{R}^m$ is *convex*, for $m \geq 2$, iff, for any pair of points $p, q \in S$, all straight line segments pq are fully contained in S.

For example, a circular or rectangular region in $\mathbb{R}^2$ is convex, but an *annulus* (i.e., the area between two concentric circles) is not. These are examples of *bounded sets*. A set is bounded iff it is contained in a disk ($m = 2$) or a sphere ($m \geq 3$) of some radius $r > 0$.

A non-convex set S does have *cavities* or *holes*; these are sets of points $r \notin S$ on straight segments pq with $p \in S$ and $q \in S$. Cavities and holes are easy to visualise for polygons in $\mathbb{R}^2$; see Fig. 4.2. However, imagine, for example, all the cavities of

F. Li, R. Klette, *Euclidean Shortest Paths*,
DOI 10.1007/978-1-4471-2256-2_4, © Springer-Verlag London Limited 2011

Fig. 4.1 A shortest path around a castle (shown as a group of buildings and walls), starting at p and ending at q, not allowed to cross the road

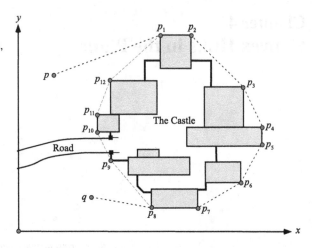

a sponge in $\mathbb{R}^3$. Let $S \subset \mathbb{R}^m$ be a bounded set. A *hole* is a bounded component in $\mathbb{R}^m \setminus S$, and a *cavity* is a subset of the unbounded component of $\mathbb{R}^m \setminus S$. See also Definition 4.4 in the next section for cavities in the case $m = 2$.

Example 4.1 A *half-plane* is unbounded. A half-plane is a set of points $(x, y) \in \mathbb{R}^2$ which satisfy a linear relation

$$ax + by \leq c$$

for a given triple a, b, and c of reals. A half-plane is also a convex set. For showing this, let $ax_1 + by_1 \leq c$ and $ax_2 + by_2 \leq c$, for two points $p_i = (x_i, y_i)$ in the half-plane, with $i = 1, 2$. We consider points on the straight segment

$$p_1 p_2 = \left\{ \left(x_1 + \lambda(x_2 - x_1), y_1 + \lambda(y_2 - y_1) \right) : 0 \leq \lambda \leq 1 \right\}.$$

Let $p_\lambda = (x_1 + \lambda(x_2 - x_1), y_1 + \lambda(y_2 - y_1))$ be on $p_1 p_2$, for $0 \leq \lambda \leq 1$. We have that

$$a\left(x_1 + \lambda(x_2 - x_1) \right) + b\left(y_1 + \lambda(y_2 - y_1) \right)$$
$$= (1 - \lambda)(ax_1 + by_1) + \lambda(ax_2 + by_2)$$
$$\leq (1 - \lambda)c + \lambda c = c.$$

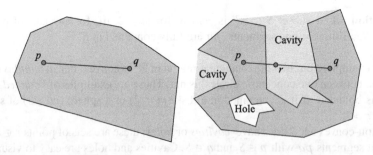

Fig. 4.2 *Left*: a convex set. *Right*: a non-convex set with two cavities and one hole

Fig. 4.3 Disk and tangential
line γ_p at point p on the
frontier of the disk. A second
point q on the line can be
used for defining the line with
an orientation; $\gamma_p = \overline{pq}$
means that the disk is left of
the oriented line $\overline{pq}$

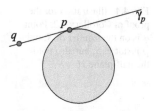

This shows that p_λ also satisfies the defining linear relation of the half-plane, for
any $0 \le \lambda \le 1$.   □

Corollary 4.1 *Let $S_1, S_2 \subset \mathbb{R}^2$ be two convex sets. Then the intersection $S_1 \cap S_2$ is
convex as well.*

Proof Consider two distinct points $p, q \in S_1 \cap S_2$. Because pq is in S_1 and also in
S_2, it is also in the intersection of both convex sets.   □

The proof of this corollary can be extended for showing that the intersection $\bigcap \mathcal{F}$
of any family $\mathcal{F} = \{S_a : a \in I\}$ of convex sets S_a, with $a \in I \subseteq \mathbb{R}$, is also a convex
set. For example, the intersection of a set of half-planes of the cardinality of $\mathbb{R}$ is
again a convex set, and such an intersection may define a disk; see the following
example.

Example 4.2 Figure 4.3 shows a disk and one of its tangential lines γ_p, defined by
a point p on the frontier of the disk. Line γ_p can be represented either as *oriented
line $\overline{pq}$* (i.e., orientation from p toward q) or oriented line $\overline{qp}$. Every straight line
defines two half-planes whose intersection is the given line. The oriented line $\overline{pq}$
and the half-plane on the left, denoted by $H_l(\overline{pq})$, contains the disk.

Now we rotate the pair of points p and q around the centre of the disk by an
angle α, defining new points p_α and q_α. Note that p_α is still on the frontier of the
disk.

Let $\mathcal{F} = \{H_l(\overline{p_\alpha q_\alpha}) : 0 \le \alpha < 2\pi)\}$. The intersection $\bigcap \mathcal{F}$ coincides with the
given disk. Family $\mathcal{F}$ has the cardinality of the real numbers.   □

Corollary 4.2 *A closed subset $S \subset \mathbb{R}^2$ is convex iff S is equal to the intersection of
all half-planes containing S.*

Proof Let $\mathcal{F} = \{H : S \subseteq H \wedge H \text{ is a half-plane}\}$. As commented after Corollary 4.1,
it follows that $\bigcap \mathcal{F}$ is convex. This already proves the sufficiency: if $S = \bigcap \mathcal{F}$ then
S is convex as well.

Obviously, we have that $S \subseteq \bigcap \mathcal{F}$. To prove the necessity, it remains to show that
both sets are equal which means that there cannot be any point p that is not in the
convex set S but in the intersection of all half-planes H containing S (i.e., p would
need to be in every half-plane $H \in \mathcal{F}$).

Fig. 4.4 Illustration for the
proof of Corollary 4.2. Point
q is on the frontier of S, point
p is not in S, but also not in
the half-plane H

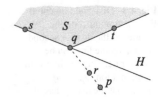

Let us assume that such a point p exists: $p \notin S$ but $p \in H$, for any $H \in \mathcal{F}$. Let q be a point in the frontier ∂S of S; because of $\partial S \subseteq S$ it follows that $q \neq p$. For any point $r \in pq$, with $r \neq q$, assume that we also have $r \notin S$; see Fig. 4.4.

Now, we rotate the ray $\vec{qp}$ clockwise around q to form two angles, at first $\sphericalangle pqs$ and then $\sphericalangle pqt$, such that qs and qt are tangents of S at s and t, respectively, where p rotates into s and later into t. Points s, q, and t are *collinear* (i.e., they are all on one straight line) iff q is on a continuous segment of the frontier of S (where 'left tangent' at q equals 'right tangent' at q).

Let $\sphericalangle sqt$ be the angle that contains point p. Because S is convex, it follows that $\sphericalangle sqt \geq 180°$. Let $H \in \mathcal{F}$ be such that the straight line $\overline{qs}$ (or line $\overline{qt}$) is its frontier and $p \notin H$. This is a contradiction to the selection of p. $\square$

Accordingly, for a convex set $S \subset \mathbb{R}^2$ we can ask for a minimum number of half-planes whose intersection defines S. For example, a rectangle is already defined by the intersection of four half-planes, but a circular region by the intersection of a family $\mathcal{F}$ of half-planes, where $\mathcal{F}$ has the cardinality of the real numbers. A simple convex polygon with n edges is defined by the intersection of n half-planes.

4.2 Convex Hull and Shortest Path; Area

A non-convex set S may be extended into a convex superset, called a *convex hull*. For example, we *could try* to define a *convex hull* $\mathrm{CH}(S)$ of a subset $S \subset \mathbb{R}^m$, for $m \geq 2$, as being the union of S with all of its cavities or holes, formally

$$S_1 = \bigcup_{p,q \in S} pq.$$

However, this will not yet define a convex set, in general. For example, let S be the set of four corners of a tetrahedron in $\mathbb{R}^3$. Then S_1 is only the set of all the six edges of that tetrahedron. We could repeat the process:

$$S_2 = \bigcup_{p,q \in S_1} pq.$$

Then we would have with S_2 the surface of the tetrahedron, formed by four triangular regions. Finally,

$$S_3 = \bigcup_{p,q \in S_2} pq.$$

would define with S_3 the whole set of all interior and frontier points of the tetra-hedron. In general, m steps of this iterative definition will define the convex hull $CH(S)$ of a set $S \subseteq \mathbb{R}^m$, for $m \geq 1$. This definition is equivalent to the following:

Definition 4.2 The *convex hull* $CH(S)$ of a subset $S \subset \mathbb{R}^m$, for $m \geq 2$, is a smallest (by contents) convex set that contains S.

The convex hull $CH(S)$ is uniquely specified this way for any $S \subset \mathbb{R}^m$: if there were two different convex hulls, then S would also be contained in the intersection of both, and the intersection would be again a convex set (see Corollary 4.1), but of smaller contents than the two assumed convex hulls. This contradicts the definition of a convex hull.

The convex hull is defined based on the measurable *content* of a set, which is the *area* for 2D, and the *volume* for 3D. We do not discuss volumes in this book, but the area of sets in a plane.

Definition 4.3 The *area* of a bounded and measurable set $S \subset \mathbb{R}^2$ is given by the following:

$$A(S) = \int_S dx \, dy.$$

The integration is for the *characteristic function* χ_S of the set S that is equal to 1 for all points in S, and equal to 0 otherwise. In a more extensive form, the definition reads as follows:

$$A(S) = \int_S \chi_S(x, y) \, dx \, dy.$$

Area is additive: If S_1 and S_2 are subsets of the Euclidean plane that have disjoint interiors, we have $A(S_1 \cup S_2) = A(S_1) + A(S_2)$. This allows us to measure the area of a set by partitioning the set into (e.g., convex) subsets and adding the areas of these subsets.

Example 4.3 Let T be a triangle pqr where $p = (x_1, y_1)$, $q = (x_2, y_2)$, and $r = (x_3, y_3)$. Then we have the following:

$$A(T) = \frac{1}{2} \cdot |D(p, q, r)| \tag{4.1}$$

where $D(p, q, r)$ is the *determinant*

$$\begin{vmatrix} x_1 & y_1 & 1 \\ x_2 & y_2 & 1 \\ x_3 & y_3 & 1 \end{vmatrix} = x_1 y_2 + x_3 y_1 + x_2 y_3 - x_3 y_2 - x_2 y_1 - x_1 y_3.$$

Note that $D(p, q, r)$ can be positive or negative; this can be used to define the orientation of the ordered triple (p, q, r).

Fig. 4.5 A *rubberband* spans around a set of (Taiwanese) *chopsticks*. Assume that the *rubberband* is in one plane, then it approximates a convex hull around the *sticks* in this plane; the *sticks* intersect this plane in a finite number of 'points'

More generally, let P be a simple polygon defined by a loop $\langle p_0, p_1, \ldots, p_n \rangle$ (i.e., $p_0 = p_n$ and $n \geq 3$). Let $p_i = (x_i, y_i)$. Then the following is true:

$$\mathcal{A}(P) = \frac{1}{2} \cdot \left| \sum_{i=1}^{n} x_i (y_{i+1} - y_{i-1}) \right| \tag{4.2}$$

where $y_{n+1} = y_1$. This general formula can be shown by complete induction, by partitioning a polygon with $n+1$ vertices into one with n vertices and one triangle, and by adding areas of both based on the additivity property of area.   □

Sets S in this book are typically (i) simple polygons P, (ii) simple or non-simple polyhedrons Π, both defined by finite sets of vertices, or (iii) finite sets of points. Figure 4.5 illustrates a convex hull of a finite set of 'points' (i.e., the intersection of the plane of the shown rubberband with the sticks). Convex hulls of such sets S are always bounded polygons.

For example, if S is a *singleton* (i.e., S only contains one point), then the convex hull is a degenerated polygon consisting only of one vertex; if S contains a finite number of points which are all collinear then the convex hull of S is a line segment. If S is non-collinear then the convex hull CH(S) is a simple polygon.

A cavity of a simple polygon is a simple polygon with three or more vertices; we may now define such a cavity precisely using the notion of a convex hull:[1]

Definition 4.4 A *cavity* CAV(P) of a simple polygon P is the topological closure of a component of CH(P) \ P.

A cavity is separated by a line segment (called the *cover* of the cavity) from the exterior $\mathbb{R}^2 \setminus$ CH(P), see Fig. 4.2. The difference CH(P) \ P does not

---

[1] Parts of this chapter have been contributed by *Gisela Klette*, at various places, starting from about here to the end of the chapter.

Fig. 4.6 For calculating the convex hull of the simple polygon P, we may embed it into a rectangle R, select in P the point p with maximal y-coordinate, add an edge pq parallel to the y-axis as shown, and calculate an ESP from p (say, starting at p to the left as indicated by the *dashed curve*) back to p in the non-simple polygon $\langle q_0, q, p, r, \ldots, s, p, q,$ $q_1, q_2, q_3, q_0 \rangle$ defined by the topological closure of $R \setminus P$ and the additional edge pq

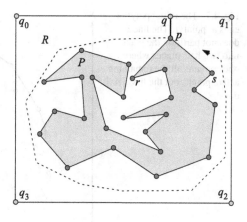

contain the two endpoints of the cover, which are in P. Thus, the difference $\mathrm{CH}(P) \setminus P$ 'disconnects' at those endpoints into different components. By taking the topological closure, the endpoints of the covers are also part of the cavities $\mathrm{CAV}_1(P), \ldots, \mathrm{CAV}_m(P)$ of a polygon P.

We are interested in calculating a Euclidean shortest path (ESP) around a finite set S of points, which might also be the set of all the vertices of a given polygon P; see Fig. 4.6. The following theorem provides the justification for this approach.

Theorem 4.1 *The convex hull of a finite non-collinear set S of points in $\mathbb{R}^2$ coincides with the ESP around S.*

Proof Let P be a polygon which contains S, and p_1 and p_2 be any two vertices of $\mathrm{CH}(S)$ defining an edge $p_1 p_2$ of this convex hull, and also a straight line γ in $\mathbb{R}^2$. Straight line γ intersects P at q_1 and q_2, and the line segment $q_1 q_2$ 'cuts off' a polygonal region from P which does not contain any point of S, thus defining a smaller polygon P_1 whose perimeter is shorter than that of P, and P_1 contains S. If P_1 does not yet coincide with $\mathrm{CH}(S)$ then continue for another edge of $\mathrm{CH}(S)$. $\square$

A polygonal path is an already sorted input for any geometric algorithm, contrary to a finite set of points which is still an unsorted input (and, thus, often more difficult to deal with).

Following Corollary 4.2, the computation of the convex hull $\mathrm{CH}(S)$, for any given finite set S of points in the plane, is equivalent to the computation of the intersection of all half-planes containing S. Algorithms for the convex hull calculation for a finite set of points are dealt with in the next section, followed by the simpler case of polygonal inputs.

The convex hull of a finite subset $S \subset \mathbb{R}^3$, or of a simple polyhedron $\Pi \subset \mathbb{R}^3$, is defined by a finite set of half-spaces. A proof can be analogous to the provided

Fig. 4.7 The initial polyline ends at point q. The line $\overline{qr}$ does not partition the plane into two half-planes where one contains all the remaining active points, but the line $\overline{qp}$ does

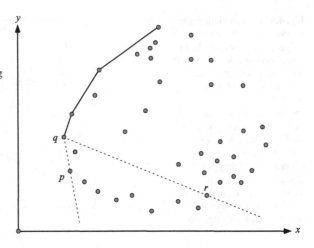

proof for Corollary 4.2; each half-space is defined by one plane in $\mathbb{R}^3$. Such a convex hull constitutes a convex polyhedron. In this book, we are interested in Euclidean shortest path problems, and there is no need for discussing algorithms for computing convex hulls of 3D sets of points.

4.3   Convex Hull of a Set of Points in the Plane

First of all, recall the throw-away principle as stated in Chap. 1: Is there a way to reduce the given set of points 'quickly' so that only points of potential impact on the calculated convex hull remain to be *active input data*?

For example, we can scan through the given set, calculate points which are extreme in directions $m \cdot \pi/4$, for $m = 0, 1, \ldots, 7$, and exclude all the points in the resulting octagon from continuing to be active. Note that these *extreme points* $p_i = (x_i, y_i)$, with $i = 1, \ldots, 8$, can easily be identified by searching for minima or maxima in x, y, $x + y$, or $x - y$ values. Note that the extreme points are also vertices of the convex hull, and they are *not* necessarily all pairwise different. If S only contains one point, then all the extreme points would coincide with this point.

It may happen that there is no point of S in the interior of the calculated octagon, but this should be a 'rare case' in general. For taking this case also into account in the worst-case sense, we stay with $n = |S|$ for describing the input complexity of convex hull algorithms in the following.

Assume that we start with identifying two subsequent vertices of the convex hull, defining a straight segment which we take as an *initial polyline*. We want to extend this polyline in counterclockwise order by more edges so that it finally becomes the frontier of the convex hull. Points that are already on the polyline are defined to be *inactive*. The current polyline ends at point q. We could apply a *straightforward algorithm* (see Fig. 4.7):

Fig. 4.8 Point p, calculated in the interior of CH(S), and an oriented reference line γ (here selected by point $p_1 \in S$) defines a polar coordinate system. All the points in S are sorted by, for example, increasing angular coordinates, here illustrated for point p_1 (i.e., angle zero) up to point p_{10}. (An application of the throw-away principle would have deactivated many points prior to this process.)

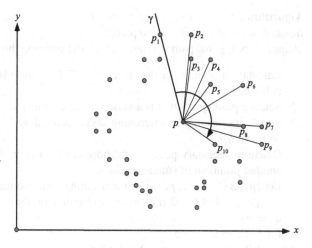

From all the remaining active points, select point p such that $\overline{qp}$ divides the plane into two half-planes where the one on the left contains all the remaining active points; extend the polyline by qp, and then p becomes the new endpoint of the extended polyline.

The time complexity of that algorithm is $\mathcal{O}(n^3)$. We can do better.

> No convex hull algorithm for n (unsorted) points in the plane can do better than in $(n \log n)$-worst-case time complexity. The *Graham algorithm* performs with upper bound $\mathcal{O}(n \log n)$, and it is thus optimal with respect to asymptotic worst-case time complexity.

Graham algorithm   The *Graham algorithm* consists of five parts:[2]

The first part identifies a point p in the interior of the convex hull CH(S), for example, by selecting the first non-collinear triple of points in S and taking p as the centroid of those three points.

The second part starts with defining an oriented line $\gamma = \overline{pp_1}$ (e.g., by connecting p with the point having a maximal y-coordinate in S; see Fig. 4.8), and expresses all points in S in polar coordinates (radius, angle) with respect to p and γ.

The third part sorts all points in S with respect to their angles (say, in increasing order, clockwise in Fig. 4.8).

In the fourth part, unnecessary points are deleted: those identical to p (i.e., radius equals zero) and those where there is another point with the same angle but a larger radius.

In the fifth part, scan through the remaining active points in order, starting with $i = 1$: if p_i, p_{i+1}, and p_{i+2} form an exterior angle on the frontier that is less than or equal to $180°$ then continue with p_{i-1}, p_i, and p_{i+2}, otherwise continue with

---

[2]Named after the mathematician *Ron Graham* (born 1935); he also popularised the Erdös numbers.

Algorithm 12 (Graham algorithm, 1972)

Input: A set S of n points in the plane.

Output: A polygonal path $\langle q_1, q_2, \ldots, q_m, q_1 \rangle$ defining the convex hull of S.

1: Calculate a point p in the interior of CH(S); stop if not possible (i.e., S is collinear).
2: Select a point $q \in S$ that is a vertex of the convex hull.
3: Sort the points in S by increasing angles defined by the polar coordinate system of p and $\overline{qp}$.
4: Delete unnecessary points (i.e., those equal to p or, if of identical angle with another point but of smaller radius).
5: Let $\langle p_1, p_2, \ldots, p_n, p_{n+1} \rangle$ be the resulting sorted sequence (e.g., $p_1 = q_1$), with $p_{n+1} = p_1$. Let $k = 2$ (i.e., number of points on the convex hull), $q_1 = p_1$, and $q_2 = p_2$.
6: **for** $i = 3$ to $n + 1$ **do**
7:     **while** angle $\sphericalangle(q_{k-1} q_k p_i) \leq 180°$ **do**
8:         **if** $k > 2$ **then**
9:             $k = k - 1$ **else** $k = i$ and $i = i + 1$
10:         **end if**
11:     **end while**
12:     $k = k + 1$ and $q_k = p_i$
13: **end for**

Fig. 4.9 Graham algorithm, where Step 2 contains a suggestion (q on the convex hull) that is not contained in the original algorithm, easy to ensure (e.g., take a point with maximal y-coordinate), and helps for defining the stop criterion

p_{i+1}, p_{i+2}, and p_{i+3}. Indices are calculated modulo the number of active points; stop after closing the loop on the frontier. The pseudocode is shown in Fig. 4.9. Figure 4.10 shows the resulting polyline of the convex hull up to p_{10}.

Parts 1, 2, 4, and 5 can all be done in linear time $\mathcal{O}(n)$, and Part 3 requires sorting, thus $\mathcal{O}(n \log n)$ in the worst-case sense.

Sklansky test The Graham algorithm depends on calculating angles used both for sorting and the test in Row 5. For increasing the speed, angles can be replaced by the cosine of angles. The test in Row 5 can be further sped-up by determinant calculations. For three points $p_i = (x_i, y_i)$, for $1 < i \leq 3$, it is sufficient to compute the determinant (compare Example 4.3)

$$D(p_1, p_2, p_3) = (y_3 - y_1)(x_2 - x_1) - (y_2 - y_1)(x_3 - x_1) = \begin{vmatrix} x_1 & y_1 & 1 \\ x_2 & y_2 & 1 \\ x_3 & y_3 & 1 \end{vmatrix} \quad (4.3)$$

to decide whether the angle $\sphericalangle(p_1 p_2 p_3)$ is less than, or equal to 180°. If less than 180°, then this is also called a *left turn*, being equal to 180° defines collinear points, and larger than 180° a *right turn*. The value of the determinant $D(p_1, p_2, p_3)$ of

Fig. 4.10 The sorted points define a simple polygon (shown by *dashed edges*). Only those vertices remain on the convex hull where the angle of the frontier is larger than 180°

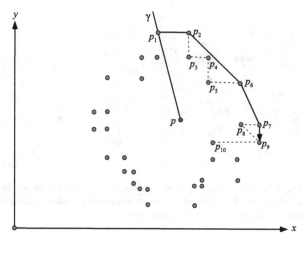

Fig. 4.11 Three points in a right-hand xy-coordinate system. The points describe a left turn in the order p_1, p_2, p_3, and a right turn in the order p_3, p_2, p_1

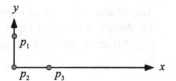

three successive points p_1, p_2, and p_3 equals the signed area of the trapezoid defined by those three points; see Eqs. (4.1) and (4.3). The use of the determinant, known as *Sklansky test*,[3] avoids the computation of trigonometric functions (and thus we stay with arithmetic algorithms as promised in Sect. 1.1).

For example, let $p_1 = (0, 1)$, $p_2 = (0, 0)$, and $p_3 = (1, 0)$; see Fig. 4.11. Then we have that $D(p_1, p_2, p_3) = 1$. This single result is already sufficient to know that a positive value of D always defines a left turn in a right-hand coordinate system, but a right turn in a left-hand coordinate system.

Assuming a right-hand coordinate system and clockwise orientation of the polygonal path, we can replace the test of the angle in Row 7 of Fig. 4.9 by the Sklansky test "$D(p_{k-1} p_k p_i) \leq 0$?". A negative value of D defines a left turn at a vertex, and this vertex is also called a *concave vertex*. A positive value of D defines a right turn at a *convex vertex*.

A quickhull algorithm The divide-and-conquer principle leads to an algorithm that determines the vertices of the convex hull of a finite set of points in the plane with $\mathcal{O}(n \log n)$ expected time complexity. The basic strategy of the algorithm (divide-and-conquer principle) was first time applied in Quicksort.[4]

---

[3] *Jack Sklansky* (born 1928) made pioneering contributions to image analysis.

[4] Quicksort was developed in 1960 by *Sir Charles Antony Richard Hoare* (born in 1934).

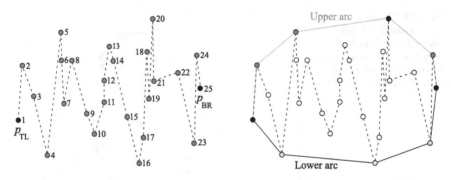

Fig. 4.12 *Left*: Sorted input with the *top-most left-most* point p_{TL} and the *bottom-most right-most* point p_{BR}. *Right*: *Upper* and *lower* arc obtained by scanning the sorted sequence from left to right, p_{TL} to p_{BR}, or vice-versa, respectively

> The *divide-and-conquer principle*: divide the given input, solve the problem for the smaller sets of input data, and merge results into a solution for the original input. Apply this principle recursively.

For a given array of points and a selected pivot, the algorithm starts by partitioning the array into two subarrays. One subarray includes all points on the left of the selected pivot, and the other subarray includes all points on the right of the selected pivot. The procedure continues for each subarray until the *base case* has been reached where the length of each subarray is 1; the convex hull of a single point is the point itself. For merging the convex hulls on the left and on the right, an upper and a lower tangential lines are calculated for defining the merging edges.

This algorithm has $\mathcal{O}(n^2)$ worst-case time complexity, but $\mathcal{O}(n \log n)$ expected time complexity.

We refrain from giving more details here for this (very efficient and popular) recursive algorithm; they can be easily found on the net also with free downloads of sources and animated demonstrations of the algorithm.

A rubberband algorithm   Figure 4.12 illustrates a rubberband algorithm for calculating the convex hull of a finite set of points in the plane.

In a first run through the given set, points are sorted left to right along the x-axis. Let p_{TL} be the point with the smallest x-coordinate. If there are more than one point with the smallest x-coordinate then we take the one with the largest y-coordinate. Analogously, p_{BR} is the one with the smallest y-coordinate of all points with the largest x-coordinate. In general, if there are points with identical x-coordinates, then they are sorted by decreasing y-coordinates.

At the beginning of the second and third run, all points (i.e., vertices) are active. During a run some of the points become inactive by deletion. This is best implemented by using a stack (see the Sklansky algorithm in the next section).

The second run considers the points from p_{TL} to p_{BR}, and the third run from p_{BR} to p_{TL}. We describe the second run; the third run is analogous but into the opposite direction.

For a current point p, let p_b be the nearest active vertex to the left ('b' from 'before'), and p_n be the nearest active vertex to the right ('n' from 'next'). At the start, we assume that $p_b = p = p_{TL}$. Now we proceed from left to right by considering the following cases:

Case A: Point p is collinear or concave in the sequence p_b, p, and p_n. We delete point p, p_b becomes the new point p, p_b is the active point before this updated p (or p_{TL} if already $p = p_{TL}$), and p_n remains.

Case B: Point p is convex in the sequence p_b, p, and p_n. Then we move p forward: p becomes p_b, p_n becomes p, and the next active vertex is the new p_n.

Note that we 'stretch the rubberband' only by local (i.e., three-point) optimisations. The sorted sequence describes a monotonously increasing (or decreasing) polyline. For such a polyline, the applied simple backtracking strategy suffices for selecting all the convex vertices. The time complexity of the first run is defined by sorting, and the second and third run are of linear time complexity. Thus, the worst-case complexity is $\mathcal{O}(n \log n)$.

Did you notice that the second and third run coincide (methodically) with the fifth part of the Graham algorithm? Thus, there is also a 'rubberband interpretation' of this fifth part.

4.4   Convex Hull of a Simple Polygon or Polyline

The convex hull of a simple polygon in 2D (see Fig. 4.13) is equal to the convex hull of its set of vertices, and we could also apply one of the $\mathcal{O}(n \log n)$ algorithms from the previous section. However, the 'sorted' input of vertices of a simple polygon allows us to calculate its convex hull actually in linear time.

Visibility has been discussed already next to Fig. 1.12. 'Visibility from the outside' defines a class of simple polygons which is often sufficient for modelling possible inputs in a given real-world application:

Definition 4.5 A polygon P is *visible from the outside* iff for any point q on the frontier of P there is a ray with q as the start point that does not intersect the polygon at any other point than q.

Part 5 of the Graham algorithm (i.e., the repeated 'local rubberband optimisation'), sped-up by using the Sklansky test, is known as the *Sklansky algorithm* for the computation of the convex hull of a simple polygon, provided that the input polygon is completely visible from the outside. It may fail for input polygons that are not visible from the outside.

Fig. 4.13 The convex hull of a shown *isothetic* (i.e., all edges parallel to coordinate axes) simple polygon. The shown polygon is also visible from the outside

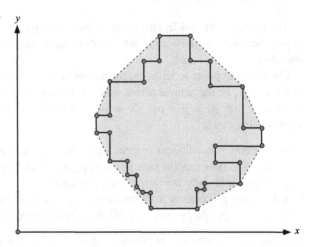

Sklansky algorithm   We provide a version of the algorithm that uses a *stack* (i.e., a first-in-last-out 1-dimensional data structure) with elements $q_0 = p_0$ (at the bottom), $q_1, \ldots, q_{top-1}, q_{top}$ at a given time assumed to be vertices of the convex hull. Operations PUSH or POP add or delete the top element, with an update top + 1 or top − 1, respectively. The use of a stack prepares for more general input cases later on.

The algorithm traces the frontier of the polygon in clockwise or counterclockwise order. For every next vertex p_k, it computes the determinant $D(q_{top-1}, q_{top}, p_k)$. The algorithm (see Fig. 4.14) works in linear time in the number of given vertices; there is no sorting included, and every vertex, once discarded, is never reconsidered.

Algorithm 13 (Sklansky algorithm, 1972)
Input: A simple polygon P, visible from the outside, defined by a loop $\langle p_0, p_1, \ldots, p_{n-1}, p_0 \rangle$.
Output: Vertices of the convex hull CH(S) in a stack.

1: Select a start vertex p_0 in the loop (modulo n) such that it is a vertex of the convex hull.
2: Let p_1 be the next vertex of P. PUSH(p_0) and PUSH(p_1) onto the stack. Let $k = 2$ and $p_n = p_0$.
3: **while** $k \leq n$ **do**
4:      **while** $D(q_{top-1}, q_{top}, p_k) \leq 0$ and top > 0 **do**
5:          POP {i.e., q_{top} is deleted from the stack and top = top − 1}
6:      **end while**
7:      PUSH p_k {i.e., top = top + 1 and $q_{top} = p_k$} and $k = k + 1$
8: **end while**
9: Vertices of the convex hull are in the stack.

Fig. 4.14 Version of the Sklansky algorithm using a stack

Fig. 4.15 A polygon with a non-convex cavity. Vertices p_2 and p_7 define the cover $p_2 p_7$ of this cavity

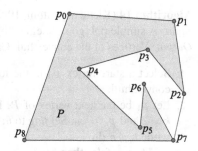

Example 4.4 The simple polygon in Fig. 4.15 is not visible from the outside. Assume we take this as an input for the Sklansky algorithm.

We start with $q_0 = p_0, \ldots, q_3 = p_3$. Then we have the first POP, remove p_3 from the stack and have $q_3 = p_4$. This happens again; we remove p_4 and have $q_3 = p_5$. Now, p_2, p_5, and p_6 form a right turn. Thus, p_5 stays in the stack, and we move on with $q_4 = p_6$.

Vertices p_5, p_6, and p_7 are again a right turn. Vertices p_5 and p_6 will not be removed from the stack.   □

Klette algorithm   The *Klette algorithm* keeps control whether the sequence of vertices is at a given time inside of a non-convex cavity or not. A cavity (see Definition 4.4) is a simple polygon defined by a cover.

The binary parameter *flag* is *true* if the algorithm is 'currently in a non-convex cavity', that is, it was crossing a cover once, and *flag* is back to *false* if the algorithm leaves the recognised non-convex cavity again.

Let q^{prev} be the vertex of P immediately preceding q in the traced polyline of all vertices of P. The vertex $q_{\text{top}}^{\text{prev}}$ in the pseudocode in Fig. 4.16 is the vertex of P immediately preceding the vertex that is currently on the top of the stack.

For example, if $q_{\text{top}} = p_7$ then $q_{\text{top}}^{\text{prev}} = p_6$. This vertex $q_{\text{top}}^{\text{prev}}$ is not necessarily an element of the stack. Vertices inside a cavity are always in the interior of the convex hull and not pushed into the stack. The Klette algorithm is correct for any simple polygon.

Example 4.5 For the polygon in Fig. 4.15, *flag* becomes *true* when the algorithm is at $k = 6$: here we have $q_0 = p_0$, $q_1 = p_1$, $q_2 = p_2$, $q_3 = p_5$, top $= 3$, vertex $q_{\text{top}}^{\text{prev}} = p_5^{\text{prev}} = p_4$, and thus with $D(p_4, q_3, p_6) = D(p_4, p_5, p_6) < 0$ a left turn.

For $k = 7$ we are back to *flag* $=$ *false* because the algorithm crosses again the line segment $p_2 p_5$. Note that a crossing of a cover was detected at $k = 6$, when $p_5 p_6$ 'turned backward into the cavity', compared to line segment $p_2 p_5$.   □

Melkman algorithm   A simple polygon is a special case of a *simple polyline*. A simple polyline is a chain of line segments in the plane such that the only common points are the end vertices and the start vertices of two consecutive line segments. Line segments do not intersect with each other. The chain does not have to form a loop.

Algorithm 14 (Klette algorithm, 1983)
Input: A simple polygon P defined by a loop $\langle p_0, p_1, \ldots, p_{n-1}, p_0 \rangle$.
Output: Vertices of the convex hull CH(S) in a stack.

1: Select a start vertex p_0 in the loop (modulo n) such that it is a vertex of the convex hull.
2: Let p_1 be the next vertex of P. PUSH(p_0) and PUSH(p_1) onto the stack. Let $k = 2$ and $p_n = p_0$. Set *flag* to *false*.
3: **while** $k \leq n$ **do**
4:   **if** *flag* $=$ *false* **then**
5:     **while** $D(q_{\text{top}-1}, q_{\text{top}}, p_k) \leq 0$ and top > 0 **do**
6:       POP
7:     **end while**
8:     **if** top $= 0$ or $D(q_{\text{top}}^{\text{prev}}, q_{\text{top}}, p_k) > 0$ **then**
9:       PUSH p_k and $k = k + 1$
10:     **else**
11:       Set *flag* to *true*.
12:     **end if**
13:   **else**
14:     **while** $D(q_{\text{top}-1}, q_{\text{top}}, p_k) \leq 0$ and top > 0 **do**
15:       POP
16:     **end while**
17:     PUSH p_k and $k = k + 1$
18:     Set *flag* to *false*.
19:   **end if**
20: **end while**
21: Vertices of the convex hull are in the stack.

Fig. 4.16 Klette algorithm

The *Melkman algorithm*[5] is a very efficient implementation for the computation of the convex hull of a simple polyline. The algorithm uses a *deque* with two ends rather than a stack. Elements can be added or removed at both ends (called the *bottom* and the *top*). Elements can be pushed (PUSH) or popped off (POP) at the top (similar to the operations of a stack), but they can also be inserted (INSERT) or deleted (DELETE) at the bottom.

Let $\{q_{\text{bot}}, q_{\text{bot}+1}, \ldots, q_{\text{top}-1}, q_{\text{top}}\}$ be the elements of the deque. INSERT p decreases bot by one; the previous element at the bottom is now at position bot $+ 1$; vertex p moves into position bot. The DELETE p operation removes the element at position bot by increasing bot by one.

The convex hull at the end of stage k (between Rows 2 an 14 in Fig. 4.17) of already processed vertices is an ordered list of elements saved in the deque $q_k = \{q_{\text{bot}}, \ldots, q_{\text{top}}\}$ with $q_{\text{bot}} = q_{\text{top}}$ at any stage.

---

[5] *Avraham Melkman* is at the Ben Gurion University of the Negev.

Algorithm 15 (Melkman algorithm, 1987)

Input: A simple polyline $\rho = \langle p_0, p_1, \ldots, p_{n-1} \rangle$ with n vertices in the plane.

Output: The vertices of the convex hull of this polyline.

1: PUSH p_0 and then p_1 onto the deque. PUSH p_2 and INSERT p_2 (i.e., at both ends). Those three vertices must constitute a triangle. Let $k = 3$.

2: **while** $k < n$ **do**

3:    **while** $D(q_{\text{bot}}, q_{\text{bot}+1}, p_k) > 0$ AND $D(q_{\text{top}-1}, q_{\text{top}}, p_k) > 0$ AND $k < n$ **do**

4:       $k = k + 1$

5:    **end while**

6:    **while** $D(q_{\text{bot}}, q_{\text{bot}+1}, p_k) \le 0$ **do**

7:       DELETE

8:    **end while**

9:    INSERT p_k

10:   **while** $D(q_{\text{top}-1}, q_{\text{top}}, p_k) \le 0$ **do**

11:      POP

12:   **end while**

13:   PUSH p_k and $k = k + 1$

14: **end while**

15: Vertices of the convex hull are in the deque.

Fig. 4.17 Melkman algorithm

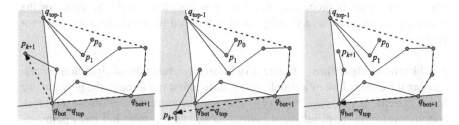

Fig. 4.18 Three different cases for positions of vertex p_{k+1}. The *dashed line* shows the convex hull calculated so far up to p_{k+1}, where the new vertex p_{k+1} becomes the new position of $q_{\text{bot}} = q_{\text{top}}$ in both cases shown on the *left* and at the *middle*

Stage $k + 1$ (see Fig. 4.18) starts with deciding whether the new vertex p_{k+1} is (already) inside the convex hull computed up to stage k. This decision requires the computation of two determinants.

If p_{k+1} is left of both straight lines $\overline{q_{\text{bot}}q_{\text{bot}+1}}$ and $\overline{q_{\text{top}-1}q_{\text{top}}}$ then p_{k+1} is inside the convex hull calculated at stage k, and the deque does not change; the algorithm moves on to the next vertex p_{k+2}.

If p_{k+1} is right of $\overline{q_{\text{bot}}q_{\text{bot}+1}}$ then the bottom of the deque will change. We delete the element q_{bot} and we check whether p_{k+1} is right of the segment defined by $\overline{q_{\text{bot}+1}q_{\text{bot}+2}}$ and so on until p_{k+1} is on the left. Then we insert p_{k+1}.

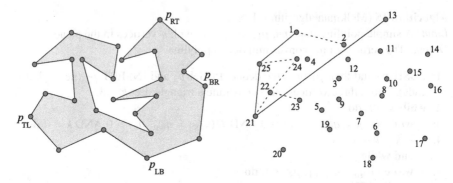

Fig. 4.19 *Left*: a simple polygon with its extreme points p_{RT}, p_{BR}, p_{LB}, and p_{TL}. The rubberband is shown with endpoints p_{TL} and p_{RT}, for local optimisation at point p. *Right*: The vertices of the polygon are numbered, and the scan through this loop is modulo 25. When testing points 21, 22, and 23, point 23 is deleted due to case B. The same case applies for the deletion of point 2 when considering points 25, 1, and 2, or for the deletion of point 15 when considering points 13, 14, and 15. Points 19 and 24 are deleted due to case A, and point 22 due to case C

If p_{k+1} is right of $\overline{q_{\text{top}-1}q_{\text{top}}}$ then the top of the deque will change. We pop off the element q_{top} and we check whether p_{k+1} is right of the segment defined by $\overline{q_{\text{top}-2}q_{\text{top}-1}}$ and so on until p_{k+1} is on the left. Then we push p_{k+1} on top of the deque.

The Melkman algorithm computes the convex hull for simple polylines in linear time. Every vertex is processed once (this was also true for the Sklansky or the Klette algorithm), and, as a novelty, there is no need to determine a special pivot vertex.

A rubberband algorithm   Figure 4.19 illustrates a rubberband algorithm for calculating the convex hull of a simple polygon. In a first run through the polygon, we detect extreme points p_{RT}, p_{BR}, p_{LB}, and p_{TL} meaning the right-most top-most, bottom-most right-most, left-most bottom-most, and top-most left-most point, respectively.

The second run starts and ends at one of those extreme points. Assume that we consider points between two subsequent extreme points q_1 and q_2 (e.g., $q_1 = p_{TL}$ and $q_2 = p_{RT}$); this situation repeats four times during the second run. Vertices are either inactive or active. At the beginning, all vertices are active. Extreme points remain always active.

The current vertex p starts at q_1 and runs through all the active vertices to q_2. For a vertex p, we also have active vertex p_b before p and active vertex p_a after p. We delete vertices (i.e., those become inactive) according to the following three cases:

Case A: Point p is concave in the sequence p_b, p, and p_a. We delete point p and p_b becomes the new point p, p_a remains, and the new p_b is the active vertex before p. (Note: this is the fifth part of the Graham algorithm again.)

Case B: Point p_a is concave in the sequence p, p_a, and q_2. We delete point p_a and continue with p (now having a new point p_a).

Fig. 4.20 The relative convex hull (*shaded area with dashed frontier*) of the shown inner and outer polygon. It is identical to the minimum-perimeter polygon, circumscribing the inner polygon and contained in the outer polygon

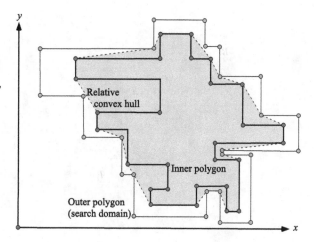

Case C: Point p_b is concave in the sequence q_1, p_b, and p. We delete point p_b and continue with p (now having a new point p_b).

When considering one triple p_b, p, and p_a, we test for cases A, B, and C in sequence. If none of the cases applies then p_a becomes the new p. We stop when reaching the same extreme point where the second run started.

The use of a doubly-linked list for storing the vertices of the given polygon ensures an efficient implementation of this linear time algorithm.

4.5 Relative Convex Hulls

The 'relative convex hull' is identical to the *minimum-perimeter polygon* (MPP). This polygon circumscribes a given discrete set (e.g., the inner polygon in Fig. 4.20) constrained by a given (polygonal) search domain (the outer polygon in Fig. 4.20), and is of shortest length.

Definition 4.6 Let $A \subseteq B \subseteq \mathbb{R}^2$. Set A is *B-convex* iff all line segments in B between any two distinct points $p, q \in A$ belong also to A.

Figure 4.21 shows three sets S, A, and B, with $S \subset A \subset B$. Set A is B-convex, and set S is neither A-convex nor B-convex.

Definition 4.7 Let $S \subseteq B \subseteq \mathbb{R}^2$. The *convex hull of S relative to B* is the intersection of all B-convex sets containing S.

The *B-convex hull of S* (see Fig. 4.22) is short for the convex hull of S relative to B, and this is also formally expressed as $\mathrm{CH}_B(S)$. The *relative convex hull of S* is a set $\mathrm{CH}_B(S)$ for some unspecified set B containing S.

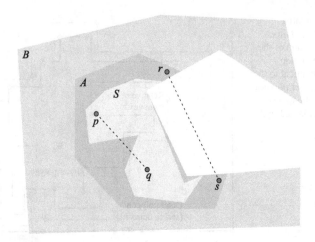

Fig. 4.21 Points p, q, r, and s are all in A. The line segment pq is in B and also in A. The line segment rs is not in B, thus not of interest for deciding about B-convexity. Points p and q are also in S, the line segment pq is in B but not in S; thus, S is not B-convex. S is also not A-convex

Fig. 4.22 The B-convex hull of S

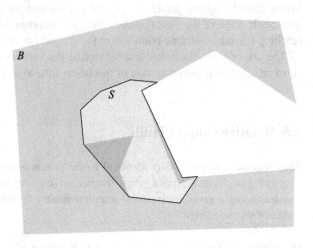

From Definition 4.4 it follows that the B-convex hull is the smallest B-convex set containing S. Any convex set S is also B-convex, only provided that S is contained in B. Any set B itself is also B-convex. If S is a simple polygon and V the set of vertices of S, then $CH_B(S) = CH_B(V)$, for any set B containing S. If $S \subseteq A \subseteq B$ then

$$S \subseteq CH_A(S) \subseteq CH_B(S) \subseteq CH(S). \tag{4.4}$$

Informally, a larger set B defines a 'more relaxed constraint' for S compared to that defined by set A, and the B-convex hull is thus 'closer' to the convex hull $CH(S)$. Analogously to Theorem 4.1, we also have:

Corollary 4.3 *Let S be a finite set of points contained in a simple polygon B. The frontier of the B-convex hull of S coincides with the ESP that circumscribes S and is contained in B.*

This ESP is uniquely defined for given sets S and P. Instead of a finite set S, we may also consider a simple polygon having S as its set of vertices. Then we call S the *inner polygon* and P the *outer polygon*.

Algorithms for calculating the P-convex hull of polygon S can make use of the following property (see Fig. 4.20 for an example):

Theorem 4.2 *Only convex vertices of polygon S and only concave vertices of polygon P are candidates for vertices of the relative convex hull* $CH_P(S)$, *for any simple polygons S and P, with $S \subseteq P$.*

4.6 Minimum-Length Polygons in Digital Pictures

The *Klette–Kovalevsky–Yip algorithm* (see pseudocode in Fig. 4.24) is based on Theorem 4.2 and was designed for estimating the perimeter of components in a digital picture.[6] Digital geometry applies adjacency relations between pixels. For example, 8-adjacency defines an *8-path* as a sequence of grid points where two subsequent vertices in this path have a d_∞-distance equal to 1 (i.e., they are diagonal or isothetic neighbours). The *4-border* of a set of pixels contains exactly all those pixels of this set which have at least one 4-adjacent pixel outside of this set.

The inner polygon is an 8-path through the 4-border of the traced component, and the outer polygon is again an 8-path, a virtual polygon along the *co-border*. The 8-path of the co-border is 'parallel' to the 8-path of the inner polygon at a distance of one pixel; see Fig. 4.23.

While tracing the 8-path of the 4-border of a given component, all its concave vertices are replaced in Line 1 (see Fig. 4.24) by corresponding concave vertices of the virtual 8-path of the co-border. In general, this is a one-to-one mapping, but there may be cases where two or three concave vertices of the inner 8-path are assigned to the same concave vertex of the outer path; see Fig. 4.23. However, we then replace those two or three concave vertices all by the same concave vertex of the outer 8-path. Identical vertices are eliminated in the algorithm because they satisfy the collinearity property.

In the resulting list CC, a vertex is marked by a plus sign if it is a convex vertex of the inner 8-path, or by a minus sign if it is a concave vertex of the outer 8-path. Collinear (e.g., repeated) vertices are ignored because they are not candidates for

---

[6]The algorithm published in [11] (see also [10]) calculates a polygonal path between inner and outer polygon whose length is multigrid convergent. However, the case discussion was incomplete; see [2] for showing an input where the calculated polygonal path actually differs from the MLP. The missed case has been taken care of in the given pseudocode.

Fig. 4.23 There is a unique mapping of concave vertices on the border of a region (i.e., the inner polygon) onto concave vertices of the virtual co-border (i.e., the outer polygon). In this example, there are two cases where two or three concave vertices map onto one concave vertex. There are also two cases where edges of the outer polygon 'are touching each-other'

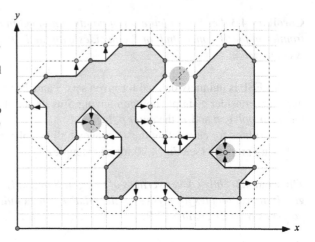

the *minimum-length polygon* (MLP), and they could already be deleted in Step 1 in Fig. 4.24.[7]

Now, the algorithm runs a second time through list CC. It starts at a vertex p_1 already known to be an MLP vertex (e.g., maximum y-coordinate). After each de-tected MLP vertex p_k, the algorithm passes all those vertices v which can still be connected by a line segment $p_k v$ contained in the difference set of outer polygon minus inner polygon; the vertex furthest away from p_k with this property defines the next MLP vertex p_{k+1}. See Fig. 4.25.

Both runs of the algorithm through the given CC list are in linear time. The pseudocode in Fig. 4.24 also contains the calculation of the length of the MLP.

4.7 Relative Convex Hulls—The General Case

Now we consider the general case where the inner and outer polygon A and B are simple polygons with $A \subseteq B$.

Toussaint algorithm   This algorithm assumes a similar cut pq as in Fig. 4.6 where the outer polygon replaces the role of the rectangle, used in this figure for embedding a given simple polygon. See Fig. 4.26. The cut connects now one ex-treme vertex p of the inner polygon (i.e., known to be on the relative convex hull) with a point q on the outer polygon having the same x-coordinate as p. The rela-tive convex hull is defined by an ESP in the generated (via the cut) simple polygon $Q = (B \setminus A)^\bullet$ that starts and ends at vertex p and circumscribes the inner polygon.

The algorithm assumes a triangulation of Q that can be done in $\mathcal{O}(n \log \log n)$ time (see Chap. 5). This is followed by finding the shortest path that can be

---

[7]The name 'minimum-length polygon' is common in digital geometry for this particular case of ESPs, defined by a border and its co-border of a component in a digital image.

Algorithm 16 (Klette–Kovalevsky–Yip algorithm, 1999)

Input: A border loop of length n of convex or concave vertices of a component in a digital picture.

Output: The vertices of the MLP saved in a list MLP $= [p_1, \ldots, p_m]$ and the length L of the MLP.

1: Compute a list CC $= [v_1, \ldots, v_n]$; it contains all convex vertices of the border loop; all concave vertices of the border loop are replaced by concave vertices of its co-border; each vertex v_j is labelled by the sign of $D(v_{j-1}, v_j, v_{j+1})$.

2: Initialise: $k = 1$ {k is the index for new MLP-vertices}, $a = 1$, $b = 1$, $j = 2$ {j is the index for all vertices in the given list CC}, $p_1 = v_1$ {p_1 is the first MLP-vertex}, $v_{n+1} = v_1$, and $L = 0$ {L is the length of the MLP}.

3: **while** $j \leq n + 1$ **do**

4:   **if** $D(p_k, v_b, v_j) > 0$ **then**

5:     **if** $D(p_k, v_a, v_j) \geq 0$ **then**

6:       **if** v_j has a negative sign {i.e., it is a concave vertex} **then**

7:         $k = k + 1$, $p_k = v_b$, $j = b$, $a = b$, $L = L + d_e(p_{k-1}, p_k)$

8:       **end if**

9:     **else**

10:       **if** v_j is labelled by a positive sign {i.e., it is a convex vertex} **then**

11:         $b = j$

12:       **else**

13:         $a = j$

14:       **end if**

15:     **end if**

16:   **else**

17:     **if** $D(p_k, v_a, v_j) \geq 0$ **then**

18:       **if** v_j is labelled by a positive sign **then**

19:         $b = j$

20:       **else**

21:         $a = j$

22:       **end if**

23:     **else**

24:       **if** v_j has a positive sign **then**

25:         $k = k + 1$, $p_k = v_a$, $j = a$, $b = a$, $L = L + d_e(p_{k-1}, p_k)$

26:       **end if**

27:     **end if**

28:   **end if**

29:   $j = j + 1$

30: **end while**

31: Now we have $m = k$ vertices in the MLP $= [p_1, \ldots, p_m]$, and it is of length L.

Fig. 4.24 Klette–Kovalevsky–Yip algorithm (revised)

Fig. 4.25 The relative
convex hull is defined by
replacing a few segments of
the *dashed polyline* (i.e., the
initial CC list) by the shown
line segments. Non-MLP
vertices are shown with *white*
interior

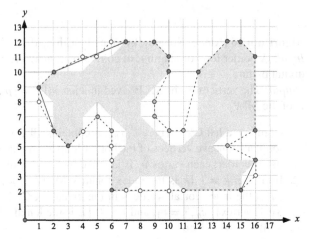

Fig. 4.26 The cut pq defines
the *light-gray shaded*
(non-simple) polygon. The
dashed line is a shortest path
in this polygon from p
(departing to the right) to p
(arriving from the left)

done in linear time on the given triangulation. The whole algorithm works thus in
$\mathcal{O}(n \log \log n)$ time.

Preliminaries for a recursive algorithm   The following properties of the relative convex hull $\mathrm{RCH}_B(A)$, for simple polygons A and B, $A \subseteq B$, prepare for the
detailed presentation of a recursive algorithm. At first we state:

Theorem 4.3 *The B-convex hull of a simple polygon A is equal to the convex hull
of A iff the convex hull of A is completely contained in B (i.e., $\mathrm{CH}(A) \subseteq B$).*

Proof If $\mathrm{CH}(A) \subseteq B$ then $\mathrm{RCH}_B(A) = \mathrm{CH}(A)$ because B is not constraining
$\mathrm{CH}(A)$ at all. On the other hand, we always have that $\mathrm{RCH}_B(A) \subseteq B$. Thus, if
$\mathrm{RCH}_B(A) = \mathrm{CH}(A)$ then also $\mathrm{CH}(A) \subseteq B$.   □

The B-convex hull of a simple polygon A is different to $\mathrm{CH}(A)$ if there exists at
least one cavity $\mathrm{CAV}_i(A)$ in A and one cavity $\mathrm{CAV}_j(B)$ in B such that the intersection $\mathrm{CAV}_i(A)^\circ \cap \mathrm{CAV}_j(B)$ is not empty. (Recall: S° is the interior of set S.)

According to the following theorem, we can determine a subset of all vertices of
the relative convex hull by computing the vertices of the convex hull for the inner
polygon.

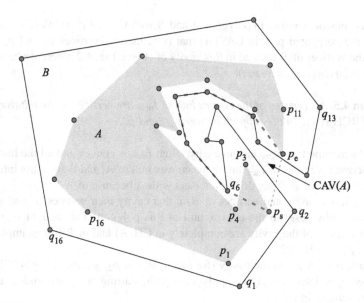

Fig. 4.27 Polygon A (*shaded*) has one cavity $CAV(A)$, with the cover $p_s p_e$.—Illustrating Theorem 4.5, I_{new} has one cavity with the cover $q_6 p_s$, and with vertex p_3 of A in its interior, and a second cavity with the cover $q_{10} p_e$, and no vertex of A in its interior

Theorem 4.4 *All vertices of the convex hull of a simple polygon A, contained in a simple polygon B, are also vertices of the B-convex hull of A.*

Proof We consider the start and end vertices p_s and p_e of the cover of a cavity $CAV(A)$; see Fig. 4.27. They are by definition vertices of the convex hull of A.

If the intersection $CAV(A)° \cap CAV(B)$ is empty, for any cavity $CAV(B)$ of B, then B has no vertices inside this cavity of A, and because $A \subseteq B$, the cover $p_s p_e$ connects vertices in A as well as points in B. Thus, p_s and p_e are vertices of $CH_B(A)$.

Now assume that $CAV(A)° \cap CAV(B)$ is not empty, for a cavity $CAV(B)$ of B. Then there is at least one vertex q_i of B in $CAV(A)$, and the line segment $p_s p_e$ crosses the frontier as well as the exterior of B. The frontier of $CH_B(A)$ contains thus a polygonal path from p_s to p_e also containing at least one vertex of B in $CAV(A)$. The segments of this polygonal path do not cross the frontier of A, and do also not cross the frontier of B. Thus, p_s and p_e are vertices of $CH_B(A)$.

Finally, consider two consecutive vertices of $CH(A)$ that are not a start or end vertex of a cavity; they are vertices of $CH_B(A)$ per definition of the relative convex hull.   □

The theorem above tells us that we can copy all vertices of the convex hull of A to the B-convex hull of A. In some cases, we have to insert additional vertices for the relative convex hull between cover endpoints p_s and p_e of a cavity in A if there are vertices of B in this cavity.

Let us consider cavities in polygons A and B with $\text{CAV}(A)^\circ \cap \text{CAV}(B) \neq \emptyset$. Let I_{new} be the polygonal path in $\text{CAV}(A)$ that is defined by vertices p_s and p_e of A and all the vertices of B located in this cavity of A (see Fig. 4.27; here we have that $I_{\text{new}} = \langle p_s, q_3, q_4, \ldots, q_{11}, p_e \rangle$).

Theorem 4.5 *All vertices of the convex hull of I_{new} are vertices of the relative convex hull* $\text{RCH}_B(A)$, *on the subpath between p_s and p_e.*

Proof We assume that I_{new} has no cavity. Then I_{new} is convex and all the line segments between vertices of I_{new} are in the convex hull of A, and they do not intersect with the exterior of B. They do not intersect with A because of $A \subseteq B$.

If I_{new} has a cavity and vertices of B in that cavity then vertices p_s and p_e on I_{new} belong obviously to the convex hull of this polygon. The straight segments between vertices of the cavity are completely in $\text{CH}(A)$ and in B (for example, see q_{10}, q_{11}, p_{10} in Fig. 4.27).

If vertices of A are in the cavity (for example, see p_2, p_3, q_6 in Fig. 4.27) then the cover $p_s p_e$ intersects A. The polygonal path, starting at p_s and ending at p_e, connects vertices in A.   □

If the new polygon I_{new} has a cavity $\text{CAV}(I_{\text{new}})$ containing vertices of A, then the start and end vertex of $\text{CAV}(I_{\text{new}})$ and the vertices of $\text{CAV}(I_{\text{new}})$ constitute again a new inner polygon O_{new}, and the start and the end vertex of $\text{CAV}(I_{\text{new}})$ and the vertices of I_{new} constitute a new outer polygon (see Fig. 4.27).

This leads us to the basic idea of a recursive algorithm for the computation of the relative convex hull.

We follow (e.g., in counterclockwise order) both frontiers of the original polygons, starting on the frontier of A at an extreme vertex that is a vertex of the relative convex hull $\text{CH}_B(A)$.

Furthermore, for the computation of the convex hull and for finding intersecting cavities in A and in B, we use the standard decision that a vertex p_i of a polygon is convex, concave, or collinear, depending on the sign or value of the determinant $D(p_{i-1}, p_i, p_{i+1})$.

Recursive algorithm   Algorithm 16 is based on a mapping of cavities of the inner polygon A onto cavities of the outer polygon B. This mapping simplifies the process of finding overlapping cavities in A and B in this special situation for MLP calculation. The following algorithm for the general case of arbitrary inner and outer simple polygons is not making use of such a mapping of cavities.

The base case of the recursion (i.e., where it stops) is a triangle. We use Theorem 4.3: The relative convex hull $\text{CH}_B(A)$ for simple polygons $A \subseteq B$ is only different from $\text{CH}(A)$ if there is at least one cavity in A and one in B such that the intersection of those cavities is not empty. The algorithm copies vertices of the convex hull of the inner polygon one by one, until it finds a cavity.

If it detects a cavity $\text{CAV}(A)$, then it finds the next cavity in B that has a nonempty intersection with $\text{CAV}(A)$, if there is any. The algorithm computes the

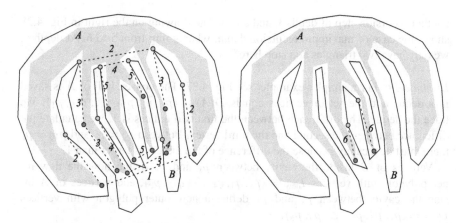

Fig. 4.28 *Left*: covers created at recursion depth m, for $m = 1$ to $m = 5$. *Right*: covers at recursion depth 6; both cavities define only a triangle each, thus the base case of the recursion

convex hull of the new inner polygon I_{new} (i.e., all vertices of B in CAV(A), including the start and end vertices of CAV(A)). For each cavity in this resulting convex hull, the algorithm computes recursively the convex hulls of the next new polygons.

If there is no cavity remaining, then the algorithm inserts the computed vertices in a cavity, between p_s and p_e, for all cavities in A, and it returns the relative convex hull of the given polygon A relatively to B.

This is the basic outline of the algorithm, and we discuss it now more in detail. We assume two lists containing all the vertices of A and B with their coordinates (in the order of polygonal paths). We compute the convex hulls of both polygons by applying the Melkman algorithm (see Algorithm 15). The computation for both polygons starts at the extreme vertices with, say, minimum y-coordinates. For a given ordered set of n vertices $A = \langle p_1, p_2, \ldots, p_n \rangle$, the Melkman algorithm delivers the convex hull in $\mathcal{O}(n)$ time in a deque $D(A)$ where the first and the last element are the same vertices.

The difference in indices of two consecutive vertices p_j and p_i in the resulting deque $D(A)$ is equal to 1 iff there is no cavity between p_j and p_i. We do not change $D(A)$. A cavity in A with the starting vertex $p_s = p_j$ and end vertex $p_e = p_i$ has been found if $(i - j) > 1$.

The next step finds a cavity in the convex hull of the outer polygon. It searches the deque $D(B)$. If it finds a cavity, it checks whether a line segment between the vertices of B in this cavity intersects the cover $p_s p_e$, and it computes the convex hull for p_s, p_e and all the vertices q of B that define a concave corner p_s, q, p_e and are also inside the cavity.

See Fig. 4.28. At recursion depth 1, one cavity of A is detected with the cover numbered "1". When calculating the convex hull of points of B in the cover, note that the sequence of points is also crossing the cover numbered "1", but convex points with respect to the cover are ignored, until we return to concave points. This can happen repeatedly (twice in the shown example). At recursion depth 2, we identify three covers of cavities, all numbered by "2". Only the middle one leads to another recursion step of depth 3, and so on. The drawing on the right in Fig. 4.28 shows the case of maximum recursion depth, when going from 5 to 6. At depth 6, we only have two triangles and stop here.

Example 4.6 For a simpler example, see Fig. 4.27. The convex hull of A is saved in a deque and the set of vertices equals $D(A) = \langle p_1, p_2, p_{10}, \ldots, p_{16}, p_1 \rangle$. We trace the deque. The difference between the first two indices is 1, we calculate the difference between the second and the third vertex. Between $p_2 = p_s$ and $p_{10} = p_e$ there must be a cavity because the difference of the indices is larger than 1.

Vertices of B inside the cavity between p_s and p_e define now the new inner polygon with vertices $I_{new} = \langle p_s, q_3, q_4, \ldots, q_{11}, p_e \rangle$, and vertices of A inside the cavity between p_s and p_e define a new outer polygon with vertices $O_{new} = \langle p_s, p_3, p_4, \ldots, p_9, p_e \rangle$.

All vertices of the convex hull $D(I_{new}) = \langle p_s, q_6, q_7, \ldots, q_{10}, p_e, p_s \rangle$ are vertices of the relative convex hull. The polygon I_{new} has one cavity p_2, p_3, q_6 containing one vertex of polygon A. This defines again a new inner polygon.

The convex hull of three vertices is the same set of three vertices; the recursion stops. We replace p_2 and q_6 by p_2, p_3, q_6 in $D(I_{new})$. We continue to check for cavities until we reach p_e. The adjusted deque replaces the start and the end vertices in $D(A)$. In our example (see Fig. 4.27), the next cavity in $D(I_{new})$ starts at q_{10} and ends at p_e. But inside the cavity there is no vertex of the outer polygon. Thus, the relative convex hull does not change.

We continue to trace the deque $D(A)$ at p_{10} until p_1 is reached, and we skip vertex by vertex because there is no other cavity in A; $D(A)$ stays unchanged.   □

See Fig. 4.29 for a pseudocode of the algorithm by Gisela Klette for calculating the relative convex hull. This algorithm applies recursively a procedure $RCH(I, O, D(I), D(O), l, t, p_s, p_e)$ also given in the same figure.

Note that the Melkman algorithm delivers the convex hull in a deque with the first element at the bottom of the deque and also at the top of the deque. We need to remove the first element from the top after the computation of the convex hull.

Note that any vertex on A or B is accessed by the algorithm only at most as often as defined by the depth of a stacked cavity. This defines this algorithm as being of linear time complexity, measured in the total number of vertices on A and B *if* this depth is limited by a constant, but the algorithm is of quadratic time in the worst case sense.

For the general case, roughly speaking, if there are "many" cavities in A, then they are all "small", and the algorithm has again linear run-time behaviour. (The number N of cavities in a simple polygon with n vertices is at most $N = \lfloor n/2 \rfloor$, for any $n > 3$.)

If there is just one "big" cavity (similar to Fig. 4.28), then the recursion only proceeds for this one cavity, and we are basically back to the upper bound for the original input, now for a reduced number of vertices plus some preprocessing in linear time. The worst case in time complexity is reached if there is a small number of "large cavities" in A and B, causing repeated recursive calls within originally

Algorithm 17 (Algorithm by Gisela Klette)
Input: Simple polygons $A = \langle p_1, p_2, \ldots, p_n, p_1 \rangle$ and $B = \langle q_1, q_2, \ldots, q_m, q_1 \rangle$, with $A \subseteq B$.
Output: Relative convex hull $\text{RCH}_B(A)$ in $D(I)$.

1: Initialise $D(I) = \emptyset$ and $D(O) = \emptyset$
2: Call Procedure $\text{RCH}(A, B, D(I), D(O), n, m, p_1, p_n)$

Procedure $\text{RCH}(A, B, D(I), D(O), n, m, p_1, p_n)$
1: Compute convex hulls of I and O in deques $D(I)$ and $D(O)$, respectively;
   l is the number of vertices in I and t the number of vertices in O;
   p_s is the start vertex and p_e the end vertex of the inner polygon;
   $k = 1$ and $j = 1$ are loop variables.
2: Remove the final element in each of the deques $D(I)$ and $D(O)$.
3: **while** $k < l$ **do**
4:    **if** there is a cavity between two consecutive vertices v_k and v_{k+1} in $D(I)$
      then
5:       $p_s = v_k$ and $p_e = v_{k+1}$
6:       **while** $j < t$ **do**
7:          **if** there is an overlapping cavity between two consecutive vertices in
            $D(O)$ **then**
8:             Update I such that I includes p_s and p_e and all vertices in O inside
               the cavity of I; L is the number of vertices.
9:             **if** $L > 3$ **then**
10:               Update O such that O includes p_s and p_e and all the vertices of I
                 in the cavity of I; T is the number of vertices.
11:               Call $\text{RCH}(I, O, D(I), D(O), L, T, p_s, p_e)$
12:             **end if**
13:             Insert q between p_s and p_e in $D(I)$.
14:          **end if**
15:       **end while**
16:       Return $D(I)$.
17:    **end if**
18: **end while**
19: Return $D(I)$.

Fig. 4.29 The recursive algorithm by Gisela Klette

large cavities of A, and the number of recursive calls defines the *depth* of those stacked cavities. However, such cases are often unlikely in real-world applications.

4.8 Problems

Problem 4.1 Show the correctness of Eq. (4.3).

Problem 4.2 Provide a correctness proof for the Graham algorithm (i.e., correct calculation of the convex hull, for any finite set of points in the plane).

Problem 4.3 Figures 4.14 and 4.21 illustrate rubberband algorithms for convex hull calculations, either for a set of points or for a simple polygon. In the latter case, we suggested the use of four extreme points, but only two for a set of points. Why?

Problem 4.4 Prove that (4.4) is true, for any subsets S, A, and B of $\mathbb{R}^2$ satisfying $S \subseteq A \subseteq B$.

Problem 4.5 (Programming exercise) Implement the rubberband algorithms for calculating the convex hull of a set of n points in the plane, or of a simple polygon. Compare the actual run-time of your rubberband algorithm with that of the Graham algorithm (i.e., input data are a set) and that of the Melkman algorithm (i.e., input data are a simple polygon); there are sources for both algorithms on the net. Use randomly generated input data (see, for example, Problem 1.6).

Problem 4.6 In Fig. 4.28, we have $a = 40$ vertices for the inner polygon A, $b = 56$ vertices for the outer polygon B, and depth $r = 7$ of the resulting recursion. Simplify polygons A and B such that they have a minimum number $a + b$ of vertices but still depth $r = 7$ in the resulting recursion. Can you generalise the result by deriving a general upper bound for r for a given sum $a + b$?

Problem 4.7 (Programming exercise) Implement a program which does the following:

(a) Read an arbitrary gray-level image (in a format such as *bmp*, *tiff*, *png*, or *raw*, which supports accurate image value representation, as opposed to, for example, *jpg* format which is defined by a lossy image encoding scheme) and apply a threshold, for generating a binary image containing black *object pixels* and white *non-object pixels*.

(b) Identify a component (e.g., defined by 4-adjacency) of object pixels by a mouse click (i.e., by clicking somewhere inside of this component). Identify this region on screen by changing all pixel values within this region into a particular colour (note: this step may apply a common graphic routine for 'filling' a connected region with some constant value).

(c) Estimate the area of the selected component by counting its number of pixels. (This is a sound way for estimating the area.)

(d) Estimate its perimeter by calculating the perimeter of its minimum-length circumscribing polygon (MLP). For doing so, implement first a *border tracing algorithm* which allows 'walking around a connected region' exactly once, consider this as being the *inner border*, and now include the calculation of an *outer border* into your tracing algorithm being one pixel away (in a region of non-object pixels) from the traced inner border. Finally, the MLP is calculated between inner and outer border. (Again, the use of the MLP defines a sound way for estimating the perimeter.)

(e) Output the *shape factor* of the selected component, defined by the area divided by the square of the perimeter. Discuss the importance of this shape factor for characterising the shape of connected image regions.

Problem 4.8 A polygon P is called *monotone* (*strictly monotone*) with respect to a straight line γ if every line orthogonal to γ intersects the frontier of P at most twice (at most in two points). Simplify the Sklansky algorithm so that it calculates the convex hull of any monotone polygon correctly without containing unnecessary instructions.

Problem 4.9 (Research problem) Implement Gisela's algorithm described in Fig. 4.29 and provide experimentally measured run-times especially for polygons having 'stacked' cavities, for identifying those shapes which cause maximum run times.

Problem 4.10 (Research problem) Design a rubberband algorithm for computing relative convex hulls, and compare its performance with that of Gisela's algorithm.

Problem 4.11 Prove that the rubberband algorithm described at the end of Sect. 4.3 can be computed in time $\mathcal{O}(n \log n)$.

4.9 Notes

For the geometry of convex sets in the context of discrete geometry, see, for example, [6].

The Graham algorithm was published in 1972 (on less than two pages) in [5]. A different strategy has been used by *William F. Eddy* [4] for the computation of the convex hull for any finite set of points in the plane. Also in 1972, *Jack Sklansky* published in [16] the first linear-time algorithm for calculating the convex hull of a simple polygon using the determinant $D(p_1, p_2, p_3)$ for testing for a 'left turn' or 'right turn' (actually, there introduced as the cross product of two vectors). However, it turned out a few years later that this algorithm is not correctly calculating the convex hull for any simple polygon.

Part 5 of the Graham algorithm and the Sklansky algorithm, as well as several convex hull algorithms published after 1972, apply a 'three steps local optimisation' approach what is a defining feature of rubberband algorithms in general. In these convex hull algorithms, steps (of the RBA) are single points only, and the RBA is just about eliminating some of these steps. For convex-hull algorithms in general, see also textbooks in computational geometry, for example, [13].

The Klette algorithm was published in [9] and is one of the various convex-hull algorithms for simple polygons which fixed this issue. The Melkman algorithm, published in [12], may have simple polygons as input, but also just simple (i.e., no self-intersection) polygonal chains.

The papers [15–17] proposed *minimum-perimeter polygons* (i.e., constrained or relative convex hulls in the regular orthogonal grid) for measuring the perimeter of sets digitised in the plane. The relative convex hull was identified as a way to estimate the length of curves [18] with convergence to the true value. For Theorem 4.2, see [18]. The relative-convex-hull algorithm in [11] applies this property. It computes the ESP for components in a digital picture in time linear with respect to the number of pixels on the border of a component.

The book [10] discusses in detail the issue of multigrid convergence of property estimators (e.g., length, area, curvature, volume, or surface area); the paper [1] discusses multigrid convergence for the particular case of length estimation in digital images. Algorithms for relative convex-hull calculations are also published in [8, 11, 14, 19]. The material on relative convex hull in this chapter is partially authored by *Gisela Klette*; see also her DGCI-conference paper [8].

For ant-colony optimisation (ACO), as referred to for the introductory figure of Part II, see the book [3]. Concepts of local optimisation are also a subject in biologically-inspired concepts of computing, such as swarm intelligence; see the book [7]. However, we cite those optimisation concepts here only for some conceptual similarity (say, accumulating local knowledge for global optimisation), and they are not further discussed in this book.

References

1. Coeurjolly, D., Klette, R.: A comparative evaluation of length estimators of digital curves. IEEE Trans. Pattern Anal. Mach. Intell. **26**, 252–258 (2004)
2. de Vieilleville, F., Lachaud, J.-O.: Digital deformable model simulating active contours. In: Proc. DGCI. LNCS, vol. 5810, pp. 203–216. Springer, Heidelberg (2009)
3. Dorigo, M., Stützle, T.: Ant Colony Optimization. MIT Press, Cambridge (2004)
4. Eddy, W.F.: A new convex hull algorithm for planar sets. ACM Trans. Math. Softw. **3**, 398–403 (1977)
5. Graham, R.L.: An efficient algorithm for determining the convex hull of a finite planar set. Inf. Process. Lett. **7**, 175–180 (1972)
6. Gruber, P.M.: Convex and Discrete Geometry. Springer, New York (2007)
7. Kennedy, J., Eberhardt, R.C.: Swarm Intelligence. Morgan Kaufmann, San Francisco (2001)
8. Klette, G.: Recursive calculation of relative convex hulls. In: Proc. DGCI. LNCS, vol. 6607, pp. 260–271. Springer, Heidelberg (2011)
9. Klette, R.: Mathematische Probleme der digitalen Bildverarbeitung. Bild Ton **36**, 107–113 (1983)
10. Klette, R., Rosenfeld, A.: Digital Geometry. Morgan Kaufmann, San Francisco (2004)
11. Klette, R., Kovalevsky, V.V., Yip, B.: Length estimation of digital curves. In: Vision Geometry. SPIE, vol. 3811, pp. 117–129 (1999)
12. Melkman, A.: On-line construction of the convex hull of a simple polygon. Inf. Process. Lett. **25**, 11–12 (1987)
13. O'Rourke, J.: Computational Geometry in C, 2nd edn. Cambridge Tracts in Theoretical Computer Science (1998)
14. Provençal, X., Lachaud, J.-O.: Two linear-time algorithms for computing the minimum length polygon of a digital contour. In: Proc. DGCI. LNCS, vol. 5810, pp. 104–117. Springer, Heidelberg (2009)
15. Sklansky, J.: Recognition of convex blobs. Pattern Recognit. **2**, 3–10 (1970)

16. Sklansky, J.: Measuring concavity on a rectangular mosaic. IEEE Trans. Comput. **21**, 1355–1364 (1972)
17. Sklansky, J., Chazin, R.L., Hansen, B.J.: Minimum perimeter polygons of digitized silhouettes. IEEE Trans. Comput. **C-21**, 260–268 (1972)
18. Sloboda, F., Stoer, J.: On piecewise linear approximation of planar Jordan curves. J. Comput. Appl. Math. **55**, 369–383 (1994)
19. Toussaint, G.T.: An optimal algorithm for computing the relative convex hull of a set of points in a polygon. In: EURASIP, Signal Processing III: Theories and Applications, Part 2, pp. 853–856. North-Holland, Amsterdam (1986)

Chapter 5
Partitioning a Polygon or the Plane

*Many are stubborn in pursuit of the path they have chosen, few
in pursuit of the goal.*

Friedrich Wilhelm Nietzsche (1844–1900)

The chapter describes algorithms for partitioning a simple polygon into trapezoids
or triangles (Seidel's triangulation and an algorithm using up- and down-stable vertices). Chazelle's algorithm, published in 1991 and claimed to be of linear time, is
often cited as a reference, but this algorithm was never implemented; the chapter
provides a brief presentation and discussion of this algorithm. This is followed by a
novel procedural presentation of Mitchell's continuous Dijkstra algorithm for sub-
dividing the plane into a shortest-path map for supporting queries about distances to
a fixed start point in the presence of polygonal obstacles.

5.1 Partitioning and Shape Complexity

Partitioning (or decomposition, or tessellation) of a simple polygon or other planar
sets is a common preprocessing step when solving an ESP problem in the plane (see
Chap. 6 for the general case or the Toussaint algorithm in Sect. 4.7 for a particular application). 'Triangle puzzles' are popular games (e.g., using triangles of various sizes for forming a predefined shape). Decompositions are also used in pattern
recognition or image analysis, for example, when in need of a 'shape complexity
measure'.

The *shape complexity* of a convex set S can simply be identified with the minimum number of half-planes needed for generating S by repeated intersections (i.e.,
the number of edges if S is a convex polygon). Non-convex sets can be partitioned
into convex sets. See Fig. 5.1.

F. Li, R. Klette, *Euclidean Shortest Paths*,
DOI 10.1007/978-1-4471-2256-2_5, © Springer-Verlag London Limited 2011

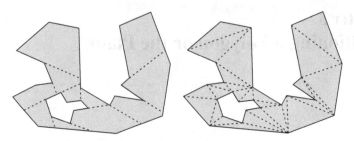

Fig. 5.1 Polygon with one *hole*. *Left*: A partitioning into convex sets by linear extension of the polygon's edges. *Right*: A partitioning into *triangles* by connecting vertices of the polygon. Those two approaches are further specified when defining algorithms

> The shape complexity of a compact set can be quantified by the minimum number of convex sets required for partitioning this set.

The complexity of the contributing convex partitioning elements can be used for a more refined specification of shape complexity.

A simple polygon (no matter whether convex or non-convex), possibly with a finite number of holes, all defined by simple polygons, can always be partitioned into triangles. The required number of triangles allows us to define the shape complexity of any simple polygon in a standard way, not depending on possible variations of the convex partition elements.

Definition 5.1 A family $\mathcal{F} = \{S_a : a \in I\}$ of sets, with index set $I \subseteq \mathbb{R}$, defines a *partitioning* of a set S iff

(i) S is the union of all the sets in $\mathcal{F}$,
(ii) any set S_a has a non-empty interior, and
(iii) the interiors of sets S_a and S_b are disjoint, for any $a, b \in I$ and $a \neq b$.

Property (ii) requires that there are no degenerated sets in $\mathcal{F}$ such as line segments or *singletons* (i.e., sets that contain only a single point). Property (iii) says that sets in $\mathcal{F}$ may share points on their frontiers. (The interior of a set was defined in Sect. 2.5.) If $S \subseteq \mathbb{R}^2$ then the elements in $\mathcal{F}$ are also called *faces*.

See Fig. 5.2 for two examples in 3D space; Listing's polyhedron on the right was already shown in Fig. 1.12. The question raised in the caption of this figure is not easy to answer in general. Listing's polyhedron is the union of four partially overlapping *toroidal polyhedrons* (i.e., each of the four is topologically equivalent to a torus). Those polyhedrons can be further subdivided by removing double copies of overlapping parts, thus defining a partitioning into pairwise disjoint polyhedrons. So far a 'brief excursus' into 3D space; this chapter discusses planar sets only.

The class of simple polygons contains 'very complex' 2D shapes. For example, consider a 'fractal tree' (see Fig. 5.3) and expand all line segments of this tree into 'thin elongated rectangles' so that their union forms a simple polygon.

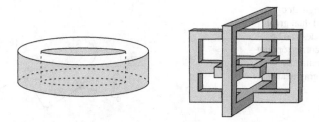

Fig. 5.2 What is the minimum number of convex sets needed for partitioning a 3D non-convex set? *Left*: Torus. *Right*: Listing's polyhedron

Fig. 5.3 Digital print of a fractal tree. Courtesy of Robert Fathauer, Tessellations Company, Phoenix, Arizona

5.2 Partitioning of Simple Polygons and Dual Graphs

We consider the partitioning of a simple polygon into triangles or trapezoids.[1] A trapezoid partitions into two triangles, and this post-process is not adding much run time to a partitioning algorithm if needed. A partitioning into triangles only is called a *triangulation*, and into triangles or trapezoids is called a *trapezoidation*.

There are many algorithms for trapezoiding a simple polygon. Sources are freely available on the net for some of them. For a start, a convex polygon is very easy to triangulate; just select a vertex and connect it by $n - 1$ line segments with all the other vertices. We apply the following

---

[1]A *trapezoid* is a quadrilateral with one pair of parallel sides. Thus, a trapezoid is always a convex polygon.

Fig. 5.4 Triangulated
polygon P and dual graph
G_P. Graph nodes can be
positioned in corresponding
triangles conveniently
anywhere for graphical
presentation

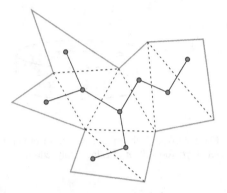

Partitioning constraint: Vertices of triangles or trapezoids of a partitioning
are restricted to be on the frontier of the given simple polygon.

Both examples in Fig. 5.1 satisfy this constraint; the example on the right even
uses only vertices of the given polygon. There are also partitioning methods which
introduce additional vertices in the interior of the polygon, but we exclude those
from our considerations.

Let P be a simple polygon and $\mathcal{T} = \{\triangle_1, \triangle_2, \ldots, \triangle_m\}$ a partitioning of P into
convex sets. Two faces $\triangle_i$ and $\triangle_j$ of $\mathcal{T}$ are either disjoint, or they share an edge e_{ij},
or they share just a single point, for $i, j = 1, \ldots, m$ and $i \neq j$.

A partitioned simple polygon defines a *dual graph* $G_P = [V, E]$. This is an undirected graph whose nodes $V = \{v_1, v_2, \ldots, v_m\}$ correspond one-to-one to the faces
of $\mathcal{T}$, and there is an edge $e = \{v_i, v_j\}$ in E iff $\triangle_i \cap \triangle_j$ is a line segment of non-zero
length (i.e., a line segment that is not just a point). See Fig. 5.4.

Lemma 5.1 *The dual graph G_P is a tree, for any partitioned simple polygon P.*

Proof By contradiction. Suppose that G_P is not a tree. Then there is a cycle
$u_1 u_2 \cdots u_k u_1$ in G_P. Consequently, there is a sequence $\triangle_1', \triangle_2', \ldots, \triangle_k'$ of sets in $\mathcal{T}$
such that $\triangle_i' \cap \triangle_j' \neq \emptyset$, for $i \neq j$ and $i, j = 1, 2, \ldots, k$. It follows that there is a
polygonal loop $\rho = w_1 w_2 \cdots w_k w_1$ with $w_i \in \triangle_i'$, for $i = 1, \ldots, k$ that is completely
contained in the interior of P. Loop ρ circumscribes a vertex of $\triangle_1'$; we denote this
by w. Vertex w cannot be on the frontier of P, which contradicts our partitioning
constraint. See Fig. 5.5.   □

Assume a fixed ESP problem where a simple polygon P defines the search space,
and end points p and q are given in P. Points p and q are in elements $\triangle_p$ and $\triangle_q$ of
a partitioning $\mathcal{T}$ of P. Any ESP from p to q has now to cross all those elements in $\mathcal{T}$
which are on a shortest path from u_p to u_p in the dual graph G_P, where u_p and u_p
are the nodes representing $\triangle_p$ and $\triangle_q$, respectively. By Lemma 5.1, we know that
G_P is a tree. By identifying the shortest path from u_p to u_p, we obtain the sequence

Fig. 5.5 Path ρ and vertex w, as discussed in the proof of Lemma 5.1

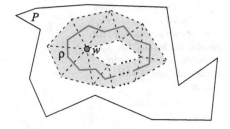

of edges e_{ij} which a solution to the ESP problem needs to cross. For example, these edges can be considered to be steps of an RBA.

Shortest path in a tree   Let T be a tree, and u_1 and u_2 be two nodes in T, $u_1 \neq u_2$. We provide a procedure for calculating a uniquely specified path from u_1 to u_2 by not allowing any returns (and thus the shortest path). There exists a linear-time algorithm for computing the shortest path between two nodes in a positive-integer weighted graph. Because we know that we have a tree, we can adapt the algorithm for this particular input; see Fig. 5.6.

Let $d(v)$ denote the *degree* of a node v in a graph (i.e., the number of adjacent nodes). In our tree, we have *leaves* v with $d(v) = 1$, and the degree of a non-leaf node can be any integer ≥ 2.

The outline of Algorithm 18 is as follows: We collect all leaves of the tree T in list S_1, not allowing u_1 and u_2 to be in this set. Then we 'prune' each branch, starting at node v in S_1, by beginning with the uniquely specified node w adjacent to v.

If $d(w) = 2$ and w is not u_1 or u_2, then we use w as the new v, insert it into V_1, and there is only one node w adjacent to v that is not yet in V_1. If $d(w) = 2$ and w is not u_1 or u_2, then we proceed as before.

Algorithm 18 (Shortest path in a tree)
Input: A tree $T = [V, E]$ and two nodes u_1 and u_2 in V.
Output: The shortest path ρ from u_1 to u_2 in T.

1: Let $S_1 = \{v : v \in V \wedge d(v) = 1 \wedge v \neq u_1 \wedge v \neq u_2\}$ and $V_1 = \emptyset$.
2: **for** each $v \in S_1$ **do**
3:      Insert v into V_1.
4:      Let w be the unique node adjacent to leaf v.
5:      **while** $d(w) = 2 \wedge w \neq u_1 \wedge w \neq u_2$ **do**
6:          $v = w$ and insert v into V_1.
7:          Let w be the unique node adjacent to v and not yet in V_1.
8:      **end while**
9:      Update T by removing V_1 from V. Reset $V_1 = \emptyset$.
10: **end for**
11: Path ρ is the remaining tree T.

Fig. 5.6 Algorithm for calculating a shortest path in a tree

Fig. 5.7 A sequence of steps
(*bold shaded edges* of faces
of the shown partitioning) for
possible use by an RBA. The
figure also shows a possible
initial path (*thin shaded
polyline*)

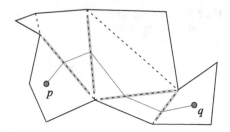

Otherwise, we go to the next node in S_1. When the program terminates, the final updated T is the desired unique path ρ from u_1 to u_2 in the originally given tree T.

Assume that we have a partitioning of a simple polygon P and its dual graph $T = G_P$. Let u_p and u_q be the vertices of T corresponding to the faces containing p and q, respectively.

We apply Algorithm 18 and interpret the obtained path from u_p to u_q in T by the sequence of corresponding faces $\{\Delta'_1, \Delta'_2, \ldots, \Delta'_k\}$.

Let $\{e_1, e_2, \ldots, e_{k-1}\}$ be a sequence of edges of faces of the partitioning such that $e_i = \Delta'_i \cap \Delta'_{i+1}$, for $i = 1, \ldots, k - 1$. When designing an RBA, we need to be prepared for degenerate cases as discussed in Sect. 3.7. Let $\{e'_1, e'_2, \ldots, e'_{k-1}\}$ be a sequence of edges such that e'_i is obtained by removing at both endpoints a sufficiently small segment of length $\varepsilon_s > 0$ from e_i, for $i = 1, \ldots, k - 1$. The set $\{e'_1, e'_2, \ldots, e'_{k-1}\}$ can then be used as a set of steps for an RBA. See Fig. 5.7, and Chap. 6 for the actual ESP algorithm.

5.3 Seidel's Algorithm for Polygon Trapezoidation

Two line segments are called *noncrossing* if the intersection of them is empty or a joint endpoint. Let S be a set of nonhorizontal, noncrossing line segments in the plane. The *left* (*right*) *horizontal extension* through an endpoint p of one of the segments in S is the line segment or ray starting at p and extending towards the left (right) until it either hits another segment in S, or it extends to infinity.

The *horizontal extension* through an endpoint p is the union of the left and right horizontal extension through p.

Definition 5.2 The *trapezoidation* $\mathcal{T}(S)$ of set S of line segments is the partitioning of the plane formed by the segments in S and the horizontal extension through endpoints in segments in S.

A face in $\mathcal{T}(S)$ is always either a trapezoid or a triangle; see Fig. 5.8.

Let T_S be a *point location query structure* for $\mathcal{T}(S)$, which is a binary decision tree with a selected node being the *root* and with one leaf node for each face in $\mathcal{T}(S)$. At this point we just assume that such a tree is given. We show first how the tree is used for a point query. Later we explain how to build this tree.

Fig. 5.8 The trapezoidation
of the shown five line
segments leads to 14 faces
(*triangles* or *trapezoids*)

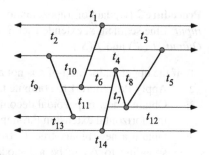

Each nonleaf node in such a tree T_S is labelled either by X or by Y. Each X-node is associated with a line segment in S. Each Y-node is associated with the y-coordinate of an endpoint of some line segment in S. The associated line segment (X-node) or y-coordinate (Y-node) is called the *key* of the node.

Let q be a *query point* in the plane; we have to identify the enclosing face in $\mathcal{T}(S)$. The query is processed as follows (see Procedure 1 in Fig. 5.9):

We start at the root and proceed along a path down to a leaf node whose corresponding face in $\mathcal{T}(S)$ encloses q. Point q can be on an edge incident with two

Procedure 1 (Query point in the plane using a binary decision tree)
Input: A point q in the plane, a partitioning $\mathcal{T}(S)$ and its binary decision tree $T(S)$.
Output: The face of $\mathcal{T}(S)$ that encloses q.

```
 1: Set the current node v to be the root of T(S).
 2: while v is not a leaf node do
 3:    if v is an X-node then
 4:       if q is to the left of the key of v then
 5:          Reset v to be the left child of itself.
 6:       else
 7:          Reset v to be the right child of itself.
 8:       end if
 9:    else
10:       if v is a Y-node then
11:          if the y-coordinate of q is greater than the key of v then
12:             Reset v to be the left child of itself.
13:          else
14:             Reset v to be the right child of itself.
15:          end if
16:       end if
17:    end if
18: end while
19: Return node v's corresponding trapezoid of T(S).
```

Fig. 5.9 Procedure for a query point and a binary decision tree $T(S)$, determining the enclosing face in $\mathcal{T}(S)$

Procedure 2 (Update of trapezoidation and binary decision tree)
Input: Line segment s, extended set $S' = S \cup \{s\}$, $\mathcal{T}(S)$, and $T(S)$.
Output: $\mathcal{T}(S')$ and $T(S')$.

1: **if** the upper endpoint a of s is not an endpoint of any line segment in S **then**
2:     Apply Procedure 1 to compute the trapezoid $\triangle_a$ of $\mathcal{T}(S)$ which contains a.
3:     Obtain a new trapezoidal decomposition $\mathcal{T}'$ from $\mathcal{T}(S)$ by splitting $\triangle_a$ with
       the horizontal extension through a, leading to two newly created trapezoids.
4:     Obtain a new binary tree T' from T by modifying the leaf of $T(S)$ corre-
       sponding to $\triangle_a$ to be a Y-node whose key is the y-coordinate of a, and
       whose two children are two new leaves corresponding to the two newly cre-
       ated trapezoids in $\mathcal{T}'$.
5: **else**
6:     Let $\mathcal{T} = \mathcal{T}'$ and $T = T'$ (i.e., a is already an endpoint of some line segment
       in S).
7: **end if**
8: For the lower endpoint b of s, we obtain analogously a new trapezoidal decomposition $\mathcal{T}''$ from $\mathcal{T}'$ and a binary tree T'' from T'.
9: Obtain $\mathcal{T}(S')$ from $\mathcal{T}''$ as follows: Find all trapezoids of $\mathcal{T}''$ that are intersected
   by s, cut each of them in two trapezoids, one on each side of s, and merge
   trapezoids that share a line segment in their frontiers.
10: Obtain tree $T(S')$ from $\mathcal{T}''$ as follows: Create a new leaf corresponding to each
    newly created trapezoid of $\mathcal{T}(S')$ (along line segment s). Each new leaf of T''
    that corresponds to a trapezoid in $\mathcal{T}''$ that was cut by s becomes an X-node
    whose key is s and whose two children are the two new leaves.

Fig. 5.10 Procedure for computing $\mathcal{T}(S')$ and $T(S')$ from previous trapezoidation and binary
decision tree

faces; in this case, one of two leaf nodes can be selected. It is suggested to use a
default decision in cases when q is on an edge (e.g., the face on the left).

At each node v along the path, the decision which of the two children of v replaces v next depends on the outcome of the comparison of q with the key of node v:
If v is an X-node, then q is either left or right of the key (where the key is a line
segment in set S in this case); at a Y-node, we consider q's y-coordinate, whether it
is smaller or greater than the key (where the key is the y-coordinate of the Y-node
in this case).

Now we consider the construction of the trapezoidation, which is done concur-
rently with the construction of the binary decision tree. Let s be a 'new' nonhorizontal line segment with upper endpoint a and lower endpoint b. Also assume that this
line segment s has either no intersection with another segment in the already given
set S of line segments, or 'just' a joint endpoint. Thus, $S' = S \cup \{s\}$ also contains
only noncrossing line segments.

We describe how to compute $\mathcal{T}(S')$ and $T(S')$ from the already given trapezoidation $\mathcal{T}(S)$ and decision tree $T(S)$. See Procedure 2 in Fig. 5.10.

Fig. 5.11 Initial steps for computing the trapezoidation $\mathcal{T}(\{s_1\})$ and the binary tree $T(\{s_1\})$. *Upper row:* Start with $S = \emptyset$. *Lower row:* Upper vertex a of first segment s_1 and its horizontal extension

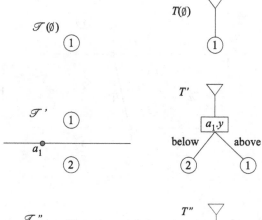

Fig. 5.12 Continuing with computing trapezoidation $\mathcal{T}(\{s_1\})$ and binary tree $T(\{s_1\})$: considering the lower vertex b of first segment s_1 and its horizontal extension

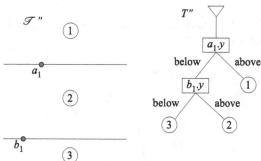

Fig. 5.13 Finalising the computation of trapezoidation $\mathcal{T}(\{s_1\})$ and binary tree $T(\{s_1\})$: segment s_1 has been taken into account

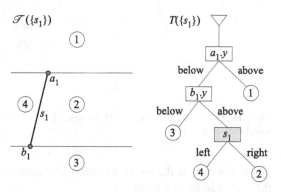

Example 5.1 Refer to Figs. 5.11, 5.12, and 5.13. This example illustrates how to calculate trapezoidation $\mathcal{T}(\{s_1\})$ and tree $T(\{s_1\})$, starting initially at $S = \emptyset$.

At the beginning, there are no segments in the set S, $\mathcal{T}(\emptyset)$ contains one face that is the whole plane (see top left in Fig. 5.11); the corresponding binary tree $T(\emptyset)$ has a root and just one leaf node (see top right in Fig. 5.11).

The second row in Fig. 5.11 shows both $\mathcal{T}'$ and T' after Lines 3 and 4 of Procedure 2. Note that we always keep the left (right) leaf node corresponding to the newly created trapezoid below (above) the horizontal extension through the current

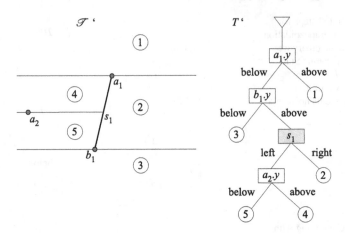

Fig. 5.14 Computation of $\mathcal{T}(\{s_1, s_2\})$ and tree $T(\{s_1, s_2\})$ from $\mathcal{T}(\{s_1\})$ and $T(\{s_1\})$: start with upper endpoint a_2

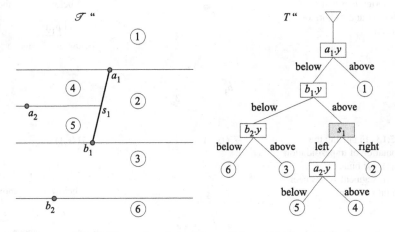

Fig. 5.15 Computation of $\mathcal{T}(\{s_1, s_2\})$ and tree $T(\{s_1, s_2\})$ from $\mathcal{T}(\{s_1\})$ and $T(\{s_1\})$: continuation with lower endpoint b_2

point; this is a_1 in the shown example. (As common in programming, $p.y$ denotes the y-coordinate of a point p.)

Figure 5.12 shows both $\mathcal{T}''$ and T'' after Line 8 of Procedure 2. Figure 5.13 shows both $\mathcal{T}(\{s_1\})$ and $T(\{s_1\})$ after Lines 9 and 10 of Procedure 2. □

Example 5.2 Refer to Figs. 5.14, 5.15, and 5.16. This example shows how to compute $\mathcal{T}(\{s_1, s_2\})$ and the binary tree $T(\{s_1, s_2\})$ from $\mathcal{T}(\{s_1\})$ and $T(\{s_1\})$ as obtained in the previous example and shown in Fig. 5.13.

This update starts with taking the upper endpoint a_2 into account, then the lower endpoint b_2, and finally the segment s_2 itself. For this final operation, we also intersect s_2 with the previous horizontal extension of b_2. □

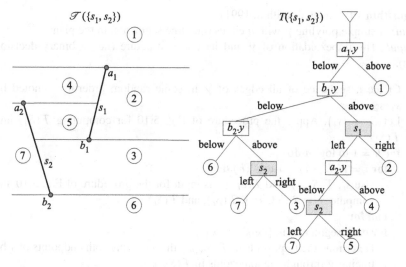

Fig. 5.16 Computation of $\mathcal{T}(\{s_1, s_2\})$ and tree $T(\{s_1, s_2\})$ from $\mathcal{T}(\{s_1\})$ and $T(\{s_1\})$—finalisation; note that the previous horizontal extension of vertex b_1 was intersected and cut by s_2

After having those two procedures at hand, we are now ready to present the *Seidel algorithm*[2] for trapezoidation of a simple polyline (i.e., a more general case than just a simple polygon). Without loss of generality, let γ be a simple polyline such that no two vertices of γ have the same y-coordinate. For a positive integer n, let

$$\log^{(0)} n = n \quad \text{and} \quad \log^{(i)} n = \log\bigl(\log^{(i-1)} n\bigr)$$

for $i \geq 1$. Let $\log^* n$ be the largest integer l such that $\log^{(l)} n \geq 1$. Finally, let $N(h) = \lceil n/\log^{(h)} n \rceil$, where $0 \leq h \leq \log^* n$ and $\lceil a \rceil$ denotes the smallest integer greater than or equal to a. For the Seidel algorithm, see now Fig. 5.17.

Line 4 starts with $i = 1$ because of $N(0) + 1 = 2$. Thus we start in Line 5 with s_2, the trapezoidation, and the tree for S_1, as available due to Line 2. Note that $N(\log^* n)$ may not be equal to n. Recall that $\log^* n$ is the largest integer l such that $\log^{(l)} n \geq 1$ (note that $\log^{(l)} n$ may be greater than 1). Thus, $N(\log^* n) = N(l) = \lceil n/\log^{(l)} n \rceil$ and $N(\log^* n) = n$ if $\log^{(l)} n = 1$.

In Line 7, we consider all the segments which have not been processed, yet. This is not only necessary but also the 'most important idea' of the algorithm: If $i > N(h-1)$ and the endpoints of s_i in $\mathcal{T}(S_{N(h-1)})$ are already known, then they can be located in $\mathcal{T}(S_i)$ via $T(S_i)$ in an expected time of at most $\mathcal{O}(\log(i/N(h-1)))$ (see Lemma 5.2 below).

If the input of Algorithm 19 is a simple polygon, then we may obtain the trapezoidation of our algorithm from the resulting trapezoidation of Algorithm 19 by simply deleting all left or right horizontal extension through an endpoint that is exterior to the input polygon.

---

[2] *Raimund Seidel* is at Saarland University.

Algorithm 19 (Seidel algorithm, 1991)

Input: A simple polyline γ with n edges (i.e., line segments) in the plane.

Output: The trapezoidation of γ and its query structure (i.e., a binary decision tree).

1: Create a sequence of all edges of γ in some random ordering, denoted by
   $s_1, s_2, \ldots, s_n$.
2: Let $S_1 = \{s_1\}$. Apply the procedure of Fig. 5.10 for computing $\mathcal{T}(S_1)$ and
   $T(S_1)$.
3: **for** $h = 1$ to $\log^* n$ **do**
4:     **for** $i = N(h-1) + 1$ to $N(h)$ **do**
5:         Use s_i, $\mathcal{T}(S_{i-1})$, and $T(S_{i-1})$ as input for the procedure of Fig. 5.10 and
           compute $S_i = S_{i-1} \cup \{s_i\}$, $\mathcal{T}(S_i)$, and $T(S_i)$.
6:     **end for**
7:     **for** each segment s in $\{s_{N(h)+1}, s_{N(h)+2}, \ldots, s_n\}$ **do**
8:         Determine the trapezoids in $\mathcal{T}(S_{N(h)})$ that contains both endpoints of s by
           tracing γ through the trapezoids in $\mathcal{T}(S_{N(h)})$.
9:     **end for**
10: **end for**
11: **for** $i = \log^* n + 1$ to n **do**
12:     Use s_i, $\mathcal{T}(S_{i-1})$, and $T(S_{i-1})$ as input for the procedure of Fig. 5.10 and
        compute $\mathcal{T}(S_i)$ and $T(S_i)$.
13: **end for**

Fig. 5.17 Seidel algorithm for triangulating a simple polyline in the plane

In order to analyse Algorithm 19, we state three lemmata. Let $S = \{s_1, s_2, \ldots, s_n\}$, where $s_1, s_2, \ldots, s_n$ are defined in Line 1 in Algorithm 19. Let $S_m = \{s_1, s_2, \ldots, s_m\}$, for $m = 1, 2, \ldots, n$.

Lemma 5.2 *Let* $1 \leq j \leq k \leq n$. *If q is a query point whose location in $\mathcal{T}(S_j)$ is already known, then q can be located in $\mathcal{T}(S_k)$ via $T(S_k)$ in an expected time of at most $\mathcal{O}(\log(k/j))$.*

Lemma 5.3 *Let* $1 < i \leq n$. *The expected number of horizontal trapezoid sides of $\mathcal{T}(S_{i-1})$ which are intersected by the interior of s_i is at most four.*

Lemma 5.4 *Let R be a random subset of S of size r. Let Z be the number of intersections between horizontal lines of $\mathcal{T}(R)$ and segments in $S \setminus R$. The expected value of Z is at most $4(n - r)$.*

The time complexity of Algorithm 19 can now be analysed as follows: Line 1 can be computed in $\mathcal{O}(n)$ time.[3] Line 2 can be computed in $\mathcal{O}(1)$ time. Time for

---

[3] See also the third paragraph in [16] for a detailed discussion about this.

Line 5 is caused by locating endpoints of s_i in $\mathcal{T}(S_{i-1})$ and "threading" s_i through $\mathcal{T}(S_{i-1})$.

By Lemma 5.2, the expected location time is $\mathcal{O}(\log(i/N(h-1))) = \mathcal{O}(\log^{(h)} n)$. By Lemma 5.3, the expected threading time is constant. Thus, Lines 4–6 can be computed in expected time $N(h) \times \mathcal{O}(\log^{(h)} n)$, that is, $\mathcal{O}(n)$.

By Lemma 5.4, the expected time of Lines 7–9 is $\mathcal{O}(n)$. Thus, the expected time of Lines 3–10 is $\mathcal{O}(n \log^* n)$. Analogously to the analysis of Lines 4–6, Lines 11–13 can be computed in expected time $\mathcal{O}(n)$.

Thus, the expected time of Algorithm 19 is $\mathcal{O}(n \log^* n)$ altogether.

5.4 Inner, Up-, Down-, or Monotone Polygons

This section provides technical definitions for presenting in the following section a worst-case $\mathcal{O}(n \log n)$ algorithm for the trapezoidation of a simple polygon.

Following the default settings in this book, we have a rectangular right-hand xy-coordinate system, the x-axis goes left to right, and the y-axis bottom to top. A simple polygon P is defined by a polygonal loop $\rho = \langle v_0, v_1, \ldots, v_{n-1}, v_0 \rangle$ in clockwise order. We recall that a polygon is always a closed set, thus containing its frontier ∂P (that is given by the polygonal loop).

Definition 5.3 Let $\varepsilon_s > 0$. Let P' be a polygon with $P' \subset P$. Let $u_0, u_1, \ldots, u_{n-1}$ be the vertices of P' in clockwise order. Then P' is called an *inner polygon of* P, denoted by $P(v_0, \varepsilon_s)$, if the following is true: $d_e(u_0, v_0) = \varepsilon_s$, edge $u_i u_{i+1}$ is parallel to edge $v_i v_{i+1}$, and $u_i v_i$ bisects the angle $\sphericalangle(v_{i-1} v_i v_{i+1})$, for $i = 0, 1, \ldots, n-1$.[4]

An inner polygon is uniquely defined by the chosen *shift distance* $\varepsilon_s > 0$ and vertex v_0. If ε_s is sufficiently small then P' is a simple polygon. In Fig. 5.18, the simple polygon $P(v_0, \varepsilon_s) = \langle u_0, u_1, \ldots, u_5, u_0 \rangle$ is the inner polygon of $P = \langle v_0, v_1, \ldots, v_5, v_0 \rangle$.

For vertices $v_1, v_2,$ and v_3 in $\{p_1, \ldots, p_n\}$, let $\rho_P(v_1, v_2)$ be the polygonal subpath from v_1 to v_2 in ρ, and $\rho_P(v_1, v_2, v_3)$ the polygonal path which goes first from v_1 to v_2 and then from v_2 to v_3, all in the same order, following the frontier of P (e.g., possibly containing v_2 twice, first time when going from v_1 to v_2).

We recall that $p.x$ and $p.y$ denote the x- and y-coordinates of a point p.

Definition 5.4 Let $u, v,$ and w be three points in ∂P with $u.y = v.y = w.y$. The path $\rho_P(u, v, w) = \langle u, \ldots, v, \ldots, w \rangle$ becomes a loop $\rho' = \langle u, \ldots, v, \ldots, w, u \rangle$ by adding the line segment wu. The polygon P', defined by loop ρ', is called an *up-(down-) polygon* with respect to point v iff $p.y \geq v.y$ $(p.y \leq v.y)$, for each point p in polygon P'.

---

[4]Addition or subtraction of indices is modulo n.

Fig. 5.18 A simple polygon
and an example of an inner
polygon, defined uniquely by
the distance ε_s between v_0
and u_0. In the shown case, the
inner polygon is again a
simple polygon

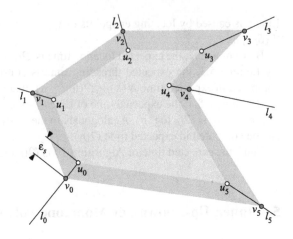

Polygon P' can be simple or non-simple. For example, P_L and P_R are up-
polygons in Fig. 5.19, both with respect to v_1. P' is a down-polygon with respect
to v_1. These three polygons are all simple. If v_3 were located higher up on the
shown line segment v_1v_{1R}, then P' would not be simple anymore, but the union of
two simple polygons.

Let v_{i-1}, v_i, v_{i+1} and v_{i+2} be four consecutive vertices of P. If

$$v_{i-1}.y < v_i.y, \qquad v_i.y = v_{i+1}.y, \quad \text{and} \quad v_i.y > v_{i+2}.y \qquad (5.1)$$

$$[\text{or} \quad v_{i-1}.y > v_i.y, \qquad v_i.y = v_{i+1}.y, \quad \text{and} \quad v_i.y < v_{i+2}.y] \qquad (5.2)$$

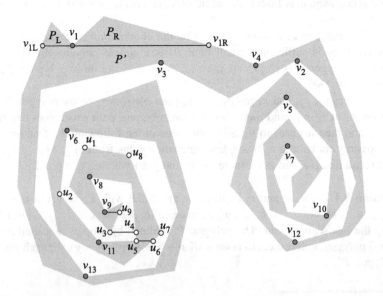

Fig. 5.19 A simple polygon and a few vertices and line segments. See the text for explanations

Fig. 5.20 The polygon on the *left* is monotone in y-direction, with one maximal edge and one minimal vertex. The polygon on the *right* is strictly monotone in x-direction, but not monotone in y-direction

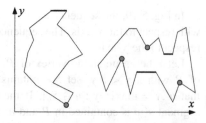

and the point $(\frac{1}{2}[v_i.x + v_{i+1}.x], v_i.y + \varepsilon_s)$ [or $(\frac{1}{2}[v_i.x + v_{i+1}.x], v_i.y - \varepsilon_s)$] is in P, for some $\varepsilon_s > 0$, then $v_i v_{i+1}$ is called an *up-* (*down-*) *stable edge* of P. Furthermore, if (5.1) is true [or (5.2) is true], and point $(\frac{1}{2}[v_i.x + v_{i+1}.x], v_i.y - \varepsilon_s)$ [or $(\frac{1}{2}[v_i.x + v_{i+1}.x], v_i.y + \varepsilon_s)$] is in P, for some $\varepsilon_s > 0$, then $v_i v_{i+1}$ is called a *maximal* (*minimal*) *edge* of P.

Let v_{i-1}, v_i, and v_{i+1} be three consecutive vertices of P. If

$$v_{i-1}.y < v_i.y \quad \text{and} \quad v_i.y > v_{i+1}.y \tag{5.3}$$

$$[\text{or} \quad v_{i-1}.y > v_i.y \quad \text{and} \quad v_i.y < v_{i+1}.y] \tag{5.4}$$

and vertex u_i of $P(v_0, \varepsilon_s)$, for some $\varepsilon_s > 0$, is not in the triangle $v_{i-1} v_i v_{i+1}$, then v_i is called an *up-* (*down-*) *stable vertex* of P, for ε_s. Furthermore, if (5.3) is true [or (5.4) is true] and vertex u_i of $P(v_0, \varepsilon_s)$, for some $\varepsilon_s > 0$, is in the triangle $v_{i-1} v_i v_{i+1}$, then v_i is a *maximal* (*minimal*) *vertex* of P.

See Fig. 5.19 for a few examples. Vertices v_2, v_3, v_5, v_6, v_7, and v_8 are up-stable. Vertices $v_1, v_4, v_9, v_{10}, v_{11}, v_{12}$, and v_{13} are down-stable. Vertex u_1 is maximal and u_3 is minimal. See also Fig. 5.20. The polygon on the right has two up-stable vertices and one up-stable edge, and one down-stable vertex and one down-stable edge.

Corollary 5.1 *A simple polygon is monotone with respect to the y-axis iff it has both a unique maximal and minimal point or edge.*

Proof The definition of a monotone polygon (see Problem 4.8) is with respect to a straight line. This is the y-axis in our case.

Assume that a simple polygon P is monotone in y-direction. A line γ orthogonal to the y-axis intersects δP at maximal or minimal points or edges once. Otherwise (if there is any intersection) a line γ intersects δP twice, and there cannot be a maximal or minimal vertex or edge below or above (in y-direction) of such an intersection, because the count would then be three or more. Now assume that two different maximal or minimal vertices or edges are incident with the same line orthogonal to the y-axis. The part on the frontier between those two vertices or edges would contradict the property that P is monotone.

Now assume that P has both a unique maximal and minimal point or edge. Then all the vertices between those two have monotonously strictly increasing (or decreasing) y-coordinates, what means that P is monotone.   □

In Fig. 5.19, the sequence $u_1, u_2, \ldots, u_8, u_1$ defines a monotone simple polygon with respect to the y-axis. The sequence $u_1, u_2, \ldots, u_8, u_9, u_1$ is not monotone with respect to the y-axis.

Let v be an up-stable vertex of P. Let S_v be the set of all minimal vertices u of P with $u.y < v.y$; set S_v contains at least two elements. Let $u' \in S_v$ be such that $u'.y = \max\{u.y : u \in S_v\}$. If there exists a point $w \in \partial P$ such that the line segment $u'w$ is contained in P and $w.y = u'.y$, then $u'w$ is called a *cut-edge* of P. The polygonal path $\rho_P(v, u')$ is called a *decreasing polygonal path from v to $u'w$*. Vertex v is called an *up-stable point with respect to the cut-edge $u'w$*, and $u'w$ is called a *cut-edge with respect to the up-stable point v*. Vertex u' is called the *nearest minimal vertex with respect to v*.

In Fig. 5.19, v_8, v_9, u_9, u_3 is a decreasing polygonal path from v_8 to edge u_3u_4, and u_3 is the nearest minimal point with respect to v_8.

Let u and w be points (i.e., not necessarily vertices) in the frontier ∂P such that $u.y = v.y = w.y$, $u.x < v.x < w.x$, and both line segments uv and vw are in P; then u and w are called the *left and right intersection points* of v, respectively. In Fig. 5.19, v_{1L} and v_{1R} are the left and right intersection points of v_1, respectively.

Let v be a maximal point of P. Let S_v be the set of all down-stable points u of P with $u.y < v.y$; set S_v can be empty. Let $u' \in S_v$ such that $u'.y = \max\{u.y : u \in S_v\}$. Vertex u' is called the *nearest* down-stable vertex with respect to v.

In Fig. 5.19, vertex v_9 is the nearest down-stable vertex with respect to the maximal vertex above the line segment v_9u_9.

5.5 Trapezoidation of a Polygon at Up- or Down-Stable Vertices

The algorithm (see Fig. 5.21) uses three subroutines, given below as Procedures 3, 4 and 5. Since the discussion of the algorithm in the case of an up- or down-stable edge is analogous to that of an up- or down-stable vertex, we will just detail the case of up- or down-stable vertices.

The algorithm partitions a given polygon into a set S of monotone polygons (Lines 1 to 23). These monotone polygons are then partitioned into trapezoids (Lines 24 to 27). If needed, each trapezoid can be further split into two triangles. Note that some of the monotone polygons in S can be triangles when arriving at Line 24.

The partitioning into monotone polygons occurs at up- or down-stable vertices. There is a sequence of figures illustrating operations in some lines of Fig. 5.21, all based on the input example of a polygon already shown in Fig. 5.19.

At the beginning, polygon P is the only element of a set $\mathcal{P}$ of polygons. Polygons in this set $\mathcal{P}$ will 'shrink' (in Line 12) or split into smaller polygons (in Line 19), and the set $\mathcal{P}$ will be empty when reaching Line 22.

See Fig. 5.22 for Line 2; up- and down-stable vertices can be identified during one scan through the given polygonal loop of P, just by applying the usual "left turn?" or "right turn?" decisions, and we skip the details.

Algorithm 20 (Partitioning of a simple polygon at up- or down-stable vertices)
Input: Let P be a simple polygon with up-stable or down-stable vertices.
Output: A set $\mathcal{T}(P)$ of trapezoids or triangles $T = \{\triangle_1, \triangle_2, \ldots, \triangle_m\}$ defining a partitioning of P.

1: Let $S = \emptyset$ and $\mathcal{P} = \{P\}$.
2: Compute the set V of all up- or down-stable vertices in P.
3: For all up-stable vertices in V, compute the left and right intersection points by applying Procedure 5.
4: Sort V for decreasing y-coordinates. Let k be the cardinality of V.
5: **for** $i = 1, 2, \ldots, k$ **do**
6:     **if** v_i is a down-stable vertex and in polygon $Q \in \mathcal{P}$ **then**
7:         Calculate two points v_{iL} and v_{iR} in ∂Q such that $v_{iL}.y = v_i.y = v_{iR}.y$ and $v_{iL}.x < v_i.x < v_{iR}.x$.
8:         Let $Q_L = v_{iL}v_i + \rho_Q(v_{iL}, v_i)$ {an up-polygon}.
9:         Let $Q_R = v_i v_{iR} + \rho_Q(v_i, v_{iL})$ {an up-polygon}.
10:        Let $Q' = v_{iL}v_{iR} + \rho_Q(v_{iL}, v_{iL})$ {the down-polygon}.
11:        Let $S = S \cup \{Q_L, Q_R\}$.
12:        Replace element Q in $\mathcal{P}$ by Q'.
13:    **else**
14:        Let v_i be an up-stable vertex in polygon $Q \in \mathcal{P}$. Let v_{iL}, v_{iR} be the left and right intersection points of v_i, respectively, as calculated in Line 3 (i.e., they are in ∂Q).
15:        Let $Q_L = v_{iL}v_i + \rho_Q(v_{iL}, v_i)$ {a down-polygon}.
16:        Let $Q_R = v_i v_{iR} + \rho_Q(v_i, v_{iR})$ {a down-polygon}.
17:        Let $Q' = v_{iL}v_{iR} + \rho_Q(v_{iL}, v_{iR})$ {the up-polygon}.
18:        Let $S = S \cup \{Q'\}$.
19:        Replace element Q in $\mathcal{P}$ by Q_L and Q_R.
20:    **end if**
21: **end for**
22: Let n be the cardinality of S and $\mathcal{T} = \emptyset$.
23: **for** $j = 1, 2, \ldots, n$ **do**
24:    Partition the monotone polygon $P_j \in S$ into a set $\mathcal{T}_j$ of trapezoids.
25:    Let $\mathcal{T} = \mathcal{T} \cup \mathcal{T}_j$.
26: **end for**
27: Output $\mathcal{T}(P) = \mathcal{T}$.

Fig. 5.21 Trapezoidation of a simple polygon using down- or up-stable points. Procedure 3 is explained below

For Line 3, see Fig. 5.23. The calculation of those intersection points needs to be organised carefully in order not to waste calculation time, and we provide with Procedure 5 a possible solution below.

In Line 4, we sort all the up- and down-stable vertices according to their decreasing y-coordinate. In the shown example, we have v_1 first, followed by v_2, v_3, and so forth—indices already show the position in the sorted set V.

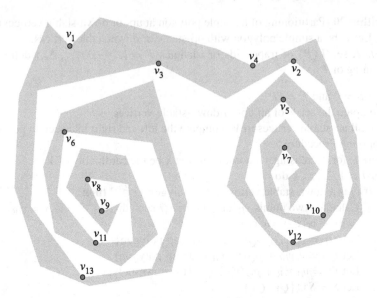

Fig. 5.22 At the end of Line 2 of Algorithm 20: all the up- and down-stable vertices of the polygon

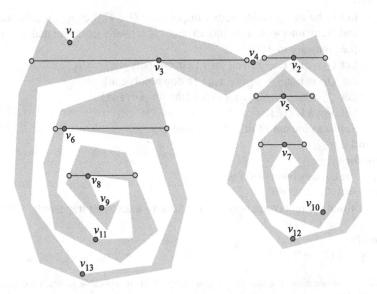

Fig. 5.23 Illustration of the final result of Line 3 in Algorithm 20, showing the calculated *left* and *right* intersection points, but also the line segments illustrating their geometric assignment to one of the up-stable vertices

If $v \in V$ is down-stable, then we still need the left-and right intersection points. Figure 5.24 illustrates all the resulting intersection points when processing down-stable vertices in Line 7. (The figure also still contains all the intersection points for up-stable vertices as calculated before in Line 3.)

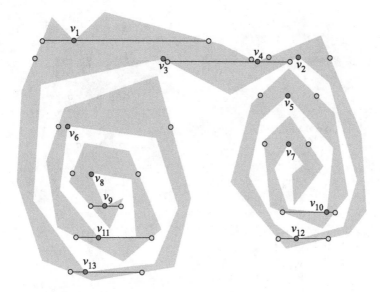

Fig. 5.24 Illustration of all the results obtained in Line 7 in Algorithm 20, showing calculated intersection points and line segments to one of the down-stable vertices

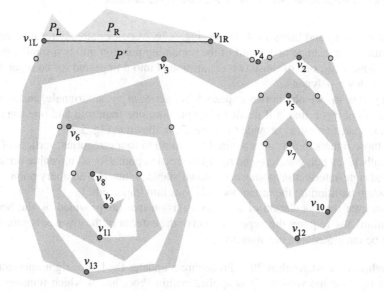

Fig. 5.25 Illustration for Lines 7–12 of Algorithm 20 when processing the first down-stable vertex v_1: two up-polygons are separated and go into S, and the remaining polygon is further processed. Set $\mathcal{P}$ still has only one element

For vertex v_1 and Lines 7 to 12, see Fig. 5.25. We have the first two monotone polygons separated and moved into S (both are triangles); the updated polygon is now without vertex v_1 but with two new vertices, the former intersection points.

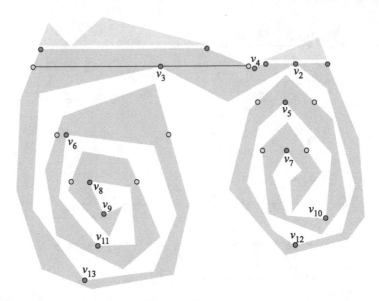

Fig. 5.26 When processing v_3 in Lines 14 to 19 of Algorithm 20, one up-polygon (a trapezoid) is separated and two down-polygons (subsets of one of the two down-polygons, created at vertex v_2) define new elements in the set $\mathcal{P}$. After v_3, $\mathcal{P}$ contains three elements

For vertex v_3 and Lines 14 to 19, see Fig. 5.26. Vertex v_2 already resulted into a separation of a triangle and a split of the remaining polygon into two smaller polygons. Now, at v_3, one of those two is partitioned into a trapezoid (to be separated) and two down-polygons.

At vertex v_4, we obtain two trapezoids for set S, at v_5 one triangle, and at v_6 a monotone polygon that still needs to be split into one trapezoid and one triangle. The final partitioning, as provided in Line 27, is shown in Fig. 5.27.

In those figures, points are coloured according to four categories: vertices of the originally given polygon, left- and right-intersection points for all up-stable vertices, left-and right-intersection points of all down-stable points, and auxiliary points created when partitioning the monotone polygons into trapezoids.

Note that the calculation of a binary decision tree (as provided by the Seidel algorithm) also requires the trapezoidation of the exterior of the given polygon. This could be integrated into the algorithm.

Procedures for Algorithm 20   Procedure 3 updates the left or right intersection points of up-stable vertices. It is applied within Procedure 4 which removes up-stable points. Procedure 5 applies Procedures 3 and 4 to compute the left and right intersection points of all up-stable vertices, from bottom to the top. This procedure is called in the main algorithm which processes both up- and down-stable vertices, from top to the bottom.

Procedure 3 is given in Fig. 5.28.

The basic idea of Procedure 3 is simple: If there is no up-stable vertex in the current simple polygon then it is easy to decompose it into some trapezoids. Each

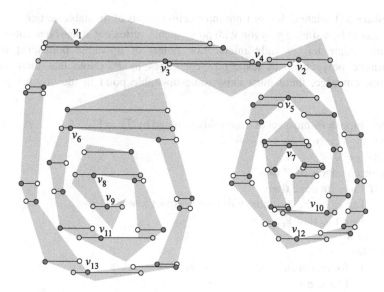

Fig. 5.27 Output of Algorithm 20 for the polygon of Fig. 5.19. If necessary, trapezoids may be partitioned into two triangles each

cut edge in this polygon corresponds to an up-stable point in the original polygon. The left or right intersection point of each up-stable vertex can be computed if we locate the trapezoid that contains the up-stable vertex.

By sorting both trapezoids (in Step 1) and up-stable vertices (in Step 9), we can select trapezoid $\triangle_i$ quickly (in linear time): if at $\triangle_i$, we search for the next trapezoid starting from index $i + 1$.

In Line 1, m is the cardinality of $\mathcal{T}_P$, $\triangle_i.y$ is the y-coordinate of endpoints of the upper edge of $\triangle_i$, for $i = 1, 2, \ldots, m$, and y_0 is the minimal y-coordinate of all vertices of P. The operations in Line 1 are straightforward because there is no up-stable point.

The original polygon is the input polygon in the main algorithm (Algorithm 20), while P is the polygon obtained from Step 7 in Procedure 4 (see Fig. 5.29) or Step 14 in Procedure 5 (see Fig. 5.30).

Strictly speaking, in Line 1 it could be $y_0 < \triangle_1.y \le \triangle_2.y \le \cdots \le \triangle_m.y$ rather than $y_0 < \triangle_1.y < \triangle_2.y < \cdots < \triangle_m.y$; see Exercise 5.10.

Each cut edge corresponds to a degenerated trapezoid whose up and down sides are identical. See Fig. 5.31 for Line 5. After the sorting, vertices are called $v_0, v_1, \ldots$ with $(\triangle_0)_y = y_0$.

In Line 13, the 'initial left and right intersection points' are as originally set in Step 12 in Procedure 5. See Fig. 5.32 for Line 13. Both v and w in this figure are originally set in Step 12 in Procedure 5. In other words, originally, v and w are left and right intersection points of u, respectively. But now, v is correctly updated to be the right intersection point of u.

The basic idea behind the procedure shown in Fig. 5.29 is as follows: There are no up-stable vertices in the input polygon when this procedure is called (Step 7) by

Procedure 3 (Update of left or right intersection points of up-stable vertices)
Input: Let P be a simple polygon with no up-stable vertex but at least one cut-edge.
Output: Update left or right intersection points of up-stable points (of original simple polygon) with respect to cut edges in P. (Note that there is at least one cut-edge, there must exist some up-stable point in the original polygon.)

1: Decompose P into a set of trapezoids, denoted by $\mathcal{T}_P = \{\triangle_1, \triangle_2, \ldots, \triangle_m\}$ such that $y_0 < \triangle_1.y < \triangle_2.y < \cdots < \triangle_m.y$.
2: Let $V_e = \emptyset$.
3: **for** each edge e of P **do**
4:     **if** e is a cut edge **then**
5:         Let v_e be the up-stable point with respect to e.
6:         Insert v_e into V_e.
7:     **end if**
8: **end for**
9: Sort V_e for increasing y-coordinates: $v_1, v_2, \ldots, v_m$.
10: **for** $i = 1$ to m **do**
11:     Select trapezoid $\triangle_i \in \mathcal{T}_P$ such that $\triangle_{i-1}.y < v_i.y < \triangle_i.y$.
12:     Compute the intersection points of line $y = v_i.y$ with the edges on the left and right of $\triangle_i$.
13:     Update the left and right intersection points of v_i by comparing the results of Line 12 with the initial left and right intersection points of v_i.
14: **end for**

Fig. 5.28 Procedure for updating left or right intersection points of up-stable vertices

Procedure 4 (Removal of maximal vertices)
Input: Let I be an interval of real numbers, and M_I be the set of maximal vertices of P such that for each element $v \in M_I$ it follows that $v.y \in I$. Assume that there is not any up-stable vertex in the input polygon P.
Output: An updated simple polygon P after removing any maximal vertex in the input polygon.

1: Let $M_I = \{v_1, v_2, \ldots, v_k\}$.
2: **for** $i = 1, 2, \ldots, k$ **do**
3:     Find a closest down-stable point with respect to v_i around the frontier of P, denoted by u_i.
4:     Find a point $w_i \in \partial P$ such that $\rho_P(u_i, v_i, w_i)$ is the shortest polygonal path in ∂P such that $u_i.y = w_i.y$.
5:     Update P by replacing $\rho_P(u_i, v_i, w_i)$ by the edge $u_i w_i$.
6:     Let P_i be the polygon created by adding edge $u_i w_i$ to $\rho_P(u_i, v_i, w_i)$.
7:     Let P_i be the input of Procedure 3 and update the left and right intersection points of all possible up-stable points.
8: **end for**

Fig. 5.29 Removal of maximal vertices in a polygon

Procedure 5 (Compute all left or right intersection points)
Input: A simple polygon P with some up-stable vertices.
Output: All left or right intersection points, for all the up-stable vertices.

1: Let $U = \{v_1, v_2, \ldots, v_k\}$ be the sorted set of up-stable vertices of P such that $v_1.y < v_2.y < \cdots < v_k.y$.
2: **for** $i = 1$ to k **do**
3:      Find closest minimal vertex u_i with respect to v_i by following the frontier of P.
4:      Let $I = [a, b]$ where $a = u_i.y$ and $b = v_i.y$.
5:      Compute M_I.
6:      **if** $M_I \neq \emptyset$ **then**
7:          Let P, I, and M_I as input, apply Procedure 4 to update P.
8:          Find a point w_i such that $\rho_P(u_i, v_i, w_i)$ is the shortest polygonal path of ∂P with $u_i.y = w_i.y$.
9:      **end if**
10:      Set initial left and right intersection points of v_i as follows: Find two points w_{iL}, w_{iR} such that $\rho_P(w_{iL}, u_i, v_i, w_{iR})$ is the shortest polygonal path in ∂P with $w_{iL}.y = v_i.y = w_{iR}.y$.
11:      Let the initial left and right intersection points of v_i be w_{iL} and w_{iR}, respectively.
12:      Update P by replacing path $\rho_P(u_i, v_i, w_i)$ by edge $u_i w_i$.
13:      Now let P be the input for Procedure 3; update the left and right intersection points of all possible (analogous to Step 7 in Procedure 4) up-stable vertices.
14: **end for**

Fig. 5.30 Computation of all left or right intersection points for a given simple polygon

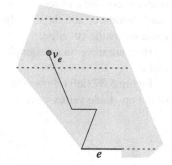

Fig. 5.31 Example for Line 5 of Procedure 1. Here, e is a cut edge. There is a decreasing polygonal path from up-stable vertex v_e to edge e. Each cut edge corresponds to an up-stable vertex via a decreasing polygonal path

the next procedure. Now, each maximal vertex (i.e., v_i in the procedure) corresponds to a down-stable vertex (i.e., u_i in Line 3) that again corresponds to a cut edge (i.e., $u_i w_i$ in Line 5) by the assumption. Thus, each maximal vertex and the cut edge corresponds a simple polygon (i.e., P_i in Line 6) that can be replaced by the cut edge in the input polygon so as to remove the maximal vertex (see Line 5).

Fig. 5.32 Example for
Line 13 of Procedure 3: u is
an up-stable point, v is the
result of Line 12, w is the
initial right intersection point
of u

For Line 1, see Fig. 5.33, first on the left: we start for this example at the specified vertex v_1.

For Line 3, see Fig. 5.33, second from the left: We find the corresponding down-stable vertex u_1 with respect v_1. In Line 3, in other words, we find a down-stable vertex u_i such that $\rho_P(v_i, u_i)$ is the shortest. There can be at most two closest down-stable points.

For Line 4, see Fig. 5.33, second from the right: We find another endpoint w_1 of the cut edge $u_1 w_1$.

For Line 5, see Fig. 5.33, first from the right: The polygon corresponding to the maximal vertex v_1 and cut edge $u_1 w_1$ is removed by replacing it by the cut edge.

For Line 7, the cut edge of P_i does not necessary correspond to an up-stable vertex of the original polygon. For example, in Fig. 5.33, the cut edge $u_1 w_1$ does not correspond to any up-stable vertex of the original polygon. In Fig. 5.31, if the cut edge is the bottom edge (i.e., the edge below e), then the removed polygon does contain an cut edge e that corresponds to an up-stable vertex v_e of the original polygon.

For Line 1, see Fig. 5.34. We sort six up-stable vertices.

For Lines 3, 8, 10, and 12, see Fig. 5.35. For Line 3, u_1 is a minimal vertex such that $\rho(u_1, v_1)$ is the shortest. For Line 8, w_1 is another endpoint of the cut edge $u_1 w_1$. For Line 10, w_{1_L} and w_{1_R} are initial left and right intersection points of v_1 respectively. We can see that w_{1_L} is correct but w_{1_R} is wrong. For Line 12, the up-stable vertex v_1 and the cut edge $u_1 w_1$ corresponds a polygon that is removed and replaced by the cut edge.

Here are three more examples. Figure 5.36 shows 4 up-stable vertices v_1, v_2, v_3, and v_4.

Figure 5.37 (left) shows the initial left and right intersection points w_{1_L} and w_{1_R} of the up-stable vertex v_1 (Line 10).

Fig. 5.33 Example for operations of Procedure 4 for the polygon in Figs. 5.19 or 5.34, only showing the lower left part of this polygon. *From left to right*: Line 1, Line 3, Line 4, and Line 5

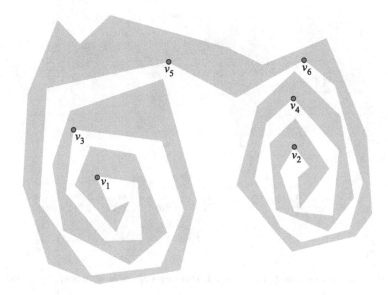

Fig. 5.34 Example for Line 1 of Procedure 5

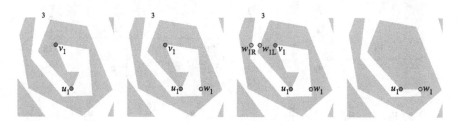

Fig. 5.35 Example for operations in Procedure 5 for the polygon in Fig. 5.34, only showing the lower left part of this polygon. *From left to right*: Line 3, Line 8, Line 10, and Line 12

Figure 5.37 (right) shows the initial left and right intersection points w_{2_L} and w_{2_R} of the up-stable vertex v_2 (for Line 10). It also shows the resulting polygon after replace the polygon $v_1 u_5 u_4$ by the cut edge $u_5 u_4$ (Line 12).

5.6 Time Complexity of Algorithm 20

Lemma 5.5 *The set of up-stable (or down-stable or maximal) vertices of a simple polygon P can be computed in $\mathcal{O}(n)$, where $n = |V(P)|$.*

Proof For a sufficiently small number $\varepsilon_s > 0$, the "start" vertex v_0' of an inner polygon $P(v_0, \varepsilon_s)$ can be computed in $\mathcal{O}(n)$; see [17]. Consider three consecutive vertices u, v, w of P with $u.x < v.x < w.x$. Let $v' = (v.x, v'.y)$ be a point of $P(v_0, \varepsilon_s)$.

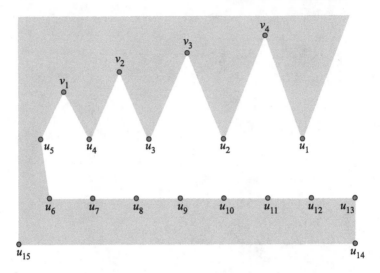

Fig. 5.36 Example 1 for Line 1 of Procedure 5: There are 4 up-stable vertices v_1, v_2, v_3, and v_4

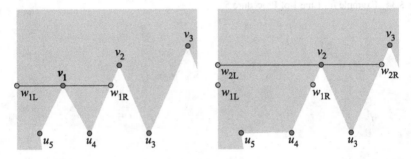

Fig. 5.37 Continuation for the polygon shown in Fig. 5.36, but only showing the upper left corner of the polygon. Examples 2 (*left*) and 3 (*right*) for Lines 10 and 12, respectively, of Procedure 5

In other words, v' is the point of $P(v_0, \varepsilon_s)$ such that its x-coordinate is identical to $v.x$.

By the definition above, if there are three consecutive vertices u, v_0, w of P with $u.x < v_0.x < w.x$, then v'_0 is the point of $P(v_0, \varepsilon_s)$ such that its x-coordinate is identical to $v_0.x$. Without loss of generality, we may assume that the coordinates of each vertex are integers, and we set $\varepsilon_s = 0.1$.

If $u.y < v.y$, $v.y > w.y$ and $v'.y > v.y$, then v is an up-stable vertex. If $u.y < v.y$, $v.y > w.y$ and $v'.y < v.y$, then v is a maximal point. If $u.y > v.y$, $v.y < w.y$ and $v'.y < v.y$, then v is a down-stable point.   □

Lemma 5.6 *Procedure 3 can be executed in $\mathcal{O}(n \log n)$ time, where $n = |V(P)|$.*

Proof If P is monotone, then it can be decomposed into a stack of trapezoids in $\mathcal{O}(|V(P)|)$. Otherwise, by assumption, P can only have a finite number of down-stable points. Analogously to Lines 7–10 in Algorithm 20, P can be decomposed into a set of trapezoids in $\mathcal{O}(|V(P)|)$ time.

Thus, Line 1 can be executed in $\mathcal{O}(|V(P)|)$. Line 2 can be performed in $\mathcal{O}(1)$. Lines 3–8 require $\mathcal{O}(|E(P)|) = \mathcal{O}(|V(P)|)$ time. Line 9 can be computed in

$$\mathcal{O}\big(|V_e|\log(|V_e|)\big) \le \mathcal{O}\big(|V(P)|\log(|V(P)|)\big)$$

time. Lines 10–14 can be executed in $\mathcal{O}(|V_e| \le \mathcal{O}(|V(P)|)$. Thus, Procedure 3 requires $\mathcal{O}(|V(P)|\log(|V(P)|))$ time. □

Lemma 5.7 *Procedure 4 requires $\mathcal{O}(n\log n)$ time, where n is the number of vertices of the original simple polygon P.*

Proof By Lemma 5.5, Line 1 can be computed in $\mathcal{O}(n\log n)$ time, where n is the number of vertices of the original simple polygon P. Line 3 can be computed in $\mathcal{O}(n_u)$ time, where n_u is the number of vertices of $\rho_P(u_i, v_i)$. Line 4 can be computed in $\mathcal{O}(n_w)$ time, where n_w is the number of vertices of $\rho_P(v_i, w_i)$. Lines 5 and 6 can be executed in constant time. By Lemma 5.6, Line 7 can be computed in $\mathcal{O}(n_i \log n_i)$ time where $n_i = |V(P_i)|$.

Thus, Procedure 4 can be computed in $\mathcal{O}(n\log n)$ altogether, where n is the number of vertices of P. □

Lemma 5.8 *Procedure 5 can be computed in $\mathcal{O}(n\log n)$ time, where n is the number of vertices of the original simple polygon P.*

Proof By Lemma 5.5, Line 1 requires time in $\mathcal{O}(n\log n)$, where n is the number of vertices of the original simple polygon P. Line 3 can be computed in $\mathcal{O}(n_u)$, where n_u is the number of vertices of $\rho(u_i, P, v_i)$. Line 4 can be performed in constant time. Line 5 can be executed in $\mathcal{O}(|M_I|)$ time. By Lemma 5.7, Line 7 require $\mathcal{O}(n_u \log n_u)$ time, where $n_u = |V(\rho(u_i, P, v_i))|$. Line 8 can be computed in $\mathcal{O}(n_w)$, where n_w is the number of vertices of $\rho_P(u_i, w_i)$. Line 10 requires $\mathcal{O}(n_i)$ time, where $n_i = |V(\rho_P(u_i, w_{il}))| + |V(\rho_P(w_i, w_{ir}))|$. Lines 11 and 12 can be computed in constant time.

By Lemma 5.6, Line 13 can be computed in $\mathcal{O}(n\log n)$ time, where n is the number of the vertices of the updated P. Therefore, Procedure 5 can be computed in $\mathcal{O}(n\log n)$, where n is the number of vertices of the original simple polygon P. □

Theorem 5.1 *Algorithm 20 has a worst-case time complexity of $\mathcal{O}(n\log n)$, where n is the number of vertices of the given simple polygon P.*

Proof Line 1 requires constant time. By Lemma 5.5, Line 2 can be computed in $\mathcal{O}(n)$, where n is the number of vertices of the given simple polygon P. By Lemma 5.8, Line 3 can be performed in $\mathcal{O}(n\log n)$ time, where n is the number of the vertices of the given simple polygon P.

Line 4 can be computed in time $\mathcal{O}(|V|\log|V|)$. Line 7 can be executed in $\mathcal{O}(n_i)$ time, where $n_i = |V(\rho_P(v_{iL}, v_i))| + |V(\rho_P(v_i, v_{iR}))|$. Lines 8–13 can be computed

Fig. 5.38 A simple polyline
ρ and its visibility map $V(C)$

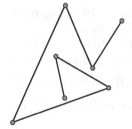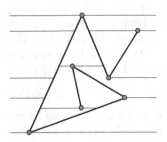

in $\mathcal{O}(1)$. Thus, Lines 6–13 require $\mathcal{O}(n_i)$ altogether, where $n_i = |V(\rho_P(v_{iL}, v_i))| + |V(\rho_P(v_i, v_{iR}))|$. Lines 15–21 only take constant time. By Lemma 5.7, Line 25 can be executed in $\mathcal{O}(|P_j| \log |P_j|)$ time.

Thus, Lines 24–27 take $\mathcal{O}(n)$ time, where n is the number of the vertices of the input polygon P. Line 28 can be computed in constant time. Therefore, the algorithm requires altogether $\mathcal{O}(n \log n)$ time.   □

5.7 Polygon Trapezoidation Method by Chazelle

The method starts with the computation of a *visibility map* for a simple polyline. The most important concept is that of *conformality* (see definition below). Restoring conformality after merging two submaps is the core of the method.

Let ρ be a simple polyline such that no two vertices of ρ have the same y-coordinate. Assume that ρ is embedded in a 'spherical plane' such that $(-\infty, y) = (+\infty, y)$, $(x, -\infty) = (-x, -\infty)$, and $(x, +\infty) = (-x, +\infty)$.

Then the horizontal extension through any vertex of ρ must hit two edges of ρ. For example, if a vertex has maximum y-coordinate, then this vertex is an endpoint of an edge. Thus, the horizontal extension through this vertex will hit this edge from both sides. Under this assumption, the *visibility map* $V(\rho)$ is identical to the trapezoidation $\mathcal{T}(S)$, where S is the set of all edges of ρ. See Fig. 5.38.

A *submap* of $V(\rho)$ is a subgraph that is obtained after removing from $V(\rho)$ some horizontal line segments, also called *chords*. Chords are not edges of ρ. A face of a submap of $V(\rho)$ is a bonded or unbounded component, where segments of ρ or chords defining the frontier of a face.

In order to make a difference between both sides of a line segment, each edge of ρ and each chord is given a sufficiently small width. See Fig. 5.39. This way, the polyline ρ is a very thin simple polygon (a subset of the space between the lines in Fig. 5.39).

The frontier of each face consists of sequences of edge segments (called an *arc*) and chords. The *weight* of a face is the maximal number of edge segments in any of its arcs. In Fig. 5.39, Faces 1, 2, and 3 have weights 5, 1, and 3, respectively.

The dual graph of $V(\rho)$ is a tree and is called the *visibility tree* of ρ. The dual graph of a submap of $V(\rho)$ is called the *visibility tree* of the submap. The weight of a node in a visibility tree is defined as the weight of the corresponding region.

Fig. 5.39 A submap of the
visibility map in Fig. 5.38.
The frontier of Face 1
consists of one (*shaded*) arc
with 5 edge segments and two
chords

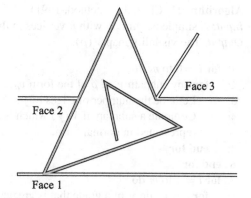

A submap is called *conformal* if the degree of each node of its visibility tree is 4 at
most. The map $V(\rho)$ is conformal.

A submap is called *k-granular* if each node of its visibility tree has a weight less
than or equal to k, and a contraction of any edge that is incident with an adjacent
node of degree 1 or 2 results in a new node of weight greater than k.

A submap is represented in *normal form* if the following information is provided
by the representation:

1. Node adjacencies in the visibility tree.
2. Each edge of the tree corresponds[5] to a chord of this submap and all arcs adjacent
   to this chord. Each arc corresponds to a node of the tree such that the node's
   corresponding face of the submap is incident with this arc.
3. Each arc is represented by an arc-structure that is stored in an *arc-sequence table*
   (not explained here).
4. If the submap is conformal, then its binary tree decomposition is given.

The *Chazelle's method* [6] is sketched in Fig. 5.40. Lines 1 to 6 have been called the
up-phase of the method, and Lines 7 to 11 the *down-phase* of the method.

The input is a simple polyline ρ with $n = 2^m + 1$ vertices. Such a number can
be obtained by padding ρ with additional vertices if necessary. A *chain of grade i*
is a subpolyline of ρ of the form $v_a, \ldots, v_b$ such that $a - 1$ is a multiple of 2^i and
$b = 2^i + a$, where $i = 0, 1, 2, \ldots, m$.

This method was published in 1991, with the claim to be a *linear-time triangulation algorithm* for any simple polyline. Until now no implementation is known to the
authors. Thus, we decided to present this algorithm here, hoping that someone could
then take the challenge of implementation. When attempting to extract a procedural presentation (i.e., an actual *algorithm*) from the original paper, we experienced
severe problems. We have two major concerns:

---

[5] Each node of the tree corresponds to a region of the submap. Two nodes are connected by an edge
if their corresponding regions share a chord. In this case, the edge "corresponds" to the chord.

[6] *Bernard Chazelle* is at Princeton University.

Algorithm 21 (Chazelle's method, 1991)
Input: A simple polyline ρ with n vertices in the plane, where $n = 2^m + 1$.
Output: The visibility map $V(\rho)$.

```
1:  for i = 0 to m do
2:      for any subpath γ of ρ of the form v_a, ..., v_b, with 2^{i-1} < b − a ≤ 2^i do
3:          Let e be the number of edges of γ.
4:          Compute a submap of V(γ) which is 2^⌈0.2⌈log e⌉⌉-granular, conformal, and
            represented in normal form.
5:      end for
6:  end for
7:  for i = 1 to m do
8:      for any chain γ of a grade that is greater than or equal to i do
9:          Compute the final V(γ) from a submap of V(γ) that is 2^⌈i/5⌉-granular,
            conformal, and represented in normal form.
10:     end for
11: end for
```

Fig. 5.40 Sketch of Chazelle's proposal for computing the visibility map of a simple polyline in the plane

(i) The method is not sufficiently specified to be called 'an algorithm': operations are often described within extensive proofs, and not explicitly with some procedures.

(ii) It appears questionable whether the method is actually of linear time: quite a few operations are described as sorting instead of the complicated, claimed to be linear operations.

The authors believe that the inclusion into our text may reactivate the consideration of this method (which gained quite some attention), being aware that we could not succeed with providing a fully comprehend presentation of an algorithm.

5.8 The Continuous Dijkstra Problem

The Dijkstra algorithm (see Fig. 1.9) solves a shortest path problem on finite weighted graphs: for a given vertex p in the graph, find shortest paths to any other vertex q of that graph.

> The *continuous Dijkstra problem*: Given is a finite set of pairwise disjoint simple polygons (the obstacles) and a start vertex $p \in \mathbb{R}^2$ which is not in any of those polygons, provide a partitioning of $\mathbb{R}^2$ which supports the calculation of ESPs to points $q \in \mathbb{R}^2$ that are not in any of those polygons.

Fig. 5.41 The continuous
Dijkstra problem: source
point p, a finite set of
polygons (obstacles), three
different destinations, and the
ESPs defined by those end
points

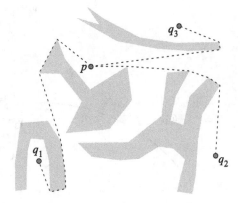

See Fig. 5.41 for an example of ESPs to three different destinations; the start
point p and obstacles remain fixed.

5.9 Wavelets and Shortest-Path Maps

Let p be the source point, d_q the length of an ESP from p to a point q, and V be the
set of vertices of all input polygons (i.e., the vertices of the obstacles). We assign a
label $l(v)$ to a vertex $v \in V$ that specifies the current upper bound for the length of
a path from p to v.

A *wavelet*[7] ω is a circular arc together with the centre $r_\omega \in V \cup \{p\}$ of the circle,
also called the *root* (of the wavelet). Let $d(r_\omega)$ be the length of a shortest path from
p to r_ω among the given obstacles. At any time we have that $d(r_\omega) \leq d_{r_\omega}$.

Example 5.3 Figure 5.42 shows a point $p = (36, 32)$ and two triangular obstacles.
We have $V = \{v_1, \ldots, v_6\}$. For demonstrating calculations further below, we have
$v_1 = (165, 64)$, $v_2 = (244, 193)$, and $v_3 = (130, 502)$ for the first triangle, and $v_4 =
(300, 600)$, $v_5 = (350, 800)$, and $v_6 = (100, 800)$ for the second triangle. It follows
that $d_{v_1} = d_e(p, v_1)$, $d_{v_2} = d_e(p, v_1) + d_e(v_1, v_2) \approx 284.178$, $d_{v_3} = d_e(p, v_3)$, and
so forth. Actually, it is $d(r_{\omega_6}) = d_{v_2} \approx 284.178$.

Initially, we have $l(p) = 0$ and $l(v_i) = +\infty$, for $i = 1, \ldots, 6$.

Figure 5.42 shows three full-circle wavelets ω_1, ω_2, and ω_3, all having p as their
root. The wavelet ω_4 also has p as its root but is not a full circle. The non-full-circle
wavelets ω_5, ω_6, and ω_7 have v_1, v_2, and v_3 as root, respectively. In other words,
we have that $r_{\omega_i} = p$, for $i = 1, 2, 3, 4$, $r_{\omega_5} = v_1$, $r_{\omega_6} = v_2$, and $r_{\omega_7} = v_3$.

We will continue with this example.   □

In the algorithm for solving the continuous Dijkstra problem, we will consider a
queue containing wavelets. A wavelet in the current queue is *active*.

---

[7]Not to be confused with wavelets in signal theory.

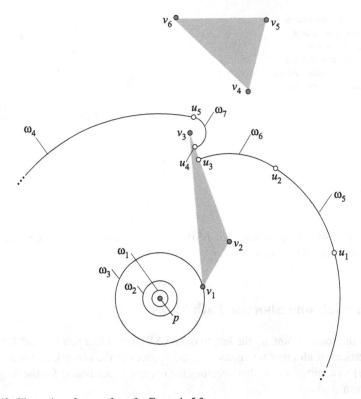

Fig. 5.42 Illustration of a wavefront for Example 5.3

An active and non-full-circle wavelet ω has a *left track* a_ω and a *right track* b_ω. Each track is the locus of the endpoints of the wavelet when it changes with the distance parameter d where

$$d = d(r_\omega) + r(\omega)$$

and $r(\omega)$ is the radius of ω. For example, in Fig. 5.42, the left track a_{ω_5} and the right track b_{ω_5} are line segments $v_1 u_2$ and $v_1 u_1$, respectively. For the radius of ω_5, we have that $r(\omega_5) = d_e(v_1, u_1) \approx 151.268$.

A full-circle wavelet is considered to be the union of two semi-circle wavelets, each having horizontal left and right tracks.

If a_ω or b_ω is not an obstacle segment then ω has a left or right adjacent wavelet $\mathrm{LW}(\omega)$ or $\mathrm{RW}(\omega)$, respectively. It follows that $b_{\mathrm{LW}(\omega)} = a_\omega$ and $a_{\mathrm{RW}(\omega)} = b_\omega$. For example, in Fig. 5.42 we have that $\mathrm{LW}(\omega_5) = \omega_6$ and $\mathrm{RW}(\omega_5) = \omega_4$.

The *bisector* $B(r_\omega, r_{\omega'})$ of two wavelets ω and ω' is the locus of points q such that

$$d(r_\omega) + d_e(r_\omega, q) = d(r_{\omega'}) + d_e(r_{\omega'}, q).$$

Below a bisector is sometimes denoted by the symbol γ of a curve in $\mathbb{R}^2$.

In Fig. 5.43, $B(v_3, v_2)$ is the bisector of two wavelets rooted at v_3 and v_2; it ends at point s. $B(v_4, v_2)$ starts at point s and is the bisector of two wavelets rooted at

Fig. 5.43 Illustration of a
shortest-path map (SPM)

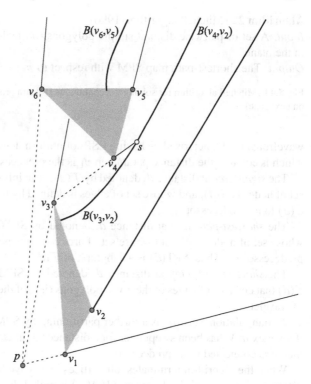

v_4 and v_2. $B(v_5, v_6)$ is the bisector of two wavelets rooted at v_5 and v_6. $B(v_3, v_2)$,
$B(v_4, v_2)$, and $B(v_5, v_6)$ are hyperbolic arcs.

In the algorithm for solving the continuous Dijkstra problem, we have different
events. A *closure event* occurs when an active wavelet comes to its end of existence
because its left and right tracks meet at a *closure point*, such as point s in Fig. 5.43.
A *vertex event* occurs when an active wavelet collides with an obstacles vertex. In
both cases, a closure point or an obstacles vertex is called the *event point*. The length
of the shortest path from p to a current event point is called the *event distance*.

For example, in Fig. 5.43, the closure point s is the intersection point of line
segment $v_4 s$ and $B(v_3, v_2)$. Line segment $v_4 s$ is the right track of the wavelet rooted
at v_4. $B(v_3, v_2)$ is the left track of the wavelet rooted at v_2. When this closure event
occurs, the event distance is $d_e(s, v_4) + d_e(v_4, v_3) + d_e(v_3, p)$.

In Fig. 5.42, wavelet ω_3 collides with obstacle vertex v_1. The event distance is
$d_e(p, v_1)$ when this event occurs.

A point q in wavelet ω is *swept over* by ω for a distance parameter $d = d(r_\omega) +
r(\omega)$. In this case, make a copy of q associated with r_ω, denoted by $q^{(r_\omega)}$. Point r_ω
is called the *predecessor* of $q^{(r_\omega)}$.

The *sweep space* $S(d)$ is the set of all copies of all points swept over by a distance
parameter less than or equal to d.

See Fig. 5.42. The set of points in $\bigcup_{i=4}^{7} \omega_i$ is a *wavefront* with distance parameter d that equals to $d_e(p, v_3) + d_e(v_3, u_5)$. Each point in the wavefront is swept over
by a corresponding wavelet for a distance parameter. It follows that all points in the

Algorithm 22 (Mitchell algorithm, 1996)
Input: A set of pairwise disjoint simple polygons (obstacles) and a source point p in the plane.
Output: The shortest-path map SPM with respect to p.

Fig. 5.44 Mitchell algorithm for solving the continuous Dijkstra problem (figure to be continued on next page)

wavefront are characterised by having ESPs of uniform length to the source point p, which is equal to the distance parameter. v_1 is the predecessor of all points in ω_5.

The *signal tree* at distance d, denoted by $\mathcal{T}(d)$, is an infinite directed graph whose set of nodes is $\mathcal{S}(d)$, and whose set of edges is defined by links from a vertex $q^{(r_\omega)} \in \mathcal{S}(d)$ to its predecessor r_ω.

The *shortest-path tree* at distance d, denoted by SPT(d), is an directed graph whose set of nodes is V, and whose set of arcs connects points $q^{(r_\omega)} \in V$ with their predecessor r_ω. Thus, SPT(d) is a subgraph of $\mathcal{T}(d)$.

The *shortest-path map* at distance d, denoted by SPM(d), is a partitioning of $\mathcal{S}(d)$ that consists of edges of the given polygons (i.e., of the obstacles) and bisectors of wavelets.

A triangulation SPM(d) is a further partitioning by SPM(d)-*triangles* such that, if a vertex in V has been swept over for a distance parameter less than or equal to d, then it is connected to its predecessor.

When the algorithm terminates, all vertices in V have been swept over by some distance parameter d'. In this case, SPM(d') is called the *shortest-path map* SPM, which is a partitioning of the free space that contains all the points outside of the given polygonal obstacles (including their frontier). For each point in the interior of the same region of the SPM, there is a unique predecessor. Figure 5.43 shows a SPM for two triangles and point p.

5.10   Mitchell's Algorithm

The *Mitchell algorithm*[8] for solving the continuous Dijkstra problem is shown in Fig. 5.44. It uses three procedures which are explained further below; the first two are shown in Figs. 5.45 and 5.46.

The input for the Mitchell algorithm is a set of pairwise disjoint simple polygons (obstacles) and a source point p in the plane.

Line 1 initialises the algorithm. We set the event distance $d = 0$; label each vertex $v \in V$ to be $l(v) = +\infty$ and $l(p) = 0$; and event queue Q has a single wavelet ω that roots at point p with radius 0.

Line 2 enters a while loop.

Line 3 takes and processes the first element in the event queue Q.

---

[8]*Joseph S.B. Mitchell* is at the State University of New York at Stony Brook.

1: Let $d = 0$. For each $v \in V$, let $l(v) = +\infty$. Let SPT(0) be a single node p and $l(p) =$
   0. Let ω be a wavelet rooted at point p with radius 0. Let event queue $Q = \{\omega\}$. In the
   rest of the algorithm, each active wavelet will be indexed by its event distance in Q.

2: **while** $Q \neq \emptyset$ **do**

3:    Remove the first event in Q. Let q be the event point, d the event distance, r_ω the
      root of the wavelet which caused this event.

4:    **if** q is a closure point **then**

5:       **if** a_ω and b_ω are not obstacle edges **then**

6:          **if** q lies to the right of the directed line segment $r_{LW(\omega)}r_{RW(\omega)}$ **then**

7:             Let q be the starting point. Call Procedure 8.

8:          **else**

9:             **if** q lies to the left of the directed line segment $r_{LW(\omega)}r_{RW(\omega)}$ **then**

10:               Update LW(ω) to have the right track $B(r_{LW(\omega)}, r_{RW(\omega)})$ and RW(ω) to
                  have the left track $B(r_{LW(\omega)}, r_{RW(\omega)})$.

11:            **end if**

12:         **end if**

13:      **else**

14:         **if** both $r_{LW(\omega)}$ and $r_{RW(\omega)}$ are obstacle edges **then**

15:            Label q with distance d.

16:         **else**

17:            **if** both a_ω [b_ω] and $b_{RW(\omega)}$ [$a_{LW(\omega)}$] are an obstacle edge e **then**

18:               Label event point $q = e \cap B(r_\omega, r_{RW(\omega)})$ [$q = e \cap B(r_\omega, r_{LW(\omega)})$] with
                  event distance d; remove both ω and RW(ω) [LW(ω)] from Q.

19:            **else**

20:               **if** a_ω [b_ω] is an obstacle edge but b_ω [a_ω] is not an obstacle edge **then**

21:                  Update RW(ω) [LW(ω)] to have left [right] track e.

22:               **end if**

23:            **end if**

24:         **end if**

25:      **end if**

26:   **else** {In this case, $q \in V$.}

27:      Rename q to be now v (note: v stands for 'vertex').

28:      **if** there exists an obstacle edge e such that $q' = e \cap r_\omega v$ **then**

29:         Let q' be the starting point. Call Procedure 8.

30:      **else**

31:         **if** $l(v) = +\infty$ **then**

32:            Let $l(v) = d(r_\omega) + d_e(r_\omega, v)$. Create a new wavelet rooted at v and its tracks
               consist of ray $r_\omega v$ and an obstacle frontier segment containing v. Let $r_v = r_\omega$. The created wavelet is inserted into Q according to its event distance.

33:         **else**

34:            Let r_ω, v, and SPM(d) be the input for Procedure 6.

35:         **end if**

36:      **end if**

37:   **end if**

38: **end while**

Fig. 5.44 (Continued)

Procedure 6 (Clip-and-merge)
Input: r_ω, v, and SPM(d), as provided in Line 33 in Algorithm 22.
Output: A *merge bisector* γ which is a straight line that passes through a start point
q and the frontier of one of the SPM(d)-triangles.

1: Call Procedure 7 for computing a start point q.
2: Call Procedure 8 for start point q.

Fig. 5.45 Pseudocode of Procedure 6 for a clip-and-merge operation

Procedure 7 (Calculate a start point)
Input: r_ω, v, and SPM(d) from Algorithm 22.
Output: A start point (that is, a "seed" point) for Procedure 8.

1: **if** line segment $r_\omega v$ intersects some edges of SPT(d) **then**
2:     Let u be that intersection point where line segment vu is minimal.
3: **else**
4:     Let $u = v$.
5: **end if**
   [Use binary search to find a "seed" point q as follows:]
6: Let ρ_u and ρ_{r_ω} be the unique paths from p to u and from p to r_ω in $\mathcal{T}(d)$,
   respectively.
7: Let ω_{r_ω} be the first active wavelet such that it is clockwise of ρ_{r_ω}; ω_u be the first
   active wavelet such that it is counterclockwise of ρ_u. Let *Search* = true.
   [In order to compute a "seed" point q, we do a binary search on a list of ordered
   active wavelets clockwise of ω_{r_ω} and counterclockwise of ω_u (see Fig. 5.49).]
8: **while** *Search* = true **do**
9:     Let ω_u be a median active wavelet on the frontier of SPM(d)-triangle $\triangle$ with
       root $r_\triangle$.
10:    **if** $\triangle \cap r_\omega u = \emptyset$ **then**
11:        Only search for q in the active wavelets which are clockwise of ω.
12:    **else**
13:        Let y be a intersection point of $\triangle$ with $r_\omega u$ such that y is closest to v.
14:        **if** $d(r_\triangle, y) + d(r_\triangle) > d(r_\omega, y) + d(r_\omega)$ **then**
15:            Only search for q in the SPM(d)-triangles which are counterclockwise
               of $\triangle$.
16:        **else**
17:            Only search for q in the SPM(d)-triangles which are clockwise of $\triangle$
               (this also includes $\triangle$ itself).
18:        **end if**
19:    **end if**
20: **end while**
21: Return the selected point q.

Fig. 5.46 A pseudocode for detecting a "seed" point q

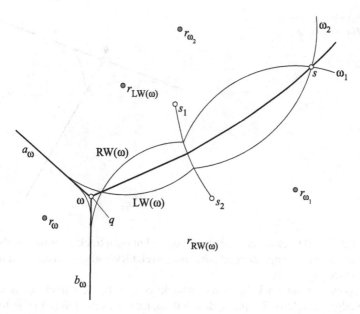

Fig. 5.47 A closure point q lies to the *right* of the directed line segment $r_{LW(\omega)}r_{RW(\omega)}$

Then the while loop divides into two main steps: Lines 4–25 handle a closure event (i.e., if q is a closure point); Lines 26–36 do a vertex event (i.e., if q is a non-closure point).

Case 1. q is a closure point (Line 4).

Case 1.1. Both left track a_ω and right track b_ω of ω are not obstacle edges (Line 5).

Case 1.1.1. q lies to the right of the directed line segment $r_{LW(\omega)}r_{RW(\omega)}$ (see Fig. 5.47), let q be the starting point. Call Procedure 8 (Lines 6–7).

Case 1.1.2. q lies to the left of the directed line segment $r_{LW(\omega)}r_{RW(\omega)}$ (e.g., in Fig. 5.43, the closure point s lies to the left of the directed line segment v_4v_2), simply update $LW(\omega)$ (i.e., the left adjacent wavelet of ω) to have the right track $B(r_{LW(\omega)}, r_{RW(\omega)})$ and $RW(\omega)$ (i.e., the right adjacent wavelet of ω) to have the left track $B(r_{LW(\omega)}, r_{RW(\omega)})$ (Lines 8–12).

Case 1.2. Left track a_ω or right track b_ω of ω is an obstacle edge (Line 13).

Case 1.2.1. Both left track a_ω and right track b_ω of ω are obstacle edges (Line 14). Simply Label q with distance d. In this case, ω dies and we remove it from the event queue Q (Line 15).

Case 1.2.2. Left track a_ω and right track $b_{RW(\omega)}$ of right adjacent wavelet of ω are an obstacle edge e (Line 17) (see Fig. 5.48). Label event point $q = e \cap B(r_\omega, r_{RW(\omega)})$ with event distance d; remove both ω and RW from the event queue Q.

Analogously, right track b_ω and left track $a_{LW(\omega)}$ of left adjacent wavelet of ω are an obstacle edge e (Line 17). Label event point $q = e \cap B(r_\omega, r_{LW(\omega)})$ with event distance d; remove both ω and LW from the event queue Q.

Fig. 5.48 A closure point q
is on an obstacle edge e

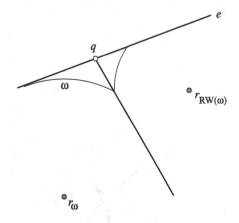

Case 1.2.3. Left track a_ω is an obstacle edge e but right track b_ω is not an obstacle edge (Line 20). Simple update right adjacent wavelet RW(ω) of ω to have left track e (Line 21) (see Fig. 5.48).

Analogously, right track b_ω is an obstacle edge e but left track a_ω is not an obstacle edge (Line 20). Simple update left adjacent wavelet LW(ω) of ω to have right track e (Line 21).

Case 2. q is a non-closure point (Line 26). In this case, q must be a vertex in V. Rename q to be v (v for vertex). Perform a ray-shooting query to test if a ray from r_ω to v hits an obstacle edge e.

Case 2.1. There exists an obstacle edge e such that $q' = e \cap r_\omega v$ (Line 28). Let q' be the starting point. Call Procedure 8 (Line 29).

Case 2.2. There does not exist an obstacle edge e such that $q' = e \cap r_\omega v$ (Line 30).

Case 2.2.1. $l(v) = +\infty$ (Line 31). That is, v is not yet hit by any wavelet. Label $l(v)$ to be $d(r_\omega) + d_e(r_\omega, v)$. Create a new wavelet rooted at v; its tracks consist of ray $r_\omega v$ and 'appreciate' (term as used by the author of this algorithm) the obstacle frontier segment containing v. Let $r_v = r_\omega$. The created wavelet is inserted into Q according to its event distance (Line 32).

Case 2.2.2. $l(v) < +\infty$ (Line 33). That is, v is already hit by a wavelet. Let r_ω, v, and SPM(d) be the input for Procedure 6 (Line 34).

In Line 33 of Algorithm 22, we have to perform a clip-and-merge operation which is done by combining Procedures 7 and 8. See Fig. 5.45.

The most difficult and important points of Algorithm 22 are Lines 7, 29, and 34 which are two procedures (Procedures 7 and 8) to be explained below.

The input for Procedure 7 are the same as the input for Procedure 6, which are r_ω, v, and SPM(d) from Algorithm 22. The output is a start point (that is, a "seed" point) for Procedure 8.

Lines 1–5 of Procedure 7 compute a point u in line segment $r_\omega v$ as follows: If line segment $r_\omega v$ intersects some edges of SPT(d), then let u be the intersection point that is closest to v. Otherwise, simply let u be v.

The rest of this procedure is applying a binary search to find a "seed" point q as follows: Let ρ_u and ρ_{r_ω} be the unique paths from p to u and from p to r_ω in $\mathcal{T}(d)$, respectively (Line 6) (see Fig. 5.49).

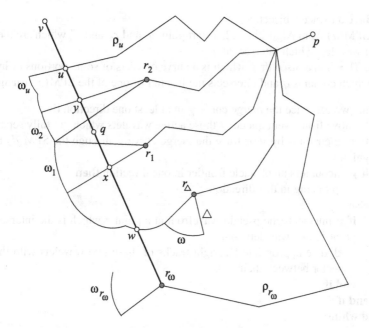

Fig. 5.49 Illustration for Procedure 7

Let ω_{r_ω} be the first active wavelet such that it is clockwise of ρ_{r_ω}; ω_u be the first active wavelet such that it is counterclockwise of ρ_u (Line 7) (see Fig. 5.49).

In order to compute a "seed" point q, we do the binary search on a list of ordered active wavelets clockwise of ω_{r_ω} and counterclockwise of ω_u.

Line 8 enters a while loop.

Let ω_u be a median active wavelet on the frontier of SPM(d)-triangle $\triangle$ with root $r_\triangle$ (Line 9).

Case 1 $\triangle \cap r_\omega u = \emptyset$ (Line 10). Only search for a "seed" point q in the active wavelets which are clockwise of ω (Line 11).

Case 2 $\triangle \cap r_\omega u \neq \emptyset$ (Line 12). Let y be a intersection point of $\triangle$ with $r_\omega u$ such that y is closest to v (Line 13).

Case 2.1 $d(r_\triangle, y) + d(r_\triangle) > d(r_\omega, y) + d(r_\omega)$ (Line 14). Only search for q in the SPM(d)-triangles which are counterclockwise of $\triangle$ (Line 15).

Case 2.2 $d(r_\triangle, y) + d(r_\triangle) \leq d(r_\omega, y) + d(r_\omega)$ (Line 16). Only search for q in the SPM(d)-triangles which are clockwise of $\triangle$ (this also includes $\triangle$ itself) (Line 17).

When can the variable *Search* be set to *false* to terminate the while loop (i.e., the binary search)? There are the following two cases:

Case 1. There is an SPM(d)-triangle $\triangle$ with root $r_\triangle$ such that $\triangle$ intersects $r_\omega u$ with two points x and y such that $d(r) + d_2(r, x) < d(r\triangle) + d_2(r\triangle, x)$ and $d(r) + d_2(r, y) > d(r\triangle) + d_2(r\triangle, y)$. This means that $\triangle$ contains a "seed" point q that can be computed approximately by another binary search.

Case 2. There are two adjacent active wavelets ω_1 and ω_2 such that the corresponding SPM(d)-triangles of them intersect $r_\omega u$ at two points x and y, re-

Procedure 8 (Trace a bisector)

Input: SPM(d) from Algorithm 22, a start point q, and r_ω and r_ω'' which are the two roots of wavelets which intersect at q.

Output: The merge bisector γ which is a curve consists of some portions of hyperbolas, which contains q and intersects the frontier of one of the SPM(d)-triangles.

1: **while** we can trace the merge curve γ in at least one direction **do**
2:      Remove from event queue Q those active wavelets which get fully separated from their roots by γ as trace the merge curve γ through the SPM(d) triangulation.
3:      **if** γ encounters an obstacle frontier in one direction **then**
4:          Stop tracing in this direction.
5:      **else**
6:          **if** γ intersects the pseudo-wavefront at a point z which is the intersection point of two wavelets **then**
7:              Update appropriate left/right tracks of these two wavelets with the bisector between their roots.
8:          **end if**
9:      **end if**
10: **end while**

Fig. 5.50 A procedure for tracing a bisector

spectively (see Fig. 5.49). Compute a SPM(d)-triangle $\triangle$ that intersects the directed line segment xy. Compute all SPM(d)-triangle in a set S' such that each triangle in S' intersects $\triangle$ with non-empty. Compute a SPM(d)-triangle $\triangle'$ in S' such that $\triangle'$ intersect SPM(d)-triangle $\triangle$ with a point q' such that $d(r_\triangle) + d_2(r_\triangle, q') = d(r_\triangle') + d_2(r_\triangle', q')$, where $r_\triangle$ and $r_\triangle'$ are the roots of $\triangle$ and $\triangle'$, respectively. Return q' as a "seed" point. S' has only constant cardinality because of the following

Lemma 5.9 *Each* SPM(d)-*triangle can be clipped by a bisector at most 6 times.*

The input for Procedure 8 (see Fig. 5.50) are SPM(d) from Algorithm 22, a start point q, and r_ω and r_ω'' which are the two roots of wavelets which intersect at q.

The output of this procedure is a merged bisector γ which is a curve consisting of some portions of hyperbolas, which contains the "seed" point q and intersects the frontier of one of the SPM(d)-triangles.

This procedure computes the curve γ following $B(r_\omega, r_\omega'')$ in two directions out of q. The curve γ lies within two SPM(d)-triangles rooted at r_ω and r_ω'', respectively.

Line 1 enters a while loop and keeps checking if we can trace in at least one direction out of q.

Line 2 removes each wavelet from event queue Q whenever this wavelet dies. For example, in the Fig. 5.43, let the "seed" point q be an interior point of $B(v_3, v_2)$. The active wavelet that roots at v_3 dies once we trace the merge curve γ through the closure point s.

γ encounters an obstacle edge $v_2 v_3$ in one direction (Line 3).

If γ intersects the pseudo-wavefront at a point z which is the intersection point of two wavelets, then update the appropriate left/right tracks of these two wavelets with the bisector between their roots (Lines 6 and 7, see Fig. 5.47). For example, in Fig. 5.47, γ (which is equal to the curve qs in this figure) intersects the pseudo-wavefront at a point $z = s$ which is the intersection point of two wavelets ω_1 and ω_2; then we update the left track of ω_1 and the right track of ω_2 with the bisector (i.e., γ) between their roots r_{ω_1} and r_{ω_2}.

5.11 Problems

Problem 5.1 Discuss differences between triangulation and trapezoidation of a simple polygon.

Problem 5.2 Show that the dual graph G_P is not a tree for any polygon P with holes.

Problem 5.3 (Programming exercise) Implement and test Algorithm 18.

Problem 5.4 Discuss the difference between an X-node and a Y-node, as introduced in Sect. 5.3.

Problem 5.5 Consider a simple polygon P. Show that a triangulation of P can be obtained from a trapezoidation of P in $\mathcal{O}(n)$, where n is the number of vertices of P.

Problem 5.6 (Programming exercise) Compute the inner polygon $P(v_0, \varepsilon_s)$ of a simple polygon P, where v_0 is a vertex of P, and $\varepsilon_s >$ is a predefined parameter.

Problem 5.7 Discuss the difference between an up- (down-) stable vertex (edge) and a maximal (minimal) vertex (edge) as introduced in Sect. 5.5.

Problem 5.8 Discuss differences between a closure event (point) versus a vertex event (point), the signal tree $\mathcal{T}(d)$ at distance d versus the shortest-path tree SPT(d) at distance d, the shortest-path map SPM(d) at distance d versus the shortest-path map SPM. See definitions in Sect. 5.9.

Problem 5.9 Can a point possibly have more than one predecessor in a shortest-path map? If a side of an SPM(d)-triangle is not a line segment, can it then be an arc and on what kind of a curve?

Problem 5.10 Consider Procedure 3. Show that we could also request $y_0 < \Delta_{1y} \leq \Delta_{2y} \leq \cdots \leq \Delta_{my}$ instead of $y_0 < \Delta_{1y} < \Delta_{2y} < \cdots < \Delta_{my}$.

Problem 5.11 (Programming exercise) Implement Algorithm 20 and use polygons as given in Figs. 5.19 and 5.36 as input for testing the correctness of your program.

Problem 5.12 What is the purpose of Lines 7 and 8 in Seidel's Algorithm (Algorithm 19)?

Problem 5.13 (Programming exercise) Implement the Seidel Algorithm (Algorithm 19) described in Sect. 5.3 and study the practical running time of this algorithm, especially with the goal of analysing the worst case complexity of your implementation. For a C-source of the algorithm, see www.cs.unc.edu/~dm/CODE/GEM/chapter.html.

Problem 5.14 (Programming exercise) Implement the Mitchell algorithm (i.e., Algorithm 22) as described in Sect. 5.9. Analyse its practical time complexity by running it on inputs of different sizes.

Problem 5.15 (Research problem) Read the article on Chazelle's triangulation method (see [3]) and determine operations which are described there as sorting instead of the complicated, claimed to be linear operations. How can those operations be implemented so that they run in linear time? Altogether, provide a linear-time implementation following the ideas of Chazelle's triangulation method—if possible at all.

Problem 5.16 (A conjecture) Can we replace the $\mathcal{O}(n \log n)$ sorting step in Algorithm 20 by incorporating some type of a scan of lesser than $n \log n$ time complexity; possibly even of linear time complexity?

5.12 Notes

Although there exists a linear algorithm for computing the shortest path between two vertices in a positive integer weighted graph [19], the provided Algorithm 18 for calculating a shortest path in a tree is much simpler.

For Fig. 5.3, see [8].

There are (at least) two ways of decomposing a simple polygon: into triangles only or also allowing trapezoids. In the first case, Theorem 4.3 in [3] says that it is possible to compute a triangulation of a simple polygon in linear time (but the method is described in [3] on 40 pages and it is "fairly complicated", as pointed out in Sect. 5.7). In the second case, Theorem 1 in [14] says a given trapezoidation given in that paper has time complexity $\mathcal{O}(n \log n)$, where n is the number of vertices of the original simple polygon Π. The chapter presented this algorithm in Sect. 5.5. This algorithm has possibly the potential to be sped-up; see Problem 5.16. The algorithm uses the calculation of tangents to two polygons as provided by [17].

The Seidel algorithm has expected time complexity $\mathcal{O}(n \log n)$ and worst case complexity $\mathcal{O}(n^2)$ and is the simplest of the three trapezoidation methods. Lemmata 5.2, 5.3, and 5.4 are Lemmata 4, 2, and 5 in [16]. See also www.cs.unc.edu/

~dm/CODE/GEM/chapter.html. There is another algorithm in [18] that was not included in this chapter.

By [4] and [9], a triangulation of a simple polygon can be derived in linear time from its visibility map.

For the Mitchell algorithm, see [15]. Lemma 5.9 is Corollary 2 in [15]. For ray-shooting query, see references [1, 5, 6, 11]. For more references on decomposition of simple polygons, see, for example, [2, 4, 7, 9, 10, 12, 13, 18, 20].

References

1. Agarwal, P.K.: Ray shooting and other applications of spanning trees and low stabbing number. In: Proc. Annu. ACM Sympos. Comput. Geom., pp. 315–325 (1989)
2. Chazelle, B.: A theorem on polygon cutting with applications. In: Proc. Annu. Sympos. on Foundations of Computer Science, pp. 339–349 (1982)
3. Chazelle, B.: Triangulating a simple polygon in linear time. Discrete Comput. Geom. **6**, 485–524 (1991)
4. Chazelle, B., Incerpi, J.: Triangulation and shape-complexity. ACM Trans. Graph. **3**, 135–152 (1984)
5. Chazelle, B., Edelsbrunner, H., Grigni, M., Guibas, L., Hershberger, J., Sharir, M., Snoeyink, J.: Ray shooting in polygons using geodesic triangulations. Algorithmica **12**(1), 54–68 (1994)
6. Cheng, S.W., Janardan, R.: Space-efficient ray-shooting and intersection searching: algorithms, dynamization, and applications. In: Proc. Annu. ACM-SIAM Sympos. Discrete Algorithms, pp. 7–16 (1991)
7. Clarkson, K.L., Tarjan, R.E., Wyk, C.J.V.: A fast Las Vegas algorithm for triangulating a simple polygon. Discrete Comput. Geom. **4**, 423–432 (1989)
8. Fathauer, R.: Website of the 'Tessellations Company'. http://members.cox.net/fathauerart/index.html (2010). Accessed July 2011
9. Fournier, A., Montuno, D.Y.: Triangulating simple polygons and equivalent problems. ACM Trans. Graph. **3**(2), 153–174 (1984)
10. Garey, M.R., Johnson, D.S., Preparata, F.P., Tarjan, R.E.: Triangulating a simple polygon. Inf. Process. Lett. **7**(4), 175–179 (1978)
11. Hershberger, J., Suri, S.: A pedestrian approach to ray shooting: shoot a ray, take a walk. J. Algorithms **18**(3), 403–431 (1995)
12. Hertel, S., Mehlhorn, K.: Fast triangulation of simple polygons. In: Proc. Conf. Found. Comput. Theory, pp. 207–218 (1983)
13. Kirkpatrick, D.G., Klawe, M.M., Tarjan, R.E.: Polygon triangulation in $\mathcal{O}(n \log \log n)$ time with simple data-structures. In: Proc. Annu. ACM Sympos. Comput. Geom., pp. 34–43 (1990)
14. Li, F., Klette, R.: Decomposing a simple polygon into trapezoids. In: Proc. CAIP. LNCS, vol. 4673, pp. 726–733 (2007)
15. Mitchell, J.S.B.: Shortest paths among obstacles in the plane. Int. J. Comput. Geom. Appl. **6**, 309–332 (1996)
16. Seidel, R.: A simple and fast incremental randomized algorithm for computing trapezoidal decompositions and for triangulating polygons. Comput. Geom. **1**, 51–64 (1991)
17. Sunday, D.: Algorithm 14: Tangents to and between polygons. http://softsurfer.com/Archive/algorithm_0201/ (2006). Accessed July 2011
18. Tarjan, R.E., Wyk, C.J.V.: An $\mathcal{O}(n \log \log n)$ algorithm for triangulating a simple polygon. SIAM J. Comput. **17**, 143–178 (1988)
19. Thorup, D.: Undirected single-source shortest paths with positive integer weights in linear time. J. ACM **3**, 362–394 (1999)
20. Toussaint, G.T., Avis, D.: On a convex hull algorithm for polygons and its application to triangulation problems. Pattern Recognit. **15**(1), 23–29 (1982)

Chapter 6
ESPs in Simple Polygons

> *Obstacles are those frightful things you see when you take your eyes off your goal.*
>
> *Henry Ford (1863–1947)*

Let p and q be two points in a simple polygon P. This chapter provides the Chazelle algorithm for computing the ESP between p and q that is contained in P. It uses triangulation of simple polygons as presented in the previous chapter as a preprocessing step, and has a time complexity that is determined by that of the prior triangulation.

This chapter also provides two rubberband algorithms for computing a shortest path between p and q that is contained in P. The two algorithms use previously known results on triangular or trapezoidal decompositions of simple polygons, and have $\kappa(\varepsilon)\mathcal{O}(n \log n)$ time complexity (where the super-linear time complexity is only due to preprocessing, i.e., for the decomposition of the simple polygon P, $\kappa(\varepsilon) = \frac{L_0 - L}{\varepsilon}$, L is the length of an optimal path and L_0 the length of the initial path, as introduced in Sect. 3.5).

6.1 Properties of ESPs in Simple Polygons

Algorithms for computing Euclidean shortest paths between two points p and q of a simple polygon P, where the path is restricted to be fully contained in P, have applications in 2-dimensional pattern recognition, picture analysis, robotics, and so forth.

The design of algorithms for calculating ESPs within a simple polygon may use one of the known partitioning algorithms as a preprocess. This chapter shows how rubberband algorithms may be used to calculate approximate or exact ESPs within simple polygons, using either decompositions into triangles or into trapezoids.

For a start, we prove a basic property of exact ESPs for such cases.

Proposition 6.1 *Each vertex ($\neq p, q$) of the shortest path is a vertex of P.*

F. Li, R. Klette, *Euclidean Shortest Paths*,
DOI 10.1007/978-1-4471-2256-2_6, © Springer-Verlag London Limited 2011

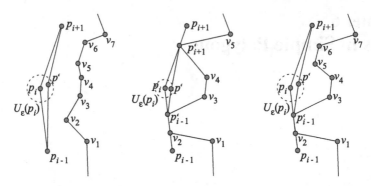

Fig. 6.1 Illustration that each vertex of a shortest path is a vertex of P, where $v_1v_2v_3v_4v_5\ldots$ is a polygonal part of the border of the simple polygon P. *Left*, *middle*, *right* illustrate Cases 1, 2, 3 as discussed in the text, respectively

Proof To see this, let $\rho = \langle p, p_1, p_2, \ldots, p_k, q \rangle$ be the shortest path from p to q completely contained in simple polygon P. Assume that at least one $p_i \in \rho$ is not a vertex of P. Also assume that each p_i is not *redundant*, which means that $p_{i-1}p_i p_{i+1}$ must be a triangle (i.e., three points p_{i-1}, p_i, and p_{i+1} are not collinear), where $i = 1, 2, \ldots, k$ and $p_0 = p$, $p_{k+1} = q$.

Case 1: Neither of the two edges $p_{i-1}p_i$ and $p_i p_{i+1}$ is on a tangent of P (see Fig. 6.1, left); then there exists a sufficiently small neighbourhood of p_i, denoted by $U(p_i)$, such that for each point $p' \in U(p_i) \cap \triangle p_{i-1}p_i p_{i+1} \subset P^\bullet$ (the topological closure of a simple polygon P), both edges $p_{i-1}p_i$ and $p_i p_{i+1}$ are completely contained in P. By elementary geometry, we have that $d_e(p_{i-1}, p') + d_e(p', p_{i+1}) < d_e(p_{i-1}, p_i) + d_e(p_i, p_{i+1})$, where d_e denotes Euclidean distance. Therefore, we may obtain a shorter path from p to q by replacing p_i by p'. This is a contraction to the assumption that p_i is a vertex of the shortest path ρ.

Case 2: Both $p_{i-1}p_i$ and $p_i p_{i+1}$ are on tangents of P (see Fig. 6.1, middle); then we can also derive a contradiction. In fact, let p'_{i-1} and p'_{i+1} be the closest vertices of P such that $p'_{i-1}p_i$ and $p_i p'_{i+1}$ are on tangents of P. Analogous to the first case, there exists a point p' such that the polygonal path $p'_{i-1}p' p'_{i+1}$ is completely contained in $P^\bullet$ and the length of $p'_{i-1}p' p'_{i+1}$ is shorter than $p'_{i-1}p_i p'_{i+1}$. This is a contradiction as well.

Case 3: Either $p_{i-1}p_i$ or $p_i p_{i+1}$ is a tangent of P (see Fig. 6.1, right); then we may arrive at the same result as in Case 2.   □

Let P be a simple polygon and u_1, u_2, v_1, and v_2 four vertices of P. Assume that v_1v_2 is a *chord* of P (i.e., line segment v_1v_2 is completely in P).

Definition 6.1 Chord v_1v_2 is called a u_1u_2-*crossing* in P if the four vertices u_1, v_1, u_2, and v_2 appear in this order on the frontier of P.

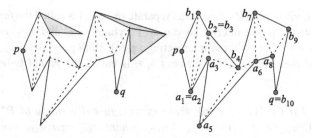

Fig. 6.2 *Left*: Triangulated simple polygon P and selected vertices p and q; all the *dashed segments* are pq-crossing and the *shaded triangles* will be removed. *Right*: Polygon P^*, also illustrating the naming scheme for a- and b-vertices; for example, we have $a_1 = a_2$ due to two edges a_1b_1 and a_2b_2. Not shown: $b_4 = b_5 = b_6$, $b_7 = b_8$, $a_3 = a_4$, $a_6 = a_7$, $a_8 = a_9$

Fig. 6.3 A funnel (*shaded polygon*) and its funnel vertex b_4. The polyline $\sigma_{b_4b_8}$ is one line segment only, and the polyline $\sigma_{b_4a_8}$ two line segments

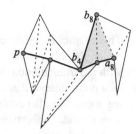

Let p and q be two vertices of P (note: not just points *in* P). Let $\mathcal{P}$ be a partitioning of P. Polygon P^* is obtained from $\mathcal{P}$ by removing at first all the edges in $\mathcal{P}$ that are pq-crossing, and by removing then all the triangles of $\mathcal{P}$ which still have all three of their edges. See Fig. 6.2 for an example. Polygon P^* can be computed from P in $\mathcal{O}(n)$ time, where n is the number of vertices of P. One has:

Lemma 6.1 *The shortest path between p and q in P is identical to the shortest path between p and q in P^*.*

Let $\mathcal{P}^\star$ be the resulting partitioning of P^* after the described removal of faces of $\mathcal{P}$. An *interior edge* of $\mathcal{P}^\star$ does only have its both endpoints on the frontier of P^*.

Lemma 6.2 *The shortest path between p and q in P^* intersects each interior edge of $\mathcal{P}^\star$ exactly once, and intersects no other edge of $\mathcal{P}^\star$.*

Let σ_{pq} be the shortest path from p to q in $P^\star$. Let $S = \{a_1b_1, \dots, a_mb_m\}$ be the set of all interior edges of P^*, indexed in the order as they are crossed by σ_{pq}. See Fig. 6.2 on the right.

Consider shortest paths σ_{pa_i} and σ_{pb_i}, for both endpoints of the interior edge a_ib_i. See Fig. 6.3 for an example, with $i = 8$. Both shortest paths are identical between p and some vertex (here $b_k = b_4$), and split then into two polylines $\sigma_{b_kb_i}$ and $\sigma_{b_ka_i}$. These two polylines and edge a_ib_i circumscribe a polygon. The shaded area is called a *funnel*. Path σ_{pq} passes through the funnel.

The vertex where both shortest paths separate (here b_4) is called the *funnel vertex* for the start vertex p and interior edge a_ib_i. Polylines $\sigma_{b_kb_i}$ and $\sigma_{b_ka_i}$ are 'turning away' from each other (without intersecting again). Both polylines $\sigma_{b_kb_i}$ and $\sigma_{b_ka_i}$ are called the *funnel sides*. The segment a_ib_i is called the *funnel base*. We state without proof:

Lemma 6.3 *For $i \in \{1, 2, \ldots, m\}$, there exists a funnel vertex v of P^* such that $\sigma_{pa_i} = \sigma_{pv} \cup \sigma_{va_i}$ and $\sigma_{pb_i} = \sigma_{pv} \cup \sigma_{vb_i}$, where σ_{va_i} and σ_{vb_i} are two non-intersecting (i.e., except at v) polylines that are both convex, one in clockwise and one in counter-clockwise order.*

Figure 6.4 shows different cases which may occur when proceeding from funnel vertex v to its interior edge a_ib_i, and then further to the next interior edge $a_{i+1}b_{b+1}$.

This chapter is organised as follows. At first we briefly recall decompositions of simple polygons, describe the Chazelle algorithm (note: not to be confused with the triangulation method discussed in the previous chapter), and specify, as a preliminary result, two approximate RBAs; we provide examples of using them. We also analyse their correctness and time complexity. These two approximate rubberband algorithms are then transformed into two exact rubberband algorithms.

6.2 Decompositions and Approximate ESPs

There are (at least) two ways of decomposing a simple polygon: into triangles or trapezoids. Step sets can be defined by selecting edges of triangles or trapezoids of those decompositions.

Triangulations   Let P be a simple polygon. Let $T_1 = \{\triangle_1, \triangle_2, \ldots, \triangle_m\}$ be such that $P = \bigcup_{i=1}^{m} \triangle_i$ and $\triangle_i \cap \triangle_j = \emptyset$ or $= e_{ij}$, where e_{ij} is an edge of both triangles $\triangle_i$ and $\triangle_j$, $i \neq j$ and $i, j = 1, 2, \ldots, m$. We construct a corresponding simple graph $G = [V, E]$ where $V = \{v_1, v_2, \ldots, v_m\}$ and each edge $e \in E$ is defined as follows: If $\triangle_i \cap \triangle_j = e_{ij} \neq \emptyset$, then let $e = v_iv_j$ (where e_{ij} is an edge of both triangles $\triangle_i$ and $\triangle_j$); and if $\triangle_i \cap \triangle_j = \emptyset$, then there is no edge between v_i and v_j, $i < j$ and $i, j = 1, 2, \ldots, m$. We say that G is a *dual graph* with respect to the triangulated simple polygon P, denoted by G_P.

Lemma 6.4 *For each triangulated simple polygon P, its dual graph G_P is a tree.*

We leave the proof of this lemma as an exercise for readers.

Let T be a tree and $p \neq q$, $p, q \in V(T)$. The following procedure will compute a unique path from p to q in T. Although there exists a linear algorithm for computing

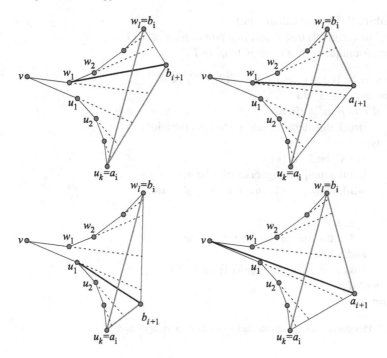

Fig. 6.4 Illustration for Algorithm 23. Interior edges are shown as *bold grey lines*. *Dashed lines* extend edges of $\sigma_{vb_i} = \langle v, w_1, \ldots, w_l = b_i \rangle$ or of $\sigma_{va_i} = \langle v, u_1, \ldots, u_k = a_i \rangle$. They define angular sectors, the next vertex a_{i+1} or b_{i+1} is in one of those, and the sector decides whether the next point is connected (see the *black bold line* segments) either to v, w_j, or u_j, where $j = 1, 2, \ldots, i - 1$. *Bottom left* illustrates the case in Line 8 of Algorithm 23. *Top left* shows the case in Line 10. *Bottom right* shows the case in Line 16, and *top right* shows the case in Line 18

the shortest path between two vertices in a positive integer weighted graph, our procedure below is much simpler because here the graph is (just) a tree.

Let v be a vertex of a graph G, $d_G(v)$ the degree of v in G.

The main idea of Procedure 9 is straightforward: We collect all the vertices of the tree T of degree 1 in an array S_1. Then we process each vertex v in S_1 as follows: We check if v equals p' or q', if so we then go to process the next vertex of degree 1 after v in S_1. Otherwise, we check the unique neighbour of v, denoted by n_v. If $d_T(n_v) = 2$ and n_v does not equal p' and q', then we update v in current T by letting $v = n_v$ and process v as before. Otherwise, we go to process the next vertex of degree 1 after v in S_1. When the program terminates, the final updated T is the desired unique path ρ from p' to q' in the original tree T.

We apply Procedure 9 (see Fig. 6.5) as follows: Let $T = G_P$ and p', q' be the vertices of T corresponding to the triangle containing p, q, respectively. Let a sequence of triangles $\{\triangle_1', \triangle_2', \ldots, \triangle_{m'}'\}$ correspond to the vertices of the path calculated by Procedure 9. Let $\{e_1, e_2, \ldots, e_{m'-1}\}$ be a sequence of edges such that $e_i = \triangle_i \cap \triangle_{i+1}$, where $i = 1, 2, \ldots, m' - 1$. Let $\{e_1', e_2', \ldots, e_{m'-1}'\}$ be a sequence of edges such that e_i' is obtained by removing a sufficiently small segment (Assume

Procedure 9 (Step set calculation)
Input: The (original) tree T and two points $p', q' \in V(T)$.
Output: A unique path ρ from p' to q' in T.

1: Let S_1 be $\{v : d(v) = 1 \wedge v \in V(T)\}$, V_1 be $\emptyset$.
2: **for** each $v \in S_1$ **do**
3:      **if** $v = p'$ or q' **then**
4:          Break this iteration and go to next iteration.
5:      **else**
6:          Let V_1 be $V_1 \cup \{v\}$.
7:          Let the unique neighbour of v be n_v.
8:          **while** $d_T(n_v) = 2 \wedge n_v \neq p' \wedge n_v \neq q'$ **do**
9:              $v = n_v$.
10:             Let V_1 be $V_1 \cup \{v\}$.
11:             Let the unique neighbour of v be n_v.
12:         **end while**
13:         Update T by removing V_1 from $V(T)$.
14:     **end if**
15: **end for**

Fig. 6.5 Procedure for step set calculation for a given triangulation

that the length of the removed segment is δ'.) from both endpoints of e_i, where $i = 1, 2, \ldots, m' - 1$. Set $\{e'_1, e'_2, \ldots, e'_{m'-1}\}$ is the approximate step set we are looking for.

Trapezoidal decomposition   Analogously to the triangulation case, let P be a simple polygon, and let $T_2 = \{t_1, t_2, \ldots, t_m\}$ be such that $P = \bigcup_{i=1}^{m} t_i$ and $t_i \cap t_j = \emptyset$ or e_{ij}, where e_{ij} is a part (a subset) of a joint edge of trapezoids t_i and t_j, $i \neq j$ and $i, j = 1, 2, \ldots, m$. We construct a corresponding simple graph $G = [V, E]$ where $V = \{v_1, v_2, \ldots, v_m\}$, and each edge $e \in E$ is defined as follows: If $t_i \cap t_j = e_{ij} \neq \emptyset$, then let $e = v_i v_j$ (where e_{ij} is a subset of a joint edge of trapezoids t_i and t_j); and if $t_i \cap t_j = \emptyset$, then there is no edge between v_i and v_j, $i < j$ and $i, j = 1, 2, \ldots, m$. We say that G is a (*corresponding*) *graph* or *dual graph* with respect to the trapezoidal decomposition of simple polygon P, denoted by G_P.

Analogously to Lemma 6.4, we also have the following:

Lemma 6.5 *For each trapezoidal decomposition of a simple polygon P, its corresponding graph G_P is a tree.*

Following the triangulation case, we apply Procedure 9 as follows: Let $T = G_P$ and let p', q' be the vertices of T corresponding to the trapezoids containing p, q, respectively. Let a sequence of trapezoids $\{t'_1, t'_2, \ldots, t'_{m'}\}$ correspond to the vertices of the path obtained by Procedure 9. Let $E' = \{e_1, e_2, \ldots, e_{m'-1}\}$ be a sequence of edges such that $e_i = t_i \cap t_{i+1}$, where $i = 1, 2, \ldots, m' - 1$. For each $i \in \{1, 2, \ldots, m' - 2\}$, if $e_i \cap e_{i+1} \neq \emptyset$, then update e_i and e_{i+1} in E' by removing sufficiently small segments from both sides of this intersection point. Then the updated set E' is the approximate step set.

6.3 Chazelle Algorithm

If p and q are not vertices of P, then P^* can also be computed from P in $\mathcal{O}(n)$ time, where n is the number vertices of P.

Let p $[q]$ be in the interior of a triangle t_p $[t_q]$ of $\mathcal{P}^\star$. Let t'_p $[t'_q]$ be the triangle whose three vertices are p, a_1, and b_1 $[q, a_m,$ and $b_m]$. Update $\mathcal{P}^*$ by replacing t_p $[t_q]$ by t'_p $[t'_q]$. Then p and q are vertices of P^*, and σ_{pq} is still the same as before. After this preprocessing we are ready to apply the *Chazelle algorithm*. See Fig. 6.6 for a pseudocode.

Figures 6.7 and 6.8 show step by step funnel vertex v, funnel base $a_i b_i$, and both funnel sides U and W of the current funnel. For current funnel base $a_i b_i$ and next interior edge $a_{i+1} b_{i+1}$, there are the following two cases:

Case 1. $a_{i+1} = a_i$ and $b_{i+1} \neq b_i$ (Line 4). Lines 5 and 6 are equivalent to computing the vertex v' such that $b_{i+1} v'$ is tangent to funnel side U or W at v'. For Line 5, see Fig. 6.4, left bottom. In Line 6, v' is the vertex of U such that the line segment $v' b_{i+1}$ is tangent of U at v'. Lines 7–11 are for updating funnel vertex, funnel sides, and funnel base.

Case 2. $b_{i+1} = b_i$ and $a_{i+1} \neq a_i$ (Line 12). Lines 13–20 are analogous to Lines 5–11.

For Line 8, see Fig. 6.4, left bottom. For Line 10, see Fig. 6.4, left top. For Line 16, see Fig. 6.4, right bottom. For Line 18, see Fig. 6.4, right top.

6.4 Two Approximate Algorithms

Figures 6.9 and 6.10 show both algorithms having decomposition, step set construction, and ESP approximation as their subprocedures.

For Step 4 in Fig. 6.9, see the description following Lemma 6.4. For Step 5 note that the approximation is not due to Algorithm 24 but due to removing small segments of length δ'.

Example 6.1 We illustrate Algorithms 24 and 25 by a few examples, using a simple polygon given by coordinates as provided in Table 6.1. Figure 6.11 shows two different triangulations of this polygon, the step sets defined by those triangulations, and also the corresponding trees.

Algorithm 23 (Chazelle algorithm, 1982)
Input: Two vertices p and q of P^*.
Output: The shortest path σ_{pq}.

1: Let p be the initial funnel vertex: $v = p$; the two sides of the initial funnel are
   $U = \langle v, a_1 \rangle$ and $W = \langle v, b_1 \rangle$.
2: **for** $i = 1$ to m **do**
3:    Let $U = \langle v, u_1, \ldots, u_k \rangle$, $W = \langle v, w_1, \ldots, w_l \rangle$, where $u_k = a_i$ and $w_l = b_i$.
4:    **if** $a_{i+1} = a_i$ and $b_{i+1} \neq b_i$ **then**
5:       Polylines U and W, and the line segments $b_{i+1}a_i$ and $b_{i+1}b_i$ form a sim-
         ple polygon; cut this by lines $u_{k-2}u_{k-1}, \ldots, u_1u_2, vu_1, vw_1, w_1w_2, \ldots,$
         $w_{l-2}w_{l-1}$ into $k + l$ faces.
6:       Determine the face F that contains b_{i+1} and that vertex v' of W or U such
         that F also contains the line segment $v'b_{i+1}$.
7:       **if** v' is a vertex of U **then**
8:          Update the current funnel as follows: reset $v = v'$, $U = \langle v, \ldots, u_k \rangle$, and
            $W = \langle v, b_{i+1} \rangle$.
9:       **else**
10:         Update the current funnel by setting $W = \langle v, w_1, \ldots, v', b_{i+1} \rangle$.
11:      **end if**
12:   **else**
13:      Polylines U and W, and line segments $a_{i+1}a_i$ and $a_{i+1}b_i$ form a sim-
         ple polygon; cut this by lines $u_{k-2}u_{k-1}, \ldots, u_1u_2, vu_1, vw_1, w_1w_2, \ldots,$
         $w_{l-2}w_{l-1}$ into $k + l$ regions.
14:      Determine the face F that contains a_{i+1} and that vertex v' of W or U such
         that F also contains the line segment $v'a_{i+1}$.
15:      **if** v' is a vertex of U **then**
16:         Update the current funnel by setting $U = \langle v, u_1, \ldots, v', a_{i+1} \rangle$.
17:      **else**
18:         Update the current funnel as follows: reset $v = v'$, $U = \langle v', a_{i+1} \rangle$, and
            $W = \langle v', \ldots, w_l \rangle$.
19:      **end if**
20:   **end if**
21: **end for**

Fig. 6.6 Chazelle algorithm for computing an ESP in a simple polygon

Tables 6.2 and 6.3 detail the resulting approximate ESPs, and those are also
visualised in Fig. 6.12. The lengths of those two paths are 1246.0330730004 and
1323.510103408.

After illustrating triangulation and Algorithm 24, we use the same polygon for
illustrating Algorithm 25 using a decomposition into trapezoids. See Fig. 6.13 for
one example of a trapezoidal decomposition, and Fig. 6.14 and Table 6.4 for results.
The length of the approximate ESP equals 1356.7016610946, thus is larger than for
both of the paths obtained for the triangulation, but the number of vertices is reduced
compared to both approximate ESPs obtained for triangulations.   □

Fig. 6.7 Illustration of Algorithm 23: *Left, top*: $v = p$, $U = \langle p, a_1 \rangle$, and $W = \langle p, b_1 \rangle$. *Right, top*: $v = p$, $U = \langle p, a_2 \rangle$, and $W = \langle p, b_2 \rangle$. *Middle, left*: $v = p$, $U = \langle p, a_3 \rangle$, and $W = \langle p, b_3 \rangle$. *Middle, right*: $v = a_4$, $U = \langle a_4 \rangle$, and $W = \langle a_4, b_4 \rangle$. *Bottom, left*: $v = a_4$, $U = \langle a_4, a_5 \rangle$, and $W = \langle a_4, b_5 \rangle$. *Bottom, right*: $v = b_6$, $U = \langle b_6, a_6 \rangle$, and $W = \langle b_6 \rangle$

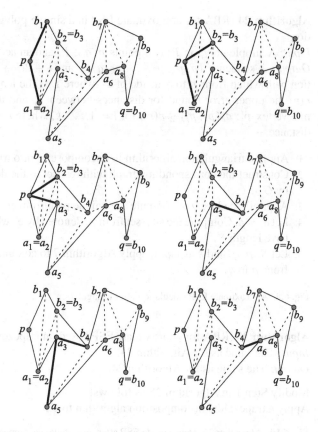

Fig. 6.8 Illustration of Algorithm 23: *Left, top*: $v = b_6$, $U = \langle b_6, a_7 \rangle$, and $W = \langle b_6, b_7 \rangle$. *Right, top*: $v = b_6$, $U = \langle b_6, a_7, a_8 \rangle$, and $W = \langle b_6, b_8 \rangle$. *Left, bottom*: $v = b_6$, $U = \langle b_6, a_7, a_9 \rangle$, and $W = \langle b_6, b_9 \rangle$. *Right, bottom*: $v = a_{10}$, $U = \langle a_{10} \rangle$, and $W = \langle a_{10}, b_{10} \rangle$

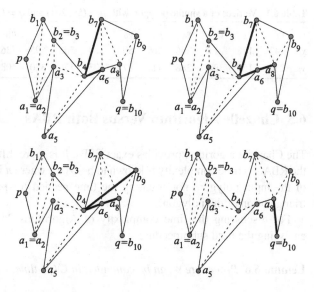

Algorithm 24 (RBA for approximate ESP in a simple polygon based on triangulation)

Input: A simple polygon P; two points $p, q \in P$, and an accuracy constant $\varepsilon > 0$.

Output: A sequence $\langle p, p_1, p_2, \ldots, p_k, q \rangle$ of an $[1 + 2(k + 1)r(\varepsilon)/L]$-approximation path which from p to q inside of P, where L is the length of an optimal path, $r(\varepsilon)$ the upper error bound for distances between p_i and the corresponding optimal vertex p_i': $d_e(p_i, p_i') \leq r(\varepsilon)$, for $i = 1, \ldots, k$, where d_e denotes the Euclidean distance.

1: Apply a triangulation algorithm to decompose P into triangles.
2: Construct the corresponding graph with respect to the decomposed P, denoted by G_P.
3: Apply Procedure 9 to compute the unique path from p' to q', denoted by ρ.
4: Let $\delta' = \varepsilon$. Compute the step set from ρ, denoted by S, where removed segments have length δ'.
5: Let S, p, q, and ε as input, apply Algorithm 7 to compute the approximate ESP from p to q.

Fig. 6.9 Approximate ESP calculation after triangulation

Algorithm 25 (RBA for approximate ESP based on trapezoidal decomposition)

Input: The same as for Algorithm 24.

Output: The same as for Algorithm 24.

Modify Step 1 in Algorithm 25 as follows:
Apply a trapezoidal decomposition algorithm to P.

Fig. 6.10 Algorithm 25: approximate ESP after trapezoidal decomposition

Table 6.1 Vertices of a simple polygon with $p = (59, 201)$ and $q = (707, 382)$

v_i	v_1	v_2	v_3	v_4	v_5	v_6	v_7	v_8	v_9	v_{10}	v_{11}	v_{12}	v_{13}	v_{14}
x_i	42	178	11	306	269	506	589	503	595	736	623	176	358	106
y_i	230	158	304	286	411	173	173	436	320	408	100	211	19	84

6.5 Chazelle Algorithm Versus Both RBAs

The Chazelle algorithm provides exact ESPs. It is more difficult to implement than the RBA. Its time complexity is known to be $\mathcal{O}(n)$, where n is the number of vertices of the input polygon, not counting the time needed for preprocessing (i.e., for a triangulation of the polygon).

For discussing the time complexity of Algorithms 24 and 25, we start with analysing the used subprocedure.

Lemma 6.6 *Procedure 9 can be computed in $\mathcal{O}(n)$ time, where $n = |V(T)|$.*

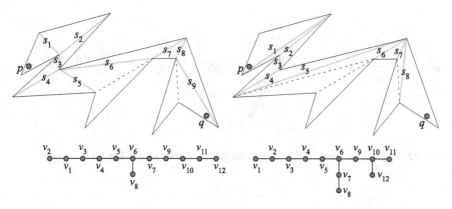

Fig. 6.11 *Top*: The step sets of two different triangulations of the same simple polygon. *Bottom*: The corresponding graphs (trees) to those two triangulations

Table 6.2 Vertices p_i calculated by Algorithm 24 for the triangulation shown in Fig. 6.11 on the left. The length of the path equals 1246.0330730004

p_i	(x_i, y_i)	p_i	(x_i, y_i)
p_1	(177.9999999928, 157.9999999926)	p_6	(374.5899740372, 188.1320635957)
p_2	(178.000000018, 157.9999999861)	p_7	(506.0000000117, 172.9999999927)
p_3	(176.9605570407, 185.5452384224)	p_8	(589.0000000034, 172.9999999927)
p_4	(175.9999999835, 211.0000000093)	p_9	(589.0000000772, 173.0000001234)
p_5	(176.000000013, 211.0000000075)		

Table 6.3 Vertices p_i calculated by Algorithm 24 for the triangulation shown in Fig. 6.11 on the right. The length of the path equals 1323.510103408

p_i	(x_i, y_i)	p_i	(x_i, y_i)
p_1	(123.3191615501, 175.7014459270)	p_5	(420.0869708340, 167.6376763887)
p_2	(178.000000018, 157.9999999861)	p_6	(510.0186257061, 170.4926523372)
p_3	(176.9605570407, 185.5452384224)	p_7	(589.0000000034, 172.9999999927)
p_4	(175.9999999835, 211.0000000093)	p_8	(609.1637118080, 208.7136929370)

Proof Line 1 can be computed in $\mathcal{O}(n)$. Inside of the for-loop, each operation outside of the while-loop can be computed in $\mathcal{O}(1)$. The while loop can also be computed in $\mathcal{O}(1)$. Thus, the for-loop can also be computed in $\mathcal{O}(n)$.   □

Again, let $\kappa(\varepsilon) = (L - L_0)/\varepsilon$, L is the true length of the ESP from p to q, L_0 that of an initial polygonal path from p to q, and n is the number of vertices of P.

Theorem 6.1 *Algorithms 24 and 25 can be computed in time $\kappa(\varepsilon) \cdot \mathcal{O}(n)$ and $\kappa(\varepsilon) \cdot \mathcal{O}(n \log n)$, respectively.*

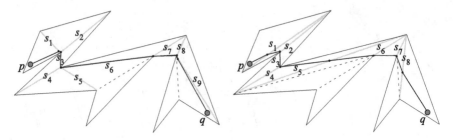

Fig. 6.12 Approximate ESPs with respect to the triangulations shown in Fig. 6.11

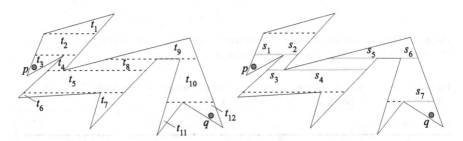

Fig. 6.13 A trapezoidal decomposition (*left*) and its step set (*right*)

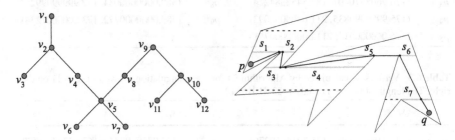

Fig. 6.14 *Left*: Corresponding graph with respect to the trapezoidal decomposition in Fig. 6.13.
Right: Resulting approximate ESP

Table 6.4 Vertices p_i calculated by Algorithm 25 for the simple polygon in Fig. 6.13. The length of the path equals 1356.7016610946	p_i	(x_i, y_i)		p_i	(x_i, y_i)
	p_1	(170.9999999999, 149)		p_5	(504, 161)
	p_2	(171.0000000001, 149)		p_6	(584, 161)
	p_3	(171.9999999999, 202)		p_7	(669.1611374407582, 312)
	p_4	(172.0000000001, 202)			

Proof Steps 2, 4 can be computed in $\mathcal{O}(n)$ time. Theorem 4.3 in [2] (Theorem 1 in [8]), Lemma 6.6, and Sect. 3.5 prove the conclusion for Algorithms 24 and 25.   □

6.6 Turning the Approximate RBA into an Exact Algorithm

We mention a possible way to improve Algorithms 24 and 25 without changing their time complexity. We may revert the removal of some segments to obtain exact vertices of P.

In Step 7 of Algorithm 26 (see Fig. 6.15), we partition the vertices of ρ into m groups so that for each group, if the number of vertices is at least 2, then the Euclidean distance between any two continuous vertices is less than ε'; and for any two continuous groups, the Euclidean distance between the last vertex of the first group and the first vertex of the second group is greater than ε'.

By Theorem 6.1, the smaller the value of the accuracy constant ε, the longer it takes for Algorithm 24 to terminate. Also, based on Proposition 6.1, the main idea of Algorithm 26 is to try a larger accuracy constant ε' ($\gg \varepsilon$) to test whether the vertices of the current path (ρ in Step 6, see also Fig. 6.17) are already close to some vertices of the simple polygon P. If this is true, then replace such vertices (which are sufficiently close to corresponding vertices) to obtain a candidate exact path (ρ in Step 8; see also Fig. 6.17, right). We check the correctness of this candidate path by testing whether an approximate path (ρ' in Step 9, see also Fig. 6.16) which is "very close" to this path is already an approximation output path of Algorithm 24.

Example 6.2 To illustrate Step 7 of Algorithm 24, see Fig. 6.17; we partition the vertices of ρ into:

$$\rho = \langle p, \{p_1\}, \{p_2, p_3\}, \{p_4\}, \{p_5\}, \{p_6\}, \{p_7\}, \{p_8\}, q \rangle$$

where $\{p_1\}$, $\{p_5\}$, and $\{p_8\}$ are redundant. The candidate exact path is

$$\rho = \langle p, p_1, p_2, p_3, p_4, q \rangle$$

in Fig. 6.17, right, which is also the true exact path.   □

This chapter provided two approximate algorithms for calculating ESPs in simple polygons. Their time complexity depends on the used preprocessing step (triangular or trapezoidal decomposition). The chapter illustrates that rubberband algorithms are of simple design, easy to implement, and can be used to solve ESP problems either approximate or in a de-facto exact way if the used accuracy threshold was chosen sufficiently small.

6.7 Problems

Problem 6.1 Give a high-level description for the Chazelle Algorithm 23. Explain what are a funnel vertex, funnel sides, and a funnel base.

Algorithm 26 (Revised RBA for exact ESP in a simple polygon based on triangulation)

Input: A simple polygon P; two points $p, q \in P$ and two accuracy constants $\varepsilon > 0$ and $\varepsilon' > 0$ such that $\varepsilon' \gg \varepsilon$.

Output: A sequence $\langle p, p_1, p_2, \ldots, p_k, q \rangle$ of an exact path from p to q inside of P.

1: Decompose P into triangles (trapezoids, respectively).
2: Construct the corresponding graph with respect to the decomposed P, denoted by G_P.
3: Apply Procedure 9 to compute the unique path from p' to q', denoted by ρ. Let *Search* = true.
4: **while** *Search* = true **do**
5:    Let $\delta' = \varepsilon'$. Compute the step set from ρ, denoted by S, where removed segments have length δ'.
6:    Let S, p, q, and ε' as input, apply Algorithm 7 to compute the approximate ESP from p to q, denoted by $\rho = \langle p, p_1, p_2, \ldots, p_k, q \rangle$.
7:    Partition the vertices of ρ into

$$\rho = \langle p, \{v_1^1, v_2^1, \ldots, v_{n_1}^1\}, \{v_1^2, v_2^2, \ldots, v_{n_2}^2\}, \ldots, \{v_1^m, v_2^m, \ldots, v_{n_m}^m\}, q \rangle$$

so that for each $i \in \{1, 2, \ldots, m\}$, if $n_i \geq 2$, then $d_e(v_j^i, v_{j+1}^i) < \varepsilon'$, where $j = 1, 2, \ldots, n_i - 1$; and for each $i \in \{1, 2, \ldots, m-1\}$, $d_e(v_{n_i}^i, v_1^{i+1}) \gg \varepsilon'$.
8:    Update the vertices of ρ as follows: For each $i \in \{1, 2, \ldots, m\}$, let v_i be the endpoint of the original step of the first vertex v_1^i, if $d_e(v_1^i, v_i) \leq \varepsilon'$, then replace each vertex in the group $\{v_1^i, v_2^i, \ldots, v_{n_i}^i\}$ by v_i. Otherwise, the group $\{v_1^i, v_2^i, \ldots, v_{n_i}^i\}$ has a single vertex, which is redundant. Still denote ρ by $\langle p, p_1, p_2, \ldots, p_k, q \rangle$.
9:    Let ρ' be $\langle p, p_1', p_2', \ldots, p_k', q \rangle$ such that p_i' is on the step of p_i and $d_e(p_i', p_i) = \varepsilon$ if p_i is a vertex of P or $p_i' = p_i$ otherwise, where $i = 1, 2, \ldots, k$. Let S' be the step set of ρ'.
10:   Let S', p, q, and ε as input, and let ρ' be an initial path, apply Algorithm 7 to compute the approximate ESP from p to q, denoted by ρ''.
11:   **if** $\rho'' = \rho'$ **then**
12:       Output ρ obtained in Line 8 as an exact path and *Search* = false.
13:   **else**
14:       $\varepsilon' = \varepsilon' \times 0.1$
15:   **end if**
16: **end while**

Fig. 6.15 Exact ESP calculation after triangulation

Problem 6.2 Show that the Chazelle Algorithm 23 is a linear-time algorithm.

Problem 6.3 Show that in Lines 5 and 13 of the Chazelle Algorithm 23 (in Fig. 6.6), the number of faces is $k + l$, where k and l are the numbers of vertices of the two current funnel sides, as defined in the algorithm.

Fig. 6.16 An example illustrating the approximation path ρ' (in *red*) in Step 9 of Algorithm 26

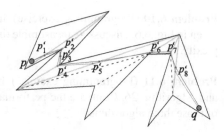

Problem 6.4 Show that Procedure 9 can be computed in linear time.

Problem 6.5 Consider Line 4 of Algorithm 24 and let $\delta' = 0$ (i.e., we do not remove any small segments from any segment in the step set S). Then, what will happen in conclusion of this?

Problem 6.6 Show that Algorithm 26 outputs an exact ESP. Why do the two accuracy constants $\varepsilon > 0$ and $\varepsilon' > 0$ have to satisfy the condition $\varepsilon' \gg \varepsilon$? Could we replace (with or without any effect?) Line 14 of the algorithm by "$\varepsilon' = \varepsilon' \times \alpha$, where α is a constant between 0 and 1."?

Problem 6.7 Prove that Algorithm 26 and Algorithm 24 (and Algorithm 25) have the same time complexity.

Problem 6.8 (Programming exercise) Implement and test Procedure 9.

Problem 6.9 In Algorithm 23 in Fig. 6.6, why are there only the following two different cases:

Case 1. $a_{i+1} = a_i$ and $b_{i+1} \neq b_i$ (Line 4), and
Case 2. $b_{i+1} = b_i$ and $a_{i+1} \neq a_i$ (Line 12)?

In Case 1, why are Lines 5 and 6 equivalent for computing a vertex v' such that $b_{i+1}v'$ is tangent to funnel side U or W at v'?

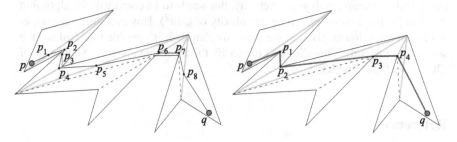

Fig. 6.17 *Left*: Example of an approximation path ρ in Line 6 of Algorithm 26. *Right*: Example of the candidate exact path ρ (in *dark red*) in Line 8 of Algorithm 26

Problem 6.10 (Programming exercise) Implement the Chazelle Algorithm 23 as given in Fig. 6.6. Discuss its measurable time complexity for inputs of varying complexity.

Problem 6.11 (Programming exercise) Implement Algorithm 24, Algorithm 25, and Algorithm 26. Compare the performance (time and resulting decompositions) of those three algorithms.

6.8 Notes

Algorithms for computing ESPs between two points p and q of a simple polygon Π have been intensively studied; see, for example, [3, 5, 6, 9].

For partitioning, there is, for example, Chazelle's [2] method (see previous chapter) proposed for triangulating a simple polygon, or an easier to describe and $\mathcal{O}(n \log n)$ algorithm for partitioning a simple polygon into trapezoids [8].

For Lemma 6.1, see [4]. For a linear-time algorithm for computing the shortest path between two vertices in a positive integer-weighted graph, see [10].

Algorithms for computing ESPs between two points p and q of a simple polygon Π, where the path is restricted to be fully contained in Π, have applications in 2-dimensional pattern recognition, picture analysis, robotics, and so forth. They have been intensively studied [3–5, 7].

This chapter provided two algorithms for calculating ESPs in simple polygons. The used preprocessing step (triangular or trapezoidal decomposition) determines their asymptotic time complexity because the subsequent ESP construction is either of $\mathcal{O}(n)$ (assuming that Chazelle's triangulation method can be turned into a linear-time algorithm) or $\kappa(\varepsilon)$-linear (RBA). The chapter illustrates that RBAs are of simple design, easy to implement, and can be used to calculate ESPs in simple polygons, either approximate or in a de-facto exact way if the used accuracy constant was chosen small enough.

Proposition 6.1 was shown in [6]. For the Chazelle algorithm for computing an ESP inside of a simple polygon, see [1]. This paper contains Lemma 6.1 as Lemma 6.1, Lemma 6.2 as Lemma 6.2, and Lemma 6.3 as Lemma 6.3.

In [8], the authors claimed to have an $\mathcal{O}(n \log n)$ rubberband algorithm for calculating ESPs in simple polygons. Actually, this needs to be corrected: the algorithm given in [8] has a worst-case time complexity of $\mathcal{O}(n^2)$. However, using the trapezoid decomposition as given in the previous chapter it is possible to calculate step sets more efficiently. This allows us to modify the algorithm given in [8] into one of $\mathcal{O}(n \log n)$ time complexity.

References

1. Chazelle, B.: A theorem on polygon cutting with applications. In: Proc. Annual IEEE Symp. Foundations Computer Science, pp. 339–349 (1982)

2. Chazelle, B.: Triangulating a simple polygon in linear time. Discrete Comput. Geom. **6**, 485–524 (1991)
3. Hershberger, J.: A new data structure for shortest path queries in a simple polygon. Inf. Process. Lett. **38**, 231–235 (1991)
4. Lee, D.T., Preparata, F.P.: Euclidean shortest paths in the presence of rectilinear barriers. Networks **14**, 393–410 (1984)
5. Guibas, L., Hershberger, J.: Optimal shortest path queries in a simple polygon. J. Comput. Syst. Sci. **39**, 126–152 (1989)
6. Guibas, L., Hershberger, J., Leven, D., Sharir, M., Tarjan, R.E.: Linear-time algorithms for visibility and shortest path problems inside triangulated simple polygons. Algorithmica **2**, 209–233 (1987)
7. Li, F., Klette, R.: Finding the shortest path between two points in a simple polygon by applying a rubberband algorithm. In: Proc. PSIVT, pp. 280–291 (2006)
8. Li, F., Klette, R.: Decomposing a simple polygon into trapezoids. In: Proc. Computer Analysis Images Patterns. LNCS, vol. 4673, pp. 726–733. Springer, Berlin (2007)
9. Mitchell, J.S.B.: Geometric shortest paths and network optimization. In: Sack J.-R., Urrutia J. (eds.) Handbook of Computational Geometry, pp. 633–701. Elsevier, Amsterdam (2000)
10. Thorup, M.: Undirected single-source shortest paths with positive integer weights in linear time. J. ACM **3**, 362–394 (1999)

Part III
Paths in 3-Dimensional Space

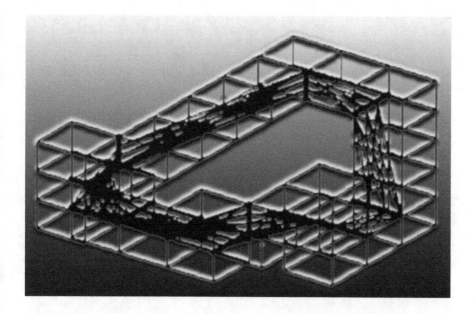

The image above was on the cover of the authors 2007 technical report at the Institute of Mathematics and its Applications (IMA), Minneapolis. It illustrates the original problem when a rubberband algorithm was studied for the first time in 1999 for solving an ESP problem: there is a simple cube-curve in 3D space, and we are interested in calculating a shortest loop which passes through all those cubes.

The third part of the book is about shortest paths in 3D space. The search domain can be the surface of a simple polyhedron, the interior of a simple polyhedron, or a union of cubes as shown in the figure above.

Chapter 7
Paths on Surfaces

Inspiration is needed in geometry, just as much as in poetry.
Alexander Pushkin (1799–1837)

This chapter presents two RBAs for the calculation of an ESP on the surface of a convex or a general polyhedron Π. Solutions are *restricted* by specified constraints. First, we consider a convex polyhedron and provide a $\kappa(\varepsilon) \cdot \mathcal{O}(kn \log n)$ RBA for computing a restricted solution. In this formula, k is the number of polygonal cuts[1] between source and target point, and n is the number of edges of Π. Second, we consider the surface of a general polyhedron Π and provide a $\kappa_1(\varepsilon) \cdot \kappa_2(\varepsilon) \cdot \mathcal{O}(n^2)$ RBA for computing a restricted solution for the surface ESP problem. In this formula, n is again the number of vertices of Π and $\kappa_i(\varepsilon) = (L_{0i} - L_i)/\varepsilon$, for $i = 1$ or $i = 2$, where L_1 is the length of a shortest path, L_{01} the length of the initial path, L_2 the length of a restricted shortest path, and L_{02} the length of an initial path for the restricted path calculation. Both proposed RBAs are easy to implement. Applications are, for example, in 3D object analysis in biomedical or industrial imaging.

7.1 Obstacle Avoidance Paths in 3D Space

In Chap. 1, we defined a polyhedron Π as a bounded volume in 3D space whose surface (i.e., its frontier) is the union of a set of simple polygons which only intersect at their frontiers (i.e., the polygons are *non-overlapping*). From here on, we also assume that this set of polygons is finite. Because each simple polygon can be triangulated, we can use the following

Definition 7.1 A *polyhedron* is a compact connected subset of $\mathbb{R}^3$ such that its frontier is a union of a finite number of non-overlapping triangles.

---

[1] See Definition 7.2 below.

F. Li, R. Klette, *Euclidean Shortest Paths*,
DOI 10.1007/978-1-4471-2256-2_7, © Springer-Verlag London Limited 2011

A polyhedron is *simple* iff it is homeomorphic (i.e., topologically equivalent) to the unit sphere. Note that this still allows that a simple polyhedron can be of 'very complex' shape. Now assume that we have a 'large' search space $\Omega \subseteq \mathbb{R}^3$ (e.g., all $\mathbb{R}^3$, or a 'large' simple polyhedron) and two points p and q in Ω such that $p \neq q$, being the start and endpoint of paths to be calculated in Ω.

An *obstacle* in the given context is a simple polyhedron contained in the set $\Omega \setminus \{p, q\}$. Assume a finite set of pairwise disjoint obstacles. (We do not assume attractions which would need to be visited.)

> *The general obstacle avoidance problem*: Calculate an ESP ρ_{pq} from p to q that does not pass through any interior point of any of the obstacles and that stays in the search space.

The problem occurs, for example, when planning optimal collision-free paths for a robot in a 3D environment. The general obstacle avoidance problem is known to be NP-hard. The problem can be simplified by also providing a simple polyhedron $\Pi \subseteq \Omega$ that contains p and q and that is disjoint with any of the given obstacles. Basically, Π defines a *corridor* for locating connecting paths.

> *The general obstacle avoidance problem with a given corridor*: Calculate an ESP ρ_{pq} from p to q, both in a given corridor Π, that does not pass through any interior point of any of the obstacles and that stays in the search space.

One option for an approximate solution for this general obstacle avoidance problem with a given corridor is that we search for a shortest path in Π that connects p with q.

> *ESP in a simple polyhedron*: Calculate an ESP ρ_{pq} from p to q, both in a given simple polyhedron Π, that stays in Π.

Another option for an approximate solution for the general obstacle avoidance problem with a given corridor is that we search for a shortest path ρ_{ab} on the surface of Π, after connecting p and q with a start and endpoint a and b on the surface of Π, respectively.

> *Surface ESP problem*: Calculate an ESP ρ_{pq} from p to q, both on the surface of a given simple polyhedron Π, that stays in the surface of Π.

This chapter provides two RBAs for specific surface ESP problems. The ESP-in-a-simple-polyhedron problem will be dealt with in the next chapter. In the rest of this chapter, Sect. 7.2 provides necessary definitions and theorems. Sections 7.3 and 7.4 present both RBAs, followed by an analysis of the time complexity of both algorithms and some concluding remarks.

7.2   Polygonal Cuts and Bands

Let Π be a convex polyhedron and $V = \{v_1, v_2, \ldots, v_n\}$ be the set of all vertices of Π. Let $\mathcal{F} = \{F_1, F_2, \ldots, F_m\}$ be the set of all faces (i.e., simple polygons) of the surface $\partial \Pi$ of Π. Let $E = \{e_1, e_2, \ldots, e_l\}$ be the set of all edges of all faces of $\partial \Pi$.

For $v \in V$, let π_v be that plane in $\mathbb{R}^3$ which contains v and is parallel to the xy-plane (i.e., π_v is defined by the equation $z = v$). Let $P_v = \pi_v \cap \Pi$.

Definition 7.2 P_v is the *polygonal cut* (of Π) with respect to v.

Because Π is a convex polyhedron, P_v is a convex polygon that is possibly degenerated into a singleton $\{v\}$. If $u.z = v.z$ and u is a vertex of P_v then $P_u = P_v$. We also write $P_u.z$ instead of $u.z$. Two polygonal cuts P_1 and P_2 are called *adjacent* iff

$$P_1.z < P_2.z \quad \text{or} \quad P_2.z < P_1.z$$

and there is no vertex $v \in V$ such that $v.z$ is between $v_1.z$ and $v_2.z$. $\mathcal{P} = \langle P_1, P_2, \ldots, P_k \rangle$ is called a *sequence of polygonal cuts* of Π iff P_i is adjacent to P_{i+1}, for $i = 1, 2, \ldots, k - 1$.

Let u and v be vertices of Π such that the polygonal cuts P_u and P_v are adjacent. Let

$$S_{uv} = \{w : w \in \partial P_u \wedge vw \subseteq \text{ some edge } e \in E\}.$$

It follows that $S_{uv} \neq S_{vu}$ for $u \neq v$. Set S_{uv} may also contain points that are not vertices of Π. Because u and v are vertices of the polyhedron, we have that $u \in S_{uv}$ iff uv is an edge of Π. Set S_{uv} defines a polyline in the frontier of P_u; see Fig. 7.1.

Definition 7.3 Let $u, v \in V$ be such that $u.z < v.z$, the line segment uv is completely contained in some edge in E, and the polygonal cuts P_u and P_v are adjacent. The set S_{uv} [S_{vu}] defines then the *downward* [*upward*] *visible polyline* of vertex v [of vertex u] in Π.

Let Π be a polyhedron (see Fig. 7.2 for an example). Let $\mathcal{F} = \{F_1, F_2, \ldots, F_m\}$ be a set of triangles such that $\partial \Pi = \bigcup_{i=1}^m F_i$ and $F_i \cap F_j = \emptyset$ or $= e_{ij}$, or $= v_{ij}$, where e_{ij} (v_{ij}) is an edge (vertex) of both F_i and F_j, $i \neq j$, respectively, with $i, j = 1, 2, \ldots, m$.

We construct a dual simple graph $G_\Pi = [V_\Pi, E_\Pi]$ where $V_\Pi = \{v_1, v_2, \ldots, v_m\}$. Each v_i is a triangle. Edges $e \in E_\Pi$ are defined as follows: If $F_i \cap F_j = e_{ij} \neq \emptyset$, then we have an edge $e = v_i v_j$ (where e_{ij} is an edge of both triangles F_i and F_j); and

Fig. 7.1 Illustration of sets $S_{uv} = \{u, w_2, w_3\}$ and $S_{vu} = \{v, w_1\}$. Those sets define the polylines $\langle v, w_1 \rangle$ and $\langle u, w_2, w_3 \rangle$ in the frontiers of P_v and P_u, respectively

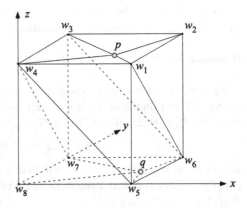

Fig. 7.2 A unit *cube* in a right-handed xyz-coordinate system, with $w_5 = (1, 0, 0)$, $w_7 = (0, 1, 0)$, $w_4 = (0, 0, 1)$, start point $p = (0.76, 0.12, 1)$, and target $q = (0.9, 0.24, 0)$

if $F_i \cap F_j = \emptyset$ or a vertex, then there is no edge between v_i and v_j, $i < j$ and $i, j = 1, 2, \ldots, m$.

In such a case, we say that G_Π is a *dual graph* with respect to the triangulated polyhedron Π. See Fig. 7.3 for an example. Analogously, we can define a dual graph for a connected surface segment (a *subsurface*) of a polyhedron. Abbreviating we may also speak about "the graph for a polyhedron" or "the graph for a subsegment of a surface".

A triangulated polyhedron Π can also be thought as being a graph such that each vertex of Π is a vertex of this graph, and each edge of a triangle is an edge of this graph. We denote this graph by G'_Π.

Let $p \neq q$, $p, q \in V(G'_\Pi)$; if ρ is a cycle of G'_Π such that $G'_\Pi \backslash \rho$ has two components, denoted by G_1 and G_2 with $p \in V(G_1)$ and $q \in V(G_2)$, then ρ is called a *cut cycle* of G'_Π or Π. For example, in Fig. 7.2, $w_1 w_2 w_3 w_4 w_1$ and $w_1 w_6 w_7 w_4 w_1$ are cut cycles of Π.

An *approximate* cycle is a graph such that it consists of a cycle plus a few more vertices, each of which is of degree 1 only, and (thus) adjacent to a vertex on the cycle. (The graph shown later in Fig. 7.12 is an approximate cycle.)

A *band* is a subsurface of a polyhedron Π such that the dual graph of it is a cycle or an approximate cycle.

Fig. 7.3 The dual graph for the triangulated *cube* in Fig. 7.2. This graph is 3-regular (i.e., each node is of degree 3)

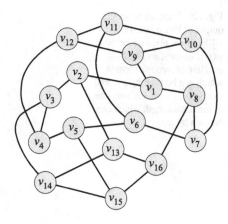

Procedure 10 (Compute visible polylines)
Input: E, V, and a vertex $u \in V$.
Output: The set of vertices of the downward (upward) visible polylines of u.

1: Let $v = \max\{w : w.z < u.z \wedge w \in V\}$.
2: Let $V_u = \{w : \exists e(e \in E \wedge w = e \cap [\text{plane } z = v.z])\}$.
3: Apply Graham's scan algorithm to compute the set of vertices of the downward visible polyline of u.
4: Analogously, compute the set of vertices of the upward visible polyline of u.

Fig. 7.4 The pseudocode of Procedure 10. This procedure is applied in Line 10 of Procedure 11

A band can also be thought as being a subgraph of G'_{Π}. Let E' be the subset of all the edges of a triangulated band such that each edge belongs to a unique triangle. Then E' consists of two cycles. Each of them is called a *frontier* of the band. For example, in Fig. 7.2, $w_1 w_2 w_3 w_4 w_1$ and $w_5 w_6 w_7 w_8 w_5$ are two frontiers of a band whose triangles are perpendicular to the xoy-plane.

If two triangulated bands share a common frontier, then they are called *continuous* (in the sense of "continuation").

7.3 ESPs on Surfaces of Convex Polyhedrons

The following algorithm uses Procedures 10 (see Fig. 7.4) and 11 (see Figs. 7.5 and 7.6). The algorithm is a simplified 3D rubberband algorithm (i.e., Algorithm 27 in Fig. 7.7), which is later used as a subprocess in our main algorithm (i.e., Algorithm 28 in Fig. 7.8). Algorithm 27 will call Procedure 11 which will call Procedure 10. Let p (q) be the start point (endpoint) on the surface of a convex polyhedron Π.

In Line 3 of Procedure 10, Graham's scan algorithm is applied, and this was given earlier in the book.

Fig. 7.5 Illustration for the
output of Procedure 11:
$P_2^* = \rho(w_2, w_3, w_4)$ because
$\rho(w_1, w_2, w_3, w_4)$ is the
polyline of maximal length
which can be seen by p_3, and
$\rho(w_2, w_3, w_4)$ is the polyline
of maximal length which can
be seen by p_1

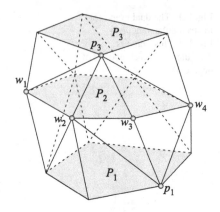

Procedure 11 (Compute a maximal visible polyline)

Input: Three consecutive polygonal cuts P_1, P_2, and P_3 (i.e., $\{P_1, P_2, P_3\}$ is a sequence of polygonal cuts) of a convex polyhedron Π, and two points $p_i \in \partial P_i$, where $i = 1, 3$, and $p_1.z < p_3.z$.

Output: The set of vertices of a maximal polyline $P_2^* \subset P_2$ such that for every point $p \in \partial P_2^*$, there is a $\triangle_i \in S$ such that the segment $p_i p \subset \partial\triangle_i$, where $i = 1, 3$. In other words, P_2^* is the polyline of maximal length between p_1 and p_3 (with respect to the z-coordinate) such that each vertex of it is visible both from p_1 and p_3 (see Fig. 7.5).

1: Let $P_i' = \emptyset$, where $i = 1, 3$.
2: **if** p_1 is not a vertex of P_1 **then**
3:    Find the unique edge e in P_2 such that p_1 and e are contained in a face of Π.
4:    Let $P_1' = \{e\}$.
5: **else**
6:    **if** p_1 is a vertex of P_1 but not a vertex of Π **then**
7:       Find the two edges e_1, e_2 in P_2 such that p_1 and e_i are contained in a face of Π, where $i = 1, 2$.
8:       Let $P_1' = \{e_1, e_2\}$.
9:    **else**
10:       Use E, V, and p_1 as input for Procedure 10; use the result for updating $V(P_1')$.
11:    **end if**
12: **end if**
13: Update $V(P_3')$ analogously.
14: Let $V(P_j') = \{w_{j_1}, w_{j_2}, \ldots, w_{j_{k_j}}\}$, where $j = 1, 3$.
15: Let $2_1 = \max\{1_1, 3_1\}$ and $2_{k_2} = \min\{1_{k_1}, 3_{k_3}\}$.
16: Output $V(P_2^*) = \{w_{2_1}, w_{2_2}, \ldots, w_{2_{k_2}}\}$.

Fig. 7.6 The pseudocode of Procedure 11. This procedure is repeatedly called in Algorithm 27

Algorithm 27 (RBA for polygonal cuts)

Input: A sequence of k pairwise disjoint polygonal cuts $P_1, P_2, \ldots, P_k$, and $P' = \{p_1, p_2, \ldots, p_k\}$ such that $p_i \in \partial P_i$, for $i = 1, 2, \ldots, k$ and $k \geq 3$; two points $p, q \notin \bigcup_{i=1}^{k} P_i$, and an accuracy constant $\varepsilon > 0$.

Output: A sequence $\langle p, p_1, p_2, \ldots, p_k, q \rangle$ of an $[1 + 2(k+1)r(\varepsilon)/L]$-approximation path which starts at p, then visits (i.e., passes through) polygonal cut P_i at p_i in the given order, and finally ends at q, where L is the length of an optimal path, $r(\varepsilon)$ the upper error bound for distances between p_i and the corresponding optimal vertex p_i': $d_e(p_i, p_i') \leq r(\varepsilon)$, for $i = 1, \ldots, k$, where d_e denotes the Euclidean distance.

1: $L_1 \leftarrow \sum_{i=0}^{k+1} L_P(p_i, p_{i+1})$ $(p = p_0 = P_0, q = p_{k+1} = P_{k+1})$; and let $L_0 \leftarrow \infty$.
2: **while** $L_0 - L_1 \geq \varepsilon$ **do**
3:    **for** each $i \in \{1, 2, \ldots, k\}$ **do**
4:       Apply Procedure 11 to compute a polyline $P_i^* \subset P_i$.
5:       Compute a point $q_i \in \partial P_i^*$ such that
        $d_e(p_{i-1}, q_i) + d_e(q_i, p_{i+1}) = \min\{d_e(p_{i-1}, p) + d_e(p, p_{i+1}) : p \in \partial P_i^*\}$
6:       Update the path $\langle p, p_1, p_2, \ldots, p_k, q \rangle$ by replacing p_i by q_i.
7:    **end for**
8:    Let $L_0 \leftarrow L_1$ and $L_1 \leftarrow \sum_{i=0}^{k+1} L_P(p_i, p_{i+1})$.
9: **end while**
10: Return $\{p, p_1, p_2, \ldots, p_k, q\}$.

Fig. 7.7 The pseudocode of Algorithm 27: A rubberband algorithm that has a set of polygonal cuts as its steps

In Line 1 of Procedure 11, the final P_1' will be the upward visible polyline of p_1, and the final P_3' will be the downward visible polyline of p_3. In Line 3, p_1 can see e. In Line 7, p_1 can see e_1 and e_2. In Line 9, p_1 is a vertex of Π.

Let $L_P(p, q)$ be the length of the shortest path, starting at p, then visiting polygonal cuts $P_1, P_2, \ldots, P_k$ in this order, and finally ending at q, where $P = \langle P_0, P_1, \ldots, P_k \rangle$.

The accuracy parameter in the input of Algorithm 27 can be chosen such that maximum possible numerical accuracy is guaranteed on a given computer. This algorithm is a simplified version of an RBA and is used in Line 14 of the main algorithm (Algorithm 28).

We note that the point q_i, computed in Line 5 in Algorithm 27, may not be unique. It is this non-uniqueness that leads our Algorithm 28 in Sect. 7.3 (i.e., this section) to produce a restricted ESP.

Now we present the main algorithm (i.e., Algorithm 28) in Fig. 7.8. Let $p, q \in \partial \Pi$ such that $p.z < q.z$.

In Line 3 of Algorithm 28, each $e \in E$ has weight equal to 1.[2]

---

[2]For Thorup's algorithm in Line 4, see [29]. *Mikkel Thorup* works for AT&T.

Algorithm 28 (RBA for surface ESP)
Input: Two points $p, q \in \partial \Pi$, and an accuracy constant $\varepsilon > 0$.
Output: An approximation path from p to q on $\partial \Pi$.

1: Compute $V_1 = \{v : p.z < v.z < q.z \wedge v \in V\}$.
2: Sort V_1 according to the z-coordinates. As a result, we have

$$V_1 = \{u_1, u_2, \ldots, u_{k'}\}$$

    with

$$u_1.z \leq u_2.z \leq \cdots \leq u_{k'}.z.$$

3: Construct a 1-weighted graph $G = [V, E]$.
4: Apply Thorup's algorithm to find the shortest path $\rho(u_1, u_{k'}) \subset G$.
5: Let $P = \{p\}$.
6: **for** each vertex $v \in V_1$ **do**
7:     Compute the polygon P_v.
8:     Find an edge $e = u_i u_{i+1}$ of $\rho(u_1, u_{k'})$ such that

$$u_i.z \leq v.z \leq u_{i+1}.z$$

    where $1 \leq i < k'$.
9:     Compute the point where P_v and e intersect, denoted by v'.
10:     Let $P = P \cup \{v'\}$.
11: **end for**
12: Let $P = P \cup \{q\}$.
13: Let $S_{\text{step}} = \{P_v : v \in V_1\}$, which is a sequence of polygonal cuts of Π from p to q.
14: Apply Algorithm 27 on S_{step} and P to compute the shortest path $\rho(p, q)$ on the surface of Π.

Fig. 7.8 The pseudocode of Algorithm 28: A rubberband algorithm surface ESP

To complete this section, we analyse (step by step) procedures and the algorithms proposed in this section.

Lemma 7.1 *Procedure 10 has time complexity $\mathcal{O}(|E| \log |E|)$, where E is the set of edges of Π.*

Proof Line 1 can be computed in $\mathcal{O}(|V|)$, where V is the set of vertices of Π. Line 2 can be computed in $\mathcal{O}(|E|)$, where E is the set of edges of Π. Line 3 can be computed in $\mathcal{O}(|V_u| \log |V_u|)$. Note that $|V_u| < |V| \leq |E|$.

It follows that Procedure 10 can be computed in $\mathcal{O}(|E| \log |E|)$. This proves the lemma. $\square$

Lemma 7.2 *Procedure 11 has time complexity $\mathcal{O}(|E| \log |E|)$, where E is the set of edges of Π.*

Proof Line 1 requires only constant time. Line 2 can be computed in $\mathcal{O}(|V(P_1)|)$. Line 3 can be computed in $\mathcal{O}(m)$, where m is the number of triangles in $\partial \Pi$. Line 6 can be computed in $\mathcal{O}(|V|)$, where V is the set of vertices of Π. Line 7 can be computed in $\mathcal{O}(m)$. By Lemma 7.1, Line 10 can be computed in $\mathcal{O}(|E|\log|E|)$, where E is the set of edges of Π. Thus, Lines 1–12 can be computed in $\mathcal{O}(|E|\log|E|)$ because of $|V(P_1)| < |V| \le |E|$. Analogously, Line 13 has the same time complexity as Lines 1–12. Lines 14, 16 can be computed in $\mathcal{O}(|E|)$. Line 15 requires constant time only. This proves the lemma. $\qquad\square$

Lemma 7.3 *The time complexity of Algorithm 27 is* $\kappa(\varepsilon) \cdot \mathcal{O}(k|E|\log|E|)$, *where* E *is the set of edges of* Π, *and* k *is the number of the polygonal cuts between* p *and* q.

Proof The difference between Algorithm 27 and Algorithm 7 is defined by Lines 4 and 5.

By Lemma 7.2, Line 4 can be computed in time $\mathcal{O}(|E|\log|E|)$, where E is the set of edges of Π. Line 5 can be computed in time $\mathcal{O}(|V(P_i^*)|)$.

Because $|V(P_i^*)| \le |V(P_i)| \le |V| + |E|$, it follows that Algorithm 27 needs $\kappa(\varepsilon) \cdot \mathcal{O}(k|E|\log|E|)$ time, where E is the set of edges of Π, and k is the number of polygonal cuts between p and q. $\qquad\square$

Theorem 7.1 *Algorithm* 28 *has a time complexity* $\kappa(\varepsilon) \cdot \mathcal{O}(kn\log n)$, *where* n *is the set of vertices of* Π, *and* k *is the number of polygonal cuts between* p *and* q.

Proof Line 1 can be computed in $\mathcal{O}(|V|)$, where V is the set of vertices of Π. Line 2 needs $\mathcal{O}(|V|\log|V|)$ time. Line 3 can be computed in time $\mathcal{O}(|E|)$, where E is the set of edges of Π. Line 4 uses $\mathcal{O}(|V|)$ time. Line 5 requires only constant time. Line 7 can be computed in time $\mathcal{O}(|E|)$, where E is the set of edges of Π. Line 8 can be computed in time $\mathcal{O}(k') = \mathcal{O}(|V_1|)$.

Line 9 requires $\mathcal{O}(|E|)$ time. Line 10 can be computed in time $\mathcal{O}(1)$. Thus, the for-loop (i.e., steps 6–11) requires $\mathcal{O}(|V_1||E|)$ time. Line 12 can be computed in time $\mathcal{O}(1)$. Line 13 consumes $\mathcal{O}(|V_1|)$ time. By Lemma 7.3, Line 14 can be computed in time $\kappa(\varepsilon) \cdot \mathcal{O}(|V_1||E|\log|E|)$. By an elementary graph, $\mathcal{O}(|E|) = \mathcal{O}(|V|)$, we have $\kappa(\varepsilon) \cdot \mathcal{O}(|V_1||E|\log|E|) = \kappa(\varepsilon) \cdot \mathcal{O}(kn\log n)$, where n is the set of vertices of Π. Altogether, this proves the theorem. $\qquad\square$

We described a simple $\kappa(\varepsilon) \cdot \mathcal{O}(kn\log n)$ algorithm.[3] Note that the provided algorithm requires the convexity of polygonal cuts. By experience of the authors, the algorithm is easy to implement.

---

[3] The algorithm may be considered as a "partial, restricted, and approximate answer" to an open problem stated in [22].

Procedure 12 (Separation)

Input: $G'_\Pi = [V(\Pi), E(\Pi)]$, and two vertices $p \neq q \in V(\Pi)$.

Output: The set of all vertices of a cycle ρ in G such that, if we cut the surface of Π along ρ into two separated parts, then p and q are on different parts.

1: Let $N_p = \{v : vp \in E(\Pi)\}$.
2: Select $u, v \in N_p$ such that $\angle upv \neq 180°$.
3: Let $V = \{p, v\}$. Let *Search* = true.
4: **while** *Search* = true **do**
5:     Let $N_v = \{w' : w'v \in E(\Pi) \wedge w' \notin V\}$.
6:     Take a vertex $w \in N_v$.
7:     **if** $w \neq u$ **then**
8:         Let $V = V \cup \{w\}$, $v = w$.
9:     **else**
10:         *Search* = false.
11:     **end if**
12: **end while**
13: **if** $q \notin V$ **then**
14:     Output V.
15: **else**
16:     Output $V \backslash \{q\}$.
17: **end if**

Fig. 7.9 The pseudocode of Procedure 12

7.4 ESPs on Surfaces of Polyhedrons

Now we move on to the case of arbitrary simple polyhedrons. Without loss of generality, we can assume that $p \neq q$, p and $q \in V(\Pi)$.

The following procedure finds a cut cycle to separate p and q such that either p or q is not a vertex of the cut cycle. [This procedure will be used in Line 1 of the main algorithm (Algorithm 29) below.]

In Line 1 of Procedure 12 (see Fig. 7.9), N_p is the set of all neighbours of p. Line 2 means, in other words, $uv \in E(\Pi)$. In Line 5, N_v is the set of all neighbours of v.

For example, in Fig. 7.2, ρ can be either $w_1 w_2 w_3 w_4 w_1$ or $w_1 w_6 w_7 w_4 w_1$, but it cannot be $w_1 w_5 w_8 w_4 w_1$.

The following procedure computes step bands (i.e., the step set for the second level RBA). It will be used in Line 2 of the main algorithm below.

In Line 3 of Procedure 13 (see Fig. 7.10), the used "minus" in graph theory can also be written as $\Pi_1 \backslash \rho_1$; in other words, we delete each vertex in ρ_1 and each edge of Π_1 which is incident with a vertex of ρ_1. In Line 5, $G_{\Pi_1}(V(\rho_1) \cup V(\rho_2))$ is the induced subgraph of G_{Π_1}.

For example, in Fig. 7.2, if a single vertex can be thought of as being a band, then we can have $S = \{B_1, B_2, B_3\}$, where $B_1 = p$, B_2 is the band such that $V(B_2) = \{w_1, w_2, w_3, w_4, w_5, w_6, w_7, w_8\}$, and $B_3 = q$.

Procedure 13 (Step set calculation)
Input: $G'_\Pi = [V(\Pi), E(\Pi)]$ and ρ, the cut cycle obtained with Procedure 12. Without loss of generality, we can assume that $p \in V(\rho)$ and $q \notin V(\rho)$.
Output: The set of the step bands $S = \{B_1, B_2, \ldots, B_m\}$ such that $p \in V(B_1)$ and $q \in V(B_m)$.

1: Let $S = \emptyset$, $\Pi_1 = \Pi$ and $\rho_1 = \rho$.
2: **while** $q \notin V(\rho_1)$ **do**
3:    Let Π_2 be the component of $\Pi_1 - \rho_1$ such that $q \in V(\Pi_2)$.
4:    Let ρ_2 be the frontier of Π_2.
5:    Let Π_1, ρ_1, and ρ_2 as the input, compute a band $B = G_{\Pi_1}(V(\rho_1) \cup V(\rho_2))$.
6:    Update ρ_1 and Π_1 by letting $\rho_1 = \rho_2$ and $\Pi_1 = \Pi_2$.
7:    Let $S = S \cup \{B\}$.
8: **end while**
9: Output S.

Fig. 7.10 The pseudocode of Procedure 13

The following Procedure 14 (see Fig. 7.11) computes step segments in a single band (i.e., a subset of the step set for the initialisation of the RBA). (It will be used in Line 2 of Procedure 15 below.)

Let F_u, F_v be two triangles such that $u \in \partial F_u$ and $v \in \partial F_v$. Let w_u and $w_v \in V(G_B)$ (Here, G_B is the dual graph of the band B; see Fig. 7.12.) be such that w_u and w_v correspond to F_u and F_v, respectively.

By the definition of a band (see Sect. 7.2), there is a cycle, denoted by ρ_B, such that either w_u (respectively, w_v) $\in V(\rho_B)$ or the unique neighbour of w_u (respectively, w_v) is in $V(\rho_B)$.

For example, in Fig. 7.13, the frontier of B consists of two cycles $uw_1w_2w_3w_4u$ and $w_5w_6w_7w_8w_5$. We have that $F_u = \triangle pw_4w_1$, $F_v = \triangle w_1w_5w_6$. $S_1 = \{w_1w_4, w_1w_5\}$ and $S_2 = \{w_4w_1, w_4w_5, w_4w_8, w_4w_7, w_3w_7, w_3w_6, w_2w_6, w_1w_6\}$, where S_1 and S_2 are the output of Procedure 14.

In Line 1 of Procedure 14, ρ_B is a cycle of the graph G_B. Cycles are defined after Fig. 7.10. In Line 4, assume that the length of the removed segment equals δ'. In Line 8, the definition of a band is in Sect. 7.2. In Line 9, Case 1 is defined in Line 1. In Line 10, only one of either w_u or w_v is not in $V(\rho_B)$. In Line 11, Cases 1 and 2 are defined in Lines 1 and 7.

The following Procedure 15 (see Fig. 7.14) is the initialisation procedure of the RBA. It will be used in Line 7 of the main algorithm (Algorithm 29 in Fig. 7.15) below.

The main algorithm defines now the iteration steps of the RBA.

In Line 3 of Algorithm 29, it is very likely that there exist further points between p_i and p_{i+1}.

We provide an analysis of run-time complexity, and an example for this algorithm. It is basically another illustration for the general comments (e.g., in [18, 19]) that the basic idea of rubberband algorithms may be applied efficiently for a diversity of shortest path problems.

Procedure 14 (Step segments in a single band)

Input: The triangulated band B and two vertices $u, v \in V(B)$ such that u and v are on two different frontiers of B, denoted by ρ_1 and ρ_2 (i.e., $u \in V(\rho_1)$ and $v \in V(\rho_2)$).

Output: Two step sets of segments (edges) S_1 and S_2 such that either S_1 or S_2 contains the vertices of an approximate surface ESP of B from u to v.

1: **if** both w_u and w_v are in $V(\rho_B)$ **then**
2:    ρ_B can be decomposed into two paths from w_u to w_v, denoted by P_1 and P_2. Let $\{F_{j_1}, F_{j_2}, \ldots, F_{j_{m_j}}\}$ be the sequence of triangles corresponding to the sequence of the vertices of P_j, where $j = 1, 2$.
3:    Let $\{e_{j_1}, e_{j_2}, \ldots, e_{j_{m_j}-1}\}$ be a sequence of edges such that $e_{j_i} = F_{j_i} \cap F_{j_{i+1}}$, where $j = 1, 2; i = 1, 2, \ldots, j_{m_j} - 1$.
4:    Let $\{e'_{j_1}, e'_{j_2}, \ldots, e'_{j_{m_j}-1}\}$ be a sequence of edges such that e'_{j_i} is obtained by removing a sufficiently small segment from both endpoints of e_{j_i}, where $j = 1, 2; i = 1, 2, \ldots, j_{m_j} - 1$.
5:    The sets $S_j = \{e'_{j_1}, e'_{j_2}, \ldots, e'_{j_{m_j}-1}\}$ (where $j = 1, 2$) are the approximate step sets we are looking for.
6: **else**
7:    **if** both w_u and w_v are not in $V(\rho_B)$ **then**
8:       Again, by the definition of a band, let w'_u (w'_v) be the unique neighbour of w_u (w_v) such that w'_u and $w'_v \notin V(\rho_B)$.
9:       In this case, ρ_B can be decomposed into two paths from w'_u to w'_v, denoted by P'_1 and P'_2. Appending w_u and w_v to both ends of P'_1 and P'_2, we obtain two paths, denoted by P_1 and P_2. Analogous to Case 1, we can compute the approximate step sets.
10:   **else**
11:      We can compute the approximate step sets, analogously to Cases 1 and 2.
12:   **end if**
13: **end if**

Fig. 7.11 The pseudocode of Procedure 14

We analyse, step by step, the time complexity for each of the procedures and the main algorithm as presented above in this section.

Lemma 7.4 *Procedure* 12 *can be computed in time* $\mathcal{O}(|V(\Pi)|^2)$.

Proof In our data structure, we identify adjacent vertices for each vertex; so Lines 1 and 5 can be computed in time $\mathcal{O}(|V(\Pi)|)$. Line 2 uses $\mathcal{O}(|N_p|)$ time. Line 3 requires $\mathcal{O}(1)$ time. Line 6 can be computed in time $\mathcal{O}(|N_v|)$. Lines 7 and 8 require $\mathcal{O}(1)$ time. The loop, from Line 4 to Line 12, needs $\mathcal{O}(|V(\Pi)|^2)$ time. Lines 13–17 can be computed in time $\mathcal{O}(|V|)$. Therefore, Procedure 12 can be computed in time $\mathcal{O}(|V(\Pi)|^2)$. $\square$

Fig. 7.12 The dual graph
with respect to B; the two
frontiers of B are
$pw_1w_2w_3w_4p$ and
$w_5w_6w_7w_8w_5$ in Fig. 7.13.
v_9 corresponds to triangle
pw_4w_1, and v_2 corresponds
to triangle $w_1w_5w_6$

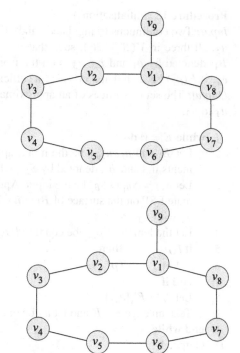

Fig. 7.13 A unit *cube* such
that $u = p$. Point v is the
midpoint of edge w_5w_6

Lemma 7.5 *Procedure* 13 *can be computed in time* $\mathcal{O}(|V(\Pi)|^2)$.

Proof Line 1 can be computed in time $\mathcal{O}(1)$. The test in Line 2 needs $\mathcal{O}(|V(\rho_1)|)$
time. Line 3 can be computed in time $\mathcal{O}(|V(\Pi_1)|)$. Line 4 can requires $\mathcal{O}(|V(\Pi_2)|)$
time. Line 5 can be computed in time $\mathcal{O}(|V(\Pi_1)|)$. Lines 6 and 7 can be com-
puted in time $\mathcal{O}(1)$. The loop, from Line 2 to Line 8, is computed in time
$\mathcal{O}(|V(\Pi_1)| \cdot |V(\rho_1)|) \le \mathcal{O}(|V(\Pi)|^2)$. Line 9 can be computed in time $\mathcal{O}(|S|)$.
Therefore, Procedure 13 can be computed in time $\mathcal{O}(|V(\Pi)|^2)$.   □

   Obviously, we have the following

Lemma 7.6 *Procedure* 14 *can be computed in time* $\mathcal{O}(|V(G_B)|)$.

Lemma 7.7 *Procedure* 15 *has time complexity* $\kappa \cdot \mathcal{O}(|V(\rho_2)| \cdot |V(B_1 \cup B_2)|)$, *where*
$\kappa = (L_0 - L)/\varepsilon$, ε *is the accuracy, and* L_0 *and* L *are the lengths of the initial and a
shortest path from* u_1 *to* u_3, *respectively.*

Proof By Lemma 7.6, Line 2 can be computed in time $\mathcal{O}(|V(B_i)|)$, where $i = 1, 2$;
see the analysis in Sect. 3.5 (see also Theorem 1.4 of [20]), Line 3 has time complexity

$$\kappa \cdot \mathcal{O}\big(|V(B_1 \cup B_2)|\big).$$

Procedure 15 (Initialisations)

Input: Two continuous triangulated bands B_1 and B_2, and three vertices u_1, u_2, and u_3, all three in $V(B_1 \cup B_2)$, such that u_1 and u_2 are on two different frontiers of B_1, denoted by ρ_1 and ρ_2; u_3 is on the frontier, denoted by ρ_3 ($\neq \rho_2$), of B_2. Let $e_{u_2} \in E(\rho_2)$ be such that $u_2 \in e_{u_2}$; l a sufficiently large integer; and $E = E(\rho_2)$.
Output: The set of vertices of an approximate ESP on the surface of $B_1 \cup B_2$ from u_1 to u_3.

1: **while** $E \neq \emptyset$ **do**
2:    Let B_i and u_i, u_{i+1} be the input; apply Procedure 14 to compute step segments in band B_i, denoted by S_{B_i}, where $i = 1, 2$.
3:    Let $S_{12} = S_{B_1} \cup S_{B_2}$ be the input. Apply Algorithm 7 to compute an approximate ESP on the surface of $B_1 \cup B_2$. This is denoted by $\rho_{e_{u_2}}$, and it connects u_1 with u_3.
4:    Let the length of $\rho_{e_{u_2}}$ be equal to $l(\rho_{e_{u_2}})$.
5:    **if** $l(\rho_{e_{u_2}}) < l$ **then**
6:       Let $V = V(\rho_{e_{u_2}})$.
7:    **end if**
8:    Let $E = E \backslash \{e_{u_2}\}$.
9:    Take an edge $e \in E$ and let u_2 be one endpoint of e; let $e_{u_2} = e$.
10: **end while**
11: Output V.

Fig. 7.14 The pseudocode of Procedure 15

Line 4 can be computed in time $\mathcal{O}(|V(\rho_{e_{u_2}})|)$. Lines 5–7 can be computed in time $\mathcal{O}(1)$. Line 8 can be computed in time $\mathcal{O}(|V(\rho_2)|)$. Line 9 can be computed in time $\mathcal{O}(1)$. The loop, from Line 1 to 13, can be computed in time

$$\kappa \cdot \mathcal{O}(|V(\rho_2)| \cdot |V(B_1 \cup B_2)|).$$

Line 11 can be computed in time $\mathcal{O}(|V|)$. Therefore, Procedure 15 can be computed in time $\kappa \cdot \mathcal{O}(|V(\rho_2)| \cdot |V(B_1 \cup B_2)|)$.   □

Theorem 7.2 *The main algorithm has time complexity $\kappa_1 \cdot \kappa_2 \cdot \mathcal{O}(|V(G_\Pi)|^2)$; $\kappa_1 = \max\{\kappa_i : i = 1, 2, \ldots, k\}$; κ_i is defined analogously to κ in Lemma 7.7, where $i = 1, 2, \ldots, k$; $\kappa_2 = (L_1 - L)/\varepsilon$, ε is the accuracy, and L_1 and L are the lengths of the initial and restricted path, respectively.*

Proof By Lemma 7.4, Line 1 can be computed in time $\mathcal{O}(|V(G_\Pi)|^2)$. According to Lemma 7.5, Line 2 can be computed in time $\mathcal{O}(|V(G_\Pi)|^2)$. Line 3 can be computed in time $\mathcal{O}(|V(\rho)|)$. Lines 4 and 10 can be computed in time $\mathcal{O}(|V(\rho)|)$. By Lemma 7.7, Line 7 can be computed in time $\kappa_i \cdot \mathcal{O}(|V(\rho_{i2})| \cdot |V(B_{i-1} \cup B_i)|)$. Line 8 can be computed in time $\mathcal{O}(|V(\rho)|)$. The loop, from Line 5 to 11, can be computed in time $\kappa_1 \cdot \kappa_2 \cdot \mathcal{O}(|V(G_\Pi)|^2)$, where $\kappa_1 = \max\{\kappa_i : i = 1, 2, \ldots, k\}$; κ_i is defined analogously to κ in Lemma 7.7, where $i = 1, 2, \ldots, k$; $\kappa_2 = (L_1 - L)/\varepsilon$, ε is the accuracy, and L_1 and L are the lengths of the initial and restricted path, respectively.

Algorithm 29 (RBA for surface ESP)

Input: $G'_\Pi = [V(\Pi), E(\Pi)]$, and two vertices $p \neq q$, $p, q \in V(\Pi)$; accuracy constant $\varepsilon > 0$.

Output: The set of vertices of an approximate and restricted ESP on the surface of Π.

1: Let G'_Π, p and q be the input; apply Procedure 12 to compute a cut cycle which separates p and q, denoted ρ_{pq}.

2: Let G'_Π and ρ_{pq} be the input; apply Procedure 13 to compute step bands $S = \{B_1, B_2, \ldots, B_{k+1}\}$ such that $p \in V(B_1)$ and $q \in V(B_{k+1})$.

3: Let p_i be a point on the frontier of B_i, where $i = 1, 2, \ldots, k+1$, $p = p_0$ and $q = p_{k+1}$. We obtain an initial path $\rho = \langle p_0, \ldots, p_1, \ldots, p_2, \ldots, p_{k+1} \rangle$.

4: Compute the length L_1 of the initial path ρ; and let $L_0 \leftarrow \infty$.

5: **while** $L_0 - L_1 \geq \varepsilon$ **do**

6:    **for** each $i \in \{1, 2, \ldots, k\}$ **do**

7:       Apply Procedure 15 to compute a point q_i on the frontier of B_i such that q_i is a vertex of an approximate ESP on the surface of $B_{i-1} \cup B_i$ from q_{i-1} to q_{i+1}.

8:       Update the path $\langle p, p_1, p_2, \ldots, p_k, q \rangle$ by replacing p_i by q_i.

9:    **end for**

10:   Let $L_0 \leftarrow L_1$ and L_1 be the length of the updated path ρ.

11: **end while**

12: Return $\{p, p_1, p_2, \ldots, p_k, q\}$.

Fig. 7.15 The pseudocode of Algorithm 29: a rubberband algorithm for surface ESP

Fig. 7.16 A unit *cube* within an xyz-coordinate system, where $p = (0.76, 0.001, 1)$, $q = (0.999, 0.001, 0)$. $pw_9w_{10}q$ is an initial surface path from p to q while $pw_{13}w_{12}w_{11}q$ is an approximate surface ESP from p to q, where $w_9 \in w_1w_2$, $w_{10}, w_{11} \in w_5w_6$, $w_{12} \in w_1w_5$ and $w_{13} \in w_1w_4$

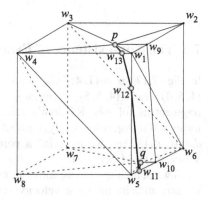

Therefore, the main algorithm can be computed in time $\kappa_1 \cdot \kappa_2 \cdot \mathcal{O}(|V(G_\Pi)|^2)$. $\square$

The following example illustrates the steps of the main algorithm. Let Π be the unit cube in Fig. 7.16.

Line 1 computes a cut cycle $\rho_{pq} = w_1w_2w_3w_4w_1$ (which may be not uniquely defined).

Line 2 computes step bands $S = \{B_1, B_2, B_3\}$, where $B_1 = p$, B_2's frontiers are two cycles $w_1w_2w_3w_4w_1$ and $w_5w_6w_7w_8w_5$, and $B_3 = q$.

Line 3 decides that we use $pw_9w_{10}q$ as an initial surface path from p to q (see Fig. 7.16).

In Line 7, the algorithm applies Procedure 15 (the initialisation procedure of the RBA) and searches each edge of the polygon $w_1w_2w_3w_4w_1$; it finds a point $w_{13} \in w_1w_4$ to update the initial point w_9, and it also inserts a new point $w'_{12} \in w_1w_5$ into the segment between w_{13} and w_{10}.

Again, in Line 7, the algorithm searches each edge of the polygon $w_5w_6w_7w_8w_5$ and finds a point $w_{11} \in w_5w_6$ for updating the initial point w_{10}; it also updates point $w'_{12} \in w_1w_5$ by point $w_{12} \in w_1w_5$ which is between w_{13} and w_{11}.

The algorithm iterates (note: the iteration steps are defined in the main algorithm) until the required accuracy is reached.

The section presented a rubberband algorithm for computing an approximate and restricted surface ESP of a polyhedron. Although it is not the most efficient, it follows a straightforward design strategy, and is thus easy to implement.

This algorithm generalised an rubberband algorithm designed for solving a 2D ESP of a simple polygon[4] to the one which solves the surface ESP of polyhedrons.

This approach is a contribution towards the exploration of efficient approximate algorithms for solving the general ESP problem.

This will allow more detailed studies of computer-represented surfaces as is typical, e.g., in biomedical or industrial 3D image analysis.

7.5 The Non-existence of Exact Algorithms for Surface ESPs

In Fig. 7.17, $p = (1, 4, 7)$, $q = (4, 7, 4)$, $w_1 = (2, 4, 5)$, $w_2 = (2, 5, 5)$, $w_3 = (4, 5, 4)$, $w_4 = (4, 5, 5)$ $(d_e(w_1, w_2) = d_e(w_3, w_4) = 1)$. We can create a surface fragment as follows: Randomly select points w_5 and w_6 on the line segments w_1w_2 and w_3w_4, respectively. Without loss of generality, assume that $d_e(w_5, w_1) \leq d_e(w_6, w_4)$. Then we may find a point w_7 on the line segment w_1w_2 such that $d_e(w_5, w_7) = d_e(w_6, w_3)$.

Now, let both w_5 and w_6 move along the line segments w_1w_2 and w_3w_4 towards w_7 and w_3 with the same velocity, respectively. We denote the locus of segment w_5w_6 by S_1. Let S_2 and S_3 be the triangles pw_5w_7 and qw_6w_3, respectively. Then we obtain a surface fragment $S = S_1 \cup S_2 \cup S_3$.

The problem of finding an ESP from p to q on S implies the problem of finding an ESP starting at p, then visiting segments w_1w_2 and w_3w_4 in this order, and finally ending at q.

---

[4]See [20].

Fig. 7.17 An example of a
surface fragment such that
there does not exist any exact
algorithm for computing a
surface ESP from p to q

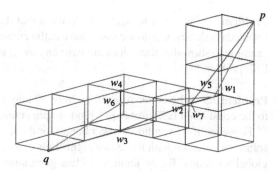

By Sect. 3.9, there does not exist any exact algorithm for the second problem.
Thus, there does not exist any exact algorithm for the first problem. Thus:

> There does not exist any exact algorithm for the general surface ESP problem.

7.6 Problems

Problem 7.1 What is the dual graph G_Π with respect to a triangulated polyhedron
Π? What is the graph G'_Π? Discuss the differences between G_Π and G'_Π.

Problem 7.2 Discuss the differences between an approximate cycle and a cycle.
When are two triangulated bands continuous?

Problem 7.3 Algorithm 27 is modified from Algorithm 7. Underline the modifications in Algorithm 27 in Fig. 7.7. Does Algorithm 27 always output a global ESP?
Why?

Problem 7.4 What are the elements (i.e., steps) in the step set S_{step} in Line 13 of
Algorithm 28 in Fig. 7.8?

Problem 7.5 Consider Line 2 of Procedure 12 in Fig. 7.9. Why can points u and v
not be collinear?

Problem 7.6 Consider Line 4 of Procedure 14 in Fig. 7.11. Why do we need to remove for each segment e_{ji} a sufficiently small segment from both endpoints? Which
lines of this procedure define Cases 1 and 2 (as mentioned in Lines 9 and 11)?

Problem 7.7 Algorithm 29 is also modified from Algorithm 7. Underline modifications in Algorithm 27 in Fig. 7.7.

Problem 7.8 We refer to Sect. 7.5. Assume that the two points w_5 and w_5 are not selected randomly. Could we then also use the constructed surface fragment S as an example to show that there does not exist any exact algorithm for the general surface ESP problem?

Problem 7.9 Why either Algorithm 28 or 29 may fail to compute a global solution to the considered ESP problems? Find simple counterexamples (one for each).

However, if we can find a lower bound for the length of the optimal path, then the solutions obtained with these two algorithms can still be thought to be approximate global solutions. Based upon this idea, generalise both Algorithms 28 and 29 to be approximate algorithms for computing global solutions to the considered ESP problems.

Problem 7.10 If the input polyhedron Π is not convex, then how will this affect the performance of Algorithm 28?

Problem 7.11 (Programming exercise)

(a) Generate a surface triangulation for a given polyhedron.
(b) Consider the generated triangulation as a labelled graph; labels are defined by the length of edges.
(c) Generate a shortest path from a selected vertex to another selected vertex by applying the Dijkstra algorithm.

Discuss the accuracy of this *discrete surface ESP algorithm* for calculating an ESP on the surface of a polyhedron.

Problem 7.12 (Programming exercise) Implement both Algorithms 28 and 29 and compare their performance.

Problem 7.13 (Programming exercise) Implement the generic RBA (i.e., Algorithm 11 for the 2.5D case) as described in Fig. 3.24, and compare its performance with that of Algorithms 28 and 29.

7.7 Notes

As a first result, [28] presented in 1984 a doubly-exponential time algorithm for solving the general obstacle avoidance problem. Reference [25] improved this by providing a singly-exponential time algorithm. The result was further improved by a PSPACE algorithm in [8].

Since the general ESP problem is known to be NP-hard [9], special cases of the problem have been studied afterwards. Reference [27] gave a polynomial time algorithm for ESP calculations for cases where all obstacles are convex and the number of obstacles is 'small'. Reference [12] solved the ESP problem with an $\mathcal{O}(n^{6k-1})$ algorithm assuming that all obstacles are vertical buildings with k different heights.

Reference [28] is the first publication considering the special case that the short-est polygonal path ρ_{pq} is constrained to stay on the surface of Π. Reference [28] presented an $\mathcal{O}(n^3 \log n)$ algorithm where Π was assumed to be convex. Reference [23] improved this result by providing an $\mathcal{O}(n^2 \log n)$ algorithm for the surface of any Π. The time complexity was even reduced to $\mathcal{O}(n^2)$ [10].

Recently, [6] presented an $\mathcal{O}(n^3)$ algorithm for the surface of any Π. Their algorithm was implemented and then evaluated on surfaces for which the correct solution (also called *the ground truth*) was known.

So far, the best known result for the surface ESP problem is due to [14]; this paper improved in 1999 the time complexity to $\mathcal{O}(n \log^2 n)$, assuming that there are $\mathcal{O}(n)$ vertices and edges on Π.

Let Π be a convex polyhedron. Let S be a set of edges (the step set) correspond-ing (by incidence of vertices of the path) to a shortest path on the surface of Π. Reference [24] shows that the cardinality $|S|$ can be calculated in $\mathcal{O}(n^4)$, where n is the number of vertices of Π. Reference [24] constructs an example such that the lower bound of $|S|$ is n^4.

Reference [1] gives an $\mathcal{O}(n^6 \beta(n) \log n)$ algorithm to compute S, where $\beta(n)$ is an extremely slowly growing function. Reference [26] proves that a lower bound for the number of maximal edge sequences (step sets which have a maximal number of edges) of shortest paths is n^3 by using the notion *star unfolding* [5]. Reference [11] solves two-point queries (i.e., given are two points p and q on the surface, find the surface ESP from p to q—after generating an auxiliary map in some preprocessing) on a (not necessarily convex) polyhedral surface in $\mathcal{O}(\log n)$ but with high complex-ities for preprocessing and high demands for available space. Reference [7] focuses on 'terrain surfaces' with various optimal path problems.

There are also already a few approximation algorithms for solving the surface ESP problem. Let n be the number of edges of a convex polyhedron. Reference [2] presents an algorithm for computing an $(1 + \varepsilon)$-shortest path on the surface of a convex polyhedron in time $\mathcal{O}(n + 1/\varepsilon^3)$. Reference [13] improves this result by an $\mathcal{O}((\log n)/\varepsilon^{1.5} + 1/\varepsilon^3)$ algorithm with a preprocessing time of $\mathcal{O}(n)$. References [3, 4, 17, 21] also discuss approximation algorithms for weighted surface ESP prob-lems.

For calculating an ESP on the surface of a convex polyhedron (in $\mathbb{R}^3$), reference [22] states on page 667 the following *open problem: Can one compute shortest paths on the surface of a convex polyhedron in $\mathbb{R}^3$ in subquadratic time? In $\mathcal{O}(n \log n)$?*

In this chapter, we generalised an RBA from solving the 2D ESP of a simple polygon (see [20] and Chap. 6 for this 2D algorithm) to a solution for the surface ESP of polyhedrons. Although this RBA is not the most time-efficient, it follows a straightforward design strategy, and the proposed algorithm is easy to implement. See [19] for results on implementing RBAs for various shortest path problems.

For shortest paths on digital surfaces (in the context of 3D picture analysis), also known as *geodesics*, see the monograph [16]. One of the earlier publications, related to the calculation of surface geodesics, is [15].

References

1. Agarwal, P.K., Aronov, B., O'Rourke, J., Schevon, C.A.: Star unfolding of a polytope with applications. SIAM J. Comput. **26**, 1689–1713 (1987)
2. Agarwal, P.K., Har-Peled, S., Sharir, M., Varadarajan, K.R.: Approximate shortest paths on a convex polytope in three dimensions. J. ACM **44**, 567–584 (1997)
3. Aleksandrov, L., Lanthier, M., Maheshwari, A., Sack, J.-R.: An ε-approximation algorithm for weighted shortest path queries on polyhedral surfaces. In: Abstracts European Workshop Comput. Geom., pp. 19–21 (1998)
4. Aleksandrov, L., Lanthier, M., Maheshwari, A., Sack, J.-R.: An ε-approximation algorithm for weighted shortest paths on polyhedral surfaces. In: Proc. Scand. Workshop Algorithm Theory. LNCS, vol. 1432, pp. 11–22. Springer, Berlin (1998)
5. Aronov, B., O'Rourke, J.: Nonoverlap of the star unfolding. Discrete Comput. Geom. **8**, 219–250 (1992)
6. Balasubramanian, M., Polimeni, J.R., Schwartz, E.L.: Exact geodesics and shortest paths on polyhedral surfaces. IEEE Trans. Pattern Anal. Mach. Intell. **31**, 1006–1016 (2009)
7. de Berg, M., van Kreveld, M.: Trekking in the alps without freezing or getting tired. Algorithmica **18**, 306–323 (1997)
8. Canny, J.: Some algebraic and geometric configurations in PSPACE. In: Proc. Annu. ACM Sympos. Theory Comput., pp. 460–467 (1988)
9. Canny, J., Reif, J.H.: New lower bound techniques for robot motion planning problems. In: Proc. IEEE Conf. Foundations Computer Science, pp. 49–60 (1987)
10. Chen, J., Han, Y.: Shortest paths on a polyhedron. In: Proc. Annu. ACM Sympos. Comput. Geom., pp. 360–369 (1990)
11. Chiang, Y.-J., Mitchell, J.S.B.: Two-point Euclidean shortest path queries in the plane. In: Proc. ACM-SIAM Sympos. Discrete Algorithms, pp. 215–224 (1999)
12. Gewali, L.P., Ntafos, S., Tollis, I.G.: Path planning in the presence of vertical obstacles. Technical report, Computer Science, University of Texas at Dallas (1989)
13. Har-Peled, S.: Approximate shortest paths and geodesic diameters on convex polytopes in three dimensions. Discrete Comput. Geom. **21**, 217–231 (1999)
14. Kapoor, S.: Efficient computation of geodesic shortest paths. In: Proc. Annu. ACM Sympos. Theory Comput., pp. 770–779 (1999)
15. Kiryati, N., Szekely, G.: Estimating shortest paths and minimal distances on digitized three dimensional surfaces. Pattern Recognit. **26**, 1623–1637 (1993)
16. Klette, R., Rosenfeld, A.: Digital Geometry. Morgan Kaufmann, San Francisco (2004)
17. Lanthier, M., Maheshwari, A., Sack, J.-R.: Approximating weighted shortest paths on polyhedral surfaces. In: Proc. Annu. ACM Sympos. Comput. Geom., pp. 274–283 (1997)
18. Li, F., Klette, R.: Exact and approximate algorithms for the calculation of shortest paths. Report 2141, IMA, The University of Minnesota, Minneapolis (2006)
19. Li, F., Klette, R.: Rubberband algorithms for solving various 2D or 3D shortest path problems. Plenary Talk. In: Proc. Computing: Theory and Applications, Platinum Jubilee Conference of The Indian Statistical Institute, pp. 9–18. IEEE, Los Alamitos (2007)
20. Li, F., Klette, R.: Euclidean shortest paths in simple polygons. Technical report CITR-202, Computer Science Department, The University of Auckland, Auckland. http://www.citr.auckland.ac.nz/techreports/2007/CITR-TR-202.pdf (2007)
21. Mata, C., Mitchell, J.S.B.: Approximation algorithms for geometric tour and network design problems. In: Proc. Annu. ACM Sympos. Comput. Geom., pp. 360–369 (1995)
22. Mitchell, J.S.B.: Geometric shortest paths and network optimization. In: Sack, J.-R., Urrutia, J. (eds.) Handbook of Computational Geometry, pp. 633–701. Elsevier, Amsterdam (2000)
23. Mitchell, J.S.B., Mount, D.M., Papadimitriou, C.H.: The discrete geodesic problem. SIAM J. Comput. **16**, 647–668 (1987)
24. Mount, D.M.: The number of shortest paths on the surface of a polyhedron. SIAM J. Comput. **19**, 593–611 (1990)
25. Reif, J.H., Storer, J.A.: A single-exponential upper bound for shortest paths in three dimensions. J. ACM **41**, 1013–1019 (1994)

26. Schevon, C., O'Rourke, J.: The number of maximal edge sequences on a convex polytope. In: Proc. Allerton Conf. Commun. Control Comput., pp. 49–57 (1988)
27. Sharir, M.: On shortest paths amidst convex polyhedra. SIAM J. Comput. **16**, 561–572 (1987)
28. Sharir, M., Schorr, A.: On shortest paths in polyhedral spaces. SIAM J. Comput. **15**, 193–215 (1986)
29. Thorup, M.: Undirected single-source shortest paths with positive integer weights in linear time. J. ACM **3**, 362–394 (1999)

Chapter 8
Paths in Simple Polyhedrons

An approximate answer to the right problem is worth a good
deal more than an exact answer to an approximate problem.
John Tukey (1915–2000)

Since the pioneering work by L. Cohen and R. Kimmel in 1997 on finding a contour as a minimal path between two endpoints, shortest paths in volume images
have raised interest in computer vision and image analysis. This chapter considers
the calculation of an ESP in a 3D polyhedral space Π. We propose an approxi-
mate $\kappa(\varepsilon) \cdot \mathcal{O}(M|V|)$ 3D ESP algorithm, not counting time for preprocessing. The
preprocessing time complexity equals $\mathcal{O}(M|E| + |\mathcal{F}| + |V| \log |V|)$ for solving a
special, but 'fairly general' case of the 3D ESP problem, where Π does not need to
be convex. V and E are the sets of vertices and edges of Π, respectively, and $\mathcal{F}$ is
the set of faces (triangles) of Π. M is the maximal number of vertices of a so-called
critical polygon, and $\kappa(\varepsilon) = (L_0 - L)/\varepsilon$ where L_0 is the length of an initial path and
L is the true (i.e., optimum) path length. The given algorithm approximately solves
three (previously known to be) NP-complete or NP-hard 3D ESP problems in time
$\kappa(\varepsilon) \cdot \mathcal{O}(k)$, where k is the number of layers in a stack, which is introduced in this
chapter as being the *problem environment*. The proposed approximation method has
straightforward applications for ESP problems when analysing polyhedral objects
(e.g., in 3D imaging), of for 'flying' over a polyhedral terrain.

8.1 Types of Polyhedrons; Strips

In this chapter, we apply a rubberband algorithm (RBA) to present an approximate

$$\kappa(\varepsilon) \cdot \mathcal{O}(M|V|) + \mathcal{O}(M|E| + |\mathcal{F}| + |V| \log |V|)$$

algorithm for ESP calculations when Π is a (type-2, see Definition 8.2 below) simply connected polyhedron which is not necessarily convex.

F. Li, R. Klette, *Euclidean Shortest Paths*,
DOI 10.1007/978-1-4471-2256-2_8, © Springer-Verlag London Limited 2011

Fig. 8.1 *Left*: A type-1
polyhedron. *Right*: Type-2
polyhedron

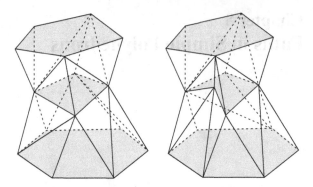

This section provides necessary definitions and theorems. Section 8.2 describes the mentioned RBA. Section 8.3 gives the time complexity of the algorithm. Section 8.4 illustrates the algorithm by some examples. Section 8.5 concludes the chapter.

We denote by Π a simple polyhedron (i.e., a compact polyhedral region which is homeomorphic to a unit ball; see Chap. 1) in the 3D Euclidean space, which is equipped with an xyz rectangular Cartesian coordinate system. Let E be the set of edges of Π; $V = \{v_1, v_2, \ldots, v_n\}$ the set of vertices of Π. For $p \in \Pi$, let π_p be the plane which is incident with p and parallel to the xy-plane. The intersection $\pi_p \cap \Pi$ is a finite set of simple polygons; a singleton (i.e., a set only containing a single point) is considered to be a degenerate polygon.

Definition 8.1 A simple polygon P, being a connected component of $\pi_p \cap \Pi$, is called a *critical polygon* of Π (with respect to p).

Any vertex p defines in general a finite set of critical polygons. The notion of a critical polygon is also generalised as follows: We assume a simply connected (possibly unbounded) polyhedron Π, and we allow that the resulting (generalised) critical polygons also be unbounded.

For example, a generalised critical polygon may have a vertex at infinity, or it can be the complement of a critical polygon, as specified in Definition 8.1. (Section 8.4 will also make use of generalised critical polygons.)

Definition 8.2 We say that a simple polyhedron Π is a *type-1* polyhedron iff any vertex p defines exactly one convex critical polygon. We say that a simple polyhedron Π is a *type-2* polyhedron iff any vertex p defines exactly one simple critical polygon.

Figure 8.1 shows a type-1 polyhedron on the left, and a type-2 polyhedron on the right. Obviously, each type-1 simple polyhedron is also a type-2 simple polyhedron. Our main algorithm below applies to type-2 simple polyhedrons.

In what follows, Π is a type-2 simple polyhedron. For a simple polygon P, we denoted by P° its topological interior, $P^\bullet$ is the closure of P°, and $\partial P = P^\bullet \backslash P^\circ$ denotes the frontier of P. Let $\rho(p, q)$ be a path from p to q.

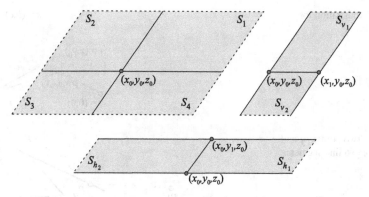

Fig. 8.2 Axis-aligned *rectangles*

Let (x_0, y_0, z_0) be a point in 3D space. Let

$$S_1 = \big\{(x, y, z_0) : x_0 \leq x < \infty \wedge y_0 \leq y < \infty\big\},$$
$$S_2 = \big\{(x, y, z_0) : -\infty < x \leq x_0 \wedge y_0 \leq y < \infty\big\},$$
$$S_3 = \big\{(x, y, z_0) : -\infty < x \leq x_0 \wedge -\infty < y \leq y_0\big\},$$
$$S_4 = \big\{(x, y, z_0) : x_0 \leq x < \infty \wedge -\infty < y \leq y_0\big\}.$$

S_i is called a *q-rectangle of type i*, where $i = 1, 2, 3, 4$. Furthermore, let (x_1, y_1, z_0) be a point in 3D space such that $x_1 > x_0$ and $y_1 > y_0$. Let

$$S_h = \big\{(x, y, z_0) : -\infty < x < \infty \wedge y_0 \leq y \leq y_1\big\},$$
$$S_v = \big\{(x, y, z_0) : x_0 \leq x \leq x_1 \wedge -\infty < y < \infty\big\}.$$

Finally, let

$$S_{h_1} = \big\{(x, y, z_0) : x_0 \leq x < \infty \wedge y_0 \leq y \leq y_1\big\},$$
$$S_{h_2} = \big\{(x, y, z_0) : -\infty < x \leq x_0 \wedge y_0 \leq y \leq y_1\big\},$$
$$S_{v_1} = \big\{(x, y, z_0) : x_0 \leq x \leq x_1 \wedge y_0 \leq y < \infty\big\},$$
$$S_{v_2} = \big\{(x, y, z_0) : x_0 \leq x \leq x_1 \wedge -\infty < y \leq y_0\big\}.$$

The sets S_h, S_v, S_{h_j}, and S_{v_j} are called *horizontal* or *vertical strips*, for $j = 1, 2$. According to their geometric shape, we notice that

(i) S_1 [S_2, S_3, S_4] is unbounded in direction $(+x, +y)$ [$(-x, +y), (-x, -y), (+x, -y)$];
(ii) S_h [S_v] is unbounded in direction $\pm x$ [$\pm y$];
(iii) S_{h_1} [$S_{h_2}, S_{v_1}, S_{v_2}$] is unbounded in direction $+x$ [$-x, +y, -y$].

The sets S_i, S_h, S_v, S_{h_j}, and S_{v_j} are also called *axis-aligned rectangles*, where $i = 1, 2, 3, 4$ and $j = 1, 2$. The stack S of axis-aligned rectangles is called *terrain-like* if, for at least one of the four directions $-x$, $+x$, $-y$, or $+y$, each rectangle in S is unbounded (see Fig. 8.2).

We recall a result in elementary geometry; see Fig. 8.3.

Fig. 8.3 Illustration for
Lemma 8.1

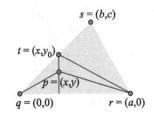

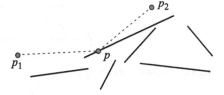

Fig. 8.4 Two points p_1 and
p_2 and $m = 6$ line segments

Lemma 8.1 *Let p be a point in $\triangle qrs$ such that p is not on any of the three line segments qr, rs, and sq. Then $d_e(p, q) + d_e(p, r) < d_e(s, q) + d_e(s, r)$.*

Let $\{s_1, s_2, \ldots, s_m\}$ be a set of m line segments and S the union of those segments. Let p_1 and p_2 be two different points not in S; see Fig. 8.4. We recall that points in $\mathbb{R}^3$ may be sorted by the lexicographic order of their coordinates.

Lemma 8.2 $d_e(p_1, p) + d_e(p_2, p) = \min\{d_e(p_1, q) + d_e(p_2, q) : q \in S\}$ *and lexicographic order define a unique point in S which can be computed in $\mathcal{O}(m)$ time.*

Proof Consider $m = 1$. For line segment s_i, there is a unique point $q_i \in s_i$ such that

$$d_e(p_1, q_i) + d_e(p_2, q_i) = \min\{d_e(p_1, q) + d_e(p_2, q) : q \in s_i\}.$$

Consider the case $m = 2$ and points q_1 and q_2. If $d_e(p_1, q_1) + d_e(p_2, q_1) < d_e(p_1, q_2) + d_e(p_2, q_2)$, then $p = q_1$. If $d_e(p_1, q_1) + d_e(p_2, q_1) > d_e(p_1, q_2) + d_e(p_2, q_2)$, then $p = q_2$. If $d_e(p_1, q_1) + d_e(p_2, q_1) = d_e(p_1, q_2) + d_e(p_2, q_2)$, then we decide for that point which comes first in lexicographic order. Cases $m > 2$ follow analogously. □

Let P be a convex critical polygon of Π, defined by the plane $z = c$. Let p_1 and p_2 be two points such that their z-coordinates satisfy $p_1.z < c < p_2.z$.

For a convex critical polygon P of Π, let $P.z$ be the z-coordinate of all points in P. Let q_1q_2 be a segment such that $q_1.z = q_2.z$. Let p_1 and p_2 be two points such that $p_1.z < q_1.z < p_2.z$ and $d_e(p_1, q_1) + d_e(p_2, q_1) = d_e(p_1, q_2) + d_e(p_2, q_2)$. Let p be a point on the line q_1q_2 such that

$$d_e(p_1, p) + d_e(p_2, p) = \min\{d_e(p_1, q) + d_e(p_2, q) : q \in q_1q_2\}.$$

Then we have the following

Lemma 8.3 p *is in between q_1 and q_2.*

Fig. 8.5 Illustration for the
proof of Lemma 8.3

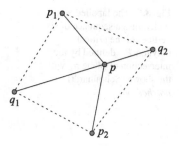

Proof Without loss of generality, suppose that q_1q_2 is parallel to the x-axis. Let the
coordinates of p_i be (a_i, b_i, c_i), where $i = 1, 2$. Let $p = (x, b, c)$ be a point on the
line q_1q_2. See Fig. 8.5. Then,

$$d_e(p_i, p) = \sqrt{(x - a_i)^2 + (b - b_i)^2 + (c - c_i)^2}$$

for $i = 1, 2$. Let $f(x) = d_e(p_1, p) + d_e(p_2, p)$. Then we have that

$$f'(x) = \frac{x - a_1}{\sqrt{(x - a_1)^2 + (b - b_1)^2 + (c - c_1)^2}}$$
$$+ \frac{x - a_2}{\sqrt{(x - a_2)^2 + (b - b_2)^2 + (c - c_2)^2}}.$$

By setting $f'(x) = 0$, we can find the unique critical point x_p of the function $f(x)$,
that is, the coordinates of p equal (x_p, b, c). Let the coordinates of q_i be equal to
(a_{q_i}, b, c), where $i = 1, 2$. Since x_p is the unique critical point of the function $f(x)$,
it follows that $f(x)$ is decreasing in the interval $(-\infty, x_p)$ and increasing in the
interval (x_p, ∞). Because

$$d_e(p_1, q_1) + d_e(p_2, q_1) = d_e(p_1, q_2) + d_e(p_2, q_2)$$

implies $f(a_{q_1}) = f(a_{q_2})$, we have that $a_{q_1} \in (-\infty, x_p)$ and $a_{q_2} \in (x_p, \infty)$. Thus,
x_p is located between a_{q_1} and a_{q_2}.   □

This lemma is used in Lemma 8.4 which will be used for justifying the main
algorithm (Algorithm 30) of this chapter; see Theorem 8.1 below.

Let P be a convex critical polygon of Π. Let e_1 and e_2 be two edges of P. Let
p_1 and p_2 be two points such that $p_1.z < P.z < p_2.z$. Let $P^\bullet$ be the closure of P.
Then we have the following

Lemma 8.4 *There is a unique point $p \in P^\bullet$ such that*

$$d_e(p_1, p) + d_e(p_2, p) = \min\{d_e(p_1, q) + d_e(p_2, q) : q \in P^\bullet\}.$$

Proof Let q be the intersection point between the plane $z = P.z$ and the segment
p_1p_2.

Case 1. $q \in P^\bullet$. Let $p = q$.

Fig. 8.6 The labelled vertex v identifies a sequence of six vertices of the critical polygon P_v, defined by the intersection of plane π_v with the shown (Schönhardt) *polyhedron*

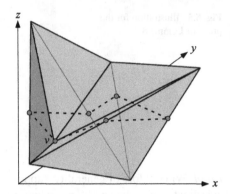

Case 2. $q \notin P^\bullet$. Let $e_1, e_2, \ldots, e_m$ be all edges of P. By Lemma 8.2, there is a unique point $q_i \in e_i$ such that $d_e(p_1, q_i) + d_e(p_2, q_i) = \min\{d_e(p_1, q) + d_e(p_2, q) : q \in e_i\}$, where $i = 1, 2, \ldots, m$. Let $d_i = d_e(p_1, q_i) + d_e(p_2, q_i)$, where $i = 1, 2, \ldots, m$. Let $d = \min\{d_i : i = 1, 2, \ldots, m\}$. Then there are no two different numbers $j \neq k \in \{1, 2, \ldots, m\}$ such that $d_j = d_k = d$. Otherwise, by Lemma 8.3, there would be a point p between q_j and q_k such that

$$d_e(p_1, p) + d_e(p_2, p) = \min\{d_e(p_1, q) + d_e(p_2, q) : q \in P^\bullet\}.$$

Because $q_j, q_k \in \partial P$, and P is convex, p must be in the interior P° of P (with $P^\circ = P^\bullet \setminus \partial P$). This contradicts Lemma 8.1.

Thus, there is a unique point $p \in \{q_i : i = 1, 2, \ldots, m\}$ such that

$$d_e(p_1, p) + d_e(p_2, p) = \min\{d_e(p_1, q_i) + d_e(p_2, q_i) : i = 1, 2, \ldots, m\}.$$

This proves the lemma.   □

8.2 ESP Computation

We start by presenting a procedure, used by a rubberband algorithm (Algorithm 30 below), and then repeatedly called in the main algorithm (Algorithm 32) of this chapter.

Let $\mathcal{F} = \{F_1, F_2, \ldots, F_m\}$ be the set of all faces of Π, and V the set of all vertices of Π. The following very basic Procedure 16 simply 'walks around' the polyhedron by tracing an intersection with a given plane. See Fig. 8.6. We do not detail this procedure; it is a fairly straightforward isoheight trace of a polyhedron, assuming that the data structure of the polyhedron links edges to faces.

In Line 5 of Procedure 16 (see Fig. 8.7), e is not parallel to the plane π_{v_1}. In Line 7, $|\pi_{v_1} \cap e| > 1$. Also note that e is parallel to the plane π_{v_1}. In Line 9, the critical polygon P_{v_1} is a triangle. In Line 15, $vw_2 \in E(S_v)$ and $vw_1 \notin E(S_v)$. Lines 15–17 are modified from Lines 12–14.

The main ideas of Algorithm 30 (see Fig. 8.8) are as follows: For a start, we randomly take a point in the closure of each critical polygon to identify an initial

Procedure 16 (Compute a sequence of vertices of the critical polygon)

Input: Set $\mathcal{F}$ and a vertex $v \in V$ such that π_v intersects Π in more than just one point.

Output: An ordered sequence of all vertices in V_v, which is the vertex set of the critical polygon P_v.

1: Set $\mathcal{F}_v \leftarrow \{F : F \in \mathcal{F} \wedge e = uw \in E(F) \wedge (u.z \leq v.z \leq w.z \vee w.z \leq v.z \leq u.z)\}$, and $E(\mathcal{F}_v) \leftarrow \{e : \exists F \in \mathcal{F}_v \wedge e \in E(F)\}$.
2: Set $V_v \leftarrow \emptyset$, and $v_1 \leftarrow v$.
3: Set $V_v \leftarrow V_v \cup \{v_1\}$.
4: Find a face $F_1 \in \mathcal{F}_v$ such that $v_1 \in V(F_1)$, and such that there exists an edge $e \in E(F_1)$ with $v_1 \notin e$ and $\pi_{v_1} \cap e \neq \emptyset$, and $\neq v_1$.
5: **if** $|\pi_{v_1} \cap e| = 1$ **then**
6:     Set $v_2 \leftarrow \pi_{v_1} \cap e$ and $V_v \leftarrow V_v \cup \{v_2\}$.
7: **else**
8:     Let w_1 and w_2 be the endpoints of edge e.
9:     **if** $v_1 w_i \in E(\mathcal{F}_v)$, for $i = 1$ and 2, **then**
10:         Set $V_v \leftarrow V_v \cup \{w_1, w_2\}$.
11:     **else**
12:         **if** $v_1 w_1 \in E(\mathcal{F}_v)$ and $v_1 w_2 \notin E(\mathcal{F}_v)$ **then**
13:             Find a face $F_2 \in \mathcal{F}_v$ such that $w_2 \in V(F_2)$, $w_1 w_2 \notin E(F_2)$, and there exists an edge $e_2 \in E(F_2)$ such that $w_2 \notin e_2$ and $\pi_{w_2} \cap e_2 \neq \emptyset$.
14:             Set $F_1 \leftarrow F_2$, $v_1 \leftarrow w_2$, $e \leftarrow e_2$, $V_v \leftarrow V_v \cup \{w_2\}$, and go to Line 5.
15:         **else**
16:             Find a face $F_2 \in \mathcal{F}_v$ such that $w_1 \in V(F_2)$, $w_1 w_2 \notin E(F_2)$ and there exists an edge $e_2 \in E(F_2)$ such that $w_1 \notin e_2$ and $\pi_{w_1} \cap e_2 \neq \emptyset$.
17:             Set $F_1 \leftarrow F_2$, $v_1 \leftarrow w_1$, $e \leftarrow e_2$, $V_v \leftarrow V_v \cup \{w_1\}$, and go to Line 5.
18:         **end if**
19:     **end if**
20: **end if**
21: Let $F_2 \in \mathcal{F}_v$ be that face which shares edge e with F_1.
22: **if** $v_2 \neq v$ **then**
23:     Set $F_1 \leftarrow F_2$ and $v_1 \leftarrow v_2$, and go to Line 3.
24: **else**
25:     Output V_v, and Stop.
26: **end if**

Fig. 8.7 The pseudocode of Procedure 16

path from p to q. Then we enter a loop; in each iteration, we optimise locally the position of point p_1 by moving it within its critical polygon, then of $p_2, \ldots$, and finally of p_k. At the end of each iteration, we check the difference between the length of the current path to that of the previous one; if it is less than a given accuracy threshold $\varepsilon > 0$ then we stop. Otherwise, we go to the next iteration. Let $p.x$ be the x-coordinate of point p, $v_1.z$ the z-coordinate of point v_1, and so forth.

Algorithm 30 (An RBA for type-1 polyhedrons)

Input: Two points p and q, a set $\{P_{v_1}^\bullet, P_{v_2}^\bullet, \ldots, P_{v_k}^\bullet\}$, where P_{v_i} is a critical polygon of a given polyhedron Π, k vertices $v_i \in \partial P_{v_i}$ such that $p.z < v_1.z < \cdots < v_k.z < q.z$, for $i = 1, 2, \ldots, k$, and there is no any other critical polygon of Π between p and q; given is also an accuracy constant $\varepsilon > 0$.

Output: The set of all vertices of an approximate shortest path which starts at p, then visits approximate optimal positions $p_1, p_2, \ldots, p_k$ in that order, and finally ends at q.

1: For each $i \in \{1, 2, \ldots, k\}$, let the initial vertex p_i be a vertex of $P_{v_i}^\bullet$.
2: Let $L_0 = \infty$. Calculate $L_1 = \sum_{i=0}^{k} d_e(p_i, p_{i+1})$, where $p_0 = p$ and $p_{k+1} = q$.
3: **while** $L_0 - L_1 \geq \varepsilon$ **do**
4:     **for** $i = 1, 2, \ldots, k$ **do**
5:         Compute a point $q_i \in P_{v_i}^\bullet$ such that
        $d_e(p_{i-1}, q_i) + d_e(q_i, p_{i+1}) = \min\{d_e(p_{i-1}, p) + d_e(p, p_{i+1}) : p \in P_{v_i}^\bullet\}$.
6:         Update the path $\langle p, p_1, p_2, \ldots, p_k, q \rangle$ by replacing p_i by q_i.
7:     **end for**
8:     Let $L_0 = L_1$ and calculate $L_1 = \sum_{i=0}^{k} d_e(p_i, p_{i+1})$.
9: **end while**
10: Return $\langle p, p_1, p_2, \ldots, p_k, q \rangle$.

Fig. 8.8 The pseudocode of Algorithm 30

A comment to Line 5 of this RBA: If $p_{i-1}p_{i+1} \cap P_{v_i}^\circ \neq \emptyset$, then assign $q_i \leftarrow p_{i-1}p_{i+1} \cap P_{v_i}^\circ$.

The set $\{P_{v_1}^\bullet, P_{v_2}^\bullet, \ldots, P_{v_k}^\bullet\}$ in Algorithm 30 is the step set of this RBA. Identifying a 'suitable' step set is normally a main issue when defining a rubberband algorithm.

Theorem 8.1 *If Π is a type-1 polyhedron in the input of Algorithm 30, then the solution obtained by Algorithm 30 is an approximate global solution to the 3D ESP problem.*

Proof Let $X = \prod_{i=1}^{k} P_{u_i}^\bullet \subset \mathbb{R}^k$, where $P_{u_i}^\bullet$ is as defined in Algorithm 30. As Π is a type-1 polyhedron, then P_{u_i} is a convex polygon, where $i = 1, 2, \ldots, k$. Let $Y \subset X$ be the set of all solutions obtained by Algorithm 30, for any initialisation in X and the given $\varepsilon > 0$.

As each P_{u_i} is a convex polygon, by Lemma 8.4, the point q_i in Line 5 of Algorithm 30 is unique, and q_i depends continuously upon p_{i-1} and p_{i+1} defined in Line 5 of Algorithm 30. Thus, Algorithm 30 defines a continuous function, denoted by f, mapping X (i.e., an initialisation) into Y, with values depending on the used accuracy $\varepsilon > 0$.[1]

---

[1] If each $P_{u_i}^\bullet$ is degenerated into a single edge, then there exists a unique solution to the ESP problem; independent of the chosen initialisation, solutions will converge to this unique solution if ε goes to zero; see [8, 23, 26].

Now let $\bar{v} = (v_1, v_2, \ldots, v_k) \in Y$. Then v_i is either located on an edge of polygon P_{u_i}, which is contained in the frontier ∂P_{u_i}, for $i = 1, 2, \ldots, k$, or v_i is located in the interior $P_{u_i}^\circ$, and v_{i-1}, v_i and v_{i+1} are collinear. Thus, Y is a finite set.

It remains to prove that Y is a singleton. Let $\bar{v}_0 \in Y$; we have that $f^{-1}(\bar{v}_0) \subset X$. For each initialisation $\bar{v} \in f^{-1}(\bar{v}_0)$, as f is a continuous function, there exists a sufficiently small open neighbourhood (with respect to the usual topology on $\mathbb{R}^k$) of $\bar{v}$, denoted by $N(\bar{v}, \delta_{\bar{v}})$, such that for each $\bar{v}' \in N(\bar{v}, \delta_{\bar{v}})$, $f(\bar{v}') = \bar{v}_0$. Thus, $N(\bar{v}, \delta_{\bar{v}}) \subseteq f^{-1}(\bar{v}_0)$ and $\bigcup_{\bar{v} \in f^{-1}(\bar{v}_0)} N(\bar{v}, \delta_{\bar{v}}) \subseteq f^{-1}(\bar{v}_0)$.

On the other hand, because (simply by definition) $f^{-1}(\bar{v}_0) = \{\bar{v} : \bar{v}_0 = f(\bar{v})\}$ and $\bar{v} \in N(\bar{v}, \delta_{\bar{v}})$, we have that $f^{-1}(\bar{v}_0) \subseteq \bigcup_{\bar{v} \in f^{-1}(\bar{v}_0)} N(\bar{v}, \delta_{\bar{v}})$. Therefore, $f^{-1}(\bar{v}_0) = \bigcup_{\bar{v} \in f^{-1}(\bar{v}_0)} N(\bar{v}, \delta_{\bar{v}})$. Because $N(\bar{v}, \delta_{\bar{v}})$ is an open set, $f^{-1}(\bar{v}_0)$ is also open. Let

$$f^{-1}(\bar{v}_0) = \bigcup_{i=1}^{k} S_i$$

where S_i is an open subset of $P_{u_i}^\bullet$, for $i = 1, 2, \ldots, k$. Recall that $f^{-1}(\bar{v}_0) \subset X$, thus there exists an S_i such that $\emptyset \subset S_i \subset P_{u_i}^\bullet$.

Without loss of generality, suppose that $\emptyset \subset S_1 \subset P_{u_1}^\bullet$. This implies that there exists a point $(x_0, y_0) \in P_{u_1}^\bullet$ such that $\emptyset \subset S_1|_{x_0} \subset P_{u_1}^\bullet|_{x_0}.$[2] Thus, $S_1|_{x_0}$ is a nonempty open subset of $P_{u_1}^\bullet|_{x_0}$. Set $S_1|_{x_0}$ is a union of a countable number of open or half-open intervals (see Proposition 5.1.4 in [24]).

Thus, there exists a point $w_1 \in P_{u_1}^\bullet|_{x_0} \setminus S_1$ such that, for every positive ε_1, there exists a point $w_1' \in N(w_1, \varepsilon_1) \cap S_1$ [again, $N(w_1, \varepsilon_1)$ is an open neighbourhood with respect to the usual topology on $P_{u_1}^\bullet$]. Therefore, there exists a point $\bar{v}_1 \in X \setminus f^{-1}(\bar{v}_0)$ such that, for each positive ε_1, there exists a point $\bar{v}_1' \in N(\bar{v}_1, \varepsilon_1) \cap f^{-1}(\bar{v}_0)$. This contradicts that f is a continuous function on X. Thus, Y is a singleton. $\qquad\square$

As an informal interpretation of this proof: If ε is sufficiently small then Y is a neighbourhood $N(\bar{v}, \delta_{\bar{v}})$ of a single point $\bar{v}$, where $\delta_{\bar{v}}$ is sufficiently small (depending on ε) such that computers regard $N(\bar{v}, \delta_{\bar{v}})$ as a single point $\bar{v}$ because of rounding.

If input Π is a type-2 polyhedron then the solution obtained by Algorithm 30 might not be an approximate global solution to the 3D ESP problem. However, following Theorem 8.1, we propose with Algorithm 31 (see Fig. 8.9) a modification of Algorithm 30 for type-2 polyhedrons, with an initial mapping of non-convex polygons on their convex hulls:

Line 2 iterates through the closures of convex hulls. The iteration through step sets $C(P_{v_i}^\bullet)$ only occurs in Line 4 (i.e., when applying Algorithm 30 for the second time, using the same ε). Algorithm 31 provides an $(1 + (L_2 - L_1)/L)$-approximate global solution for the ESP, where L is the length of an optimal path; L_1 is the length of the path obtained in Line 2; L_2 the length of the final path obtained in

---

[2] $S|_{x_0} = \{(x_0, y) : (x, y) \in S \wedge x = x_0\}$.

Algorithm 31 (An RBA for type-2 polyhedrons)
Both input and output are the same as in Algorithm 30.

1: For $i \in \{1, 2, \ldots, k\}$, apply, e.g., the Melkman algorithm for computing $C(P_{v_i})$, the convex hull of P_{v_i}.
2: Let $C(P_{v_1}^\bullet), C(P_{v_2}^\bullet), \ldots, C(P_{v_k}^\bullet)$, p, and q be the input of Algorithm 30 for computing an approximate shortest route $\langle p, p_1, \ldots, p_k, q \rangle$.
3: For $i = 1, 2, \ldots, k - 1$, find a point $q_i \in C(P_{v_i}^\bullet)$ such that $d_e(p_{i-1}, q_i) + d_e(q_i, p_{i+1}) = \min\{d_e(p_{i-1}, p) + d_e(p, p_{i+1}) : p \in C(P_{v_i}^\bullet)\}$. Update the path for each i by $p_i = q_i$.
4: Let $P_{v_1}^\bullet, P_{v_2}^\bullet, \ldots, P_{v_k}^\bullet$, p and q be the input of Algorithm 30, and points p_i as obtained in Line 3 are the initial vertices p_i in Line 1 of Algorithm 30. Continue running Algorithm 30.
5: Return $\langle p, p_1, \ldots, p_{k-1}, p_k, q \rangle$ as provided in Line 4.

Fig. 8.9 The pseudocode of Algorithm 31

Line 5. Note that $L_2 \geq L_1$, and $L_2 = L_1$ if all polygons P_i are convex. Also note that $L < L_1$, then $(1 + (L_2 - L_1)/L) \leq L_2/L_1$. Thus, Algorithm 31 provides an L_2/L_1-approximate global solution for the ESP problem.

A comment to Line 3 of this algorithm: If $p_{i-1}p_{i+1} \cap P_{p_i}^\circ \neq \emptyset$, then set $q_i \leftarrow p_{i-1}p_{i+1} \cap P_{p_i}^\circ$.

The main ideas of Algorithm 32 (see Fig. 8.10) below are as follows: We apply Procedure 16 to compute the step set of a rubberband algorithm. Then we simply apply this rubberband algorithm to compute (of course, approximately only, defined by the chosen accuracy ε) the ESP.

For the input polyhedron, we assume that it is of type-2. For example, the Schönhardt polyhedron as shown in Fig. 8.6 is of type-2, but it might be rotated so that the resulting polyhedron is not of type-2 anymore.

Line 4 of Algorithm 32 partitions the set V' into some subsets such that the points in the same subset have an identical z-coordinate. In Line 6, we have that $u_1.z < u_2.z < \cdots < u_k.z$). In Line 8, V_{u_i} is a sequence of vertices of the critical polygon P_{u_i}. In Line 13, we delete p_i if p_i is not on an edge of P_{u_i}.

By the discussions before Algorithm 31, we have the following

Theorem 8.2 *Algorithm 32 provides an L_2/L_1-approximate global solution for the ESP problem, where L_1 is the length of the path obtained in Line 2 of Algorithm 31; L_2 is the length of the final path obtained in Line 5 of Algorithm 31.*

8.3 Time Complexity

At first, we can show (calculation of upper time bounds for involved operations is fairly straightforward) that Procedure 16 can be computed in $\mathcal{O}(|V_v||E(S_v)| + |\mathcal{F}|)$ time, where $S_v = \{F : F \in \mathcal{F} \wedge e = uw \in E(F) \wedge (u.z \leq v.z \leq w.z \vee w.z \leq$

Algorithm 32 (An RBA for an approximate ESP in a polyhedron)
Input: Two points p and q in Π; sets $\mathcal{F}$ and V of faces and vertices of Π, respectively.
Output: The set of all vertices of an approximate shortest path, starting at p and ending at q, and contained in Π.

1: Initialise $V' \leftarrow \{v : p.z < v.z < q.z \wedge v \in V\}$.
2: Sort V' according to the z-coordinate.
3: We obtain $V' = \{v_1, v_2, \ldots, v_{k'}\}$ with $v_1.z \leq v_2.z \leq \cdots \leq v_{k'}.z$.
4: Partition V' into pairwise disjoint subsets $V_1, V_2, \ldots,$ and V_k such that $V_i = \{v_{i1}, v_{i2}, \ldots, v_{in_i}\}$, with $v_{ij}.z = v_{ij+1}.z$, for $j = 1, 2, \ldots, n_i - 1$, and $v_{i1}.z < v_{i+11}.z$, for $i = 1, 2, \ldots, k - 1$.
5: Set $u_i \leftarrow v_{i1}$, where $i = 1, 2, \ldots, k$.
6: Set $V'' \leftarrow \{u_1, u_2, \ldots, u_k\}$
7: **for** each $u_i \in V''$ **do**
8:     Apply Procedure 16 for computing V_{u_i}.
9: **end for**
10: Set $\mathcal{F}_{\text{step}} \leftarrow \{P_{u_1}^{\bullet}, P_{u_2}^{\bullet}, \ldots, P_{u_k}^{\bullet}\}$.
11: Set $P \leftarrow \{p\} \cup V'' \cup \{q\}$.
12: Apply Algorithm 31 on inputs $\mathcal{F}_{\text{step}}$ and P, for computing the shortest path $\rho(p, q)$ inside of Π.
13: Convert $\rho(p, q)$ into the standard form of a shortest path by deleting all vertices which are not on any edge of Π.

Fig. 8.10 The pseudocode of Algorithm 32

$v.z \leq u.z)\}$. Then we can show that the time complexity of Algorithm 30 equals $\kappa(\varepsilon) \cdot \mathcal{O}(\sum_{j=1}^{k} |V_{v_j}|)$, where $\kappa(\varepsilon)$ is the number of iterations of the while loop in Algorithm 30. By Lemma 8.2, Line 5 can be computed in $\mathcal{O}(|V_{v_j}|)$ time, where V_{v_j} is as in Algorithm 30, for $j = 1, 2, \ldots, k$. Thus, each iteration of Algorithm 30 can be computed in $\mathcal{O}(\sum_{j=1}^{k} |V_{v_j}|)$ time. Obviously, Algorithm 31 has the same time complexity as Algorithm 30. These three results allow us then to show that Algorithm 32 can be computed in

$$\kappa(\varepsilon) \cdot \mathcal{O}\left(\sum_{j=1}^{k} |V_{u_j}|\right) + \mathcal{O}\left(\sum_{j=1}^{k} |V_{u_j}| |E(S_{u_j})| + |\mathcal{F}| + |V| \log |V|\right)$$

where the second term is the time for preprocessing. This can finally be reformulated in the more compact form that Algorithm 32 is of complexity

$$\kappa(\varepsilon) \cdot \mathcal{O}(M|V|) + \mathcal{O}(M|E| + |\mathcal{F}| + |V| \log |V|)$$

for $M = \max\{|V_{u_j}| : j = 1, 2, \ldots, k\}$, where the second $\mathcal{O}(\ldots)$ term is the time for preprocessing.

In Algorithm 30, let $\kappa(\varepsilon) = \frac{L_0 - L}{\varepsilon}$ be a function which only depends upon the difference between the lengths L_0 of an initial path and L of the optimum path, and the accuracy constant ε. Let L_m be the length of the mth updated path, for

$m = 0, 1, 2, \ldots$, with $L_m - L_{m+1} \geq \varepsilon$ (otherwise the algorithm stops). It follows that

$$\kappa(\varepsilon) = \frac{L_0 - L}{\varepsilon} \geq 1 + \frac{L_1 - L}{\varepsilon} \geq \cdots \geq m + \frac{L_m - L}{\varepsilon}.$$

The sequence $\{m + \frac{L_m - L}{\varepsilon}\}$ is monotonously decreasing, lower bounded by 0, and stops at the first m_0 where $L_{m_0} - L_{m_0+1} < \varepsilon$.

We have implemented a simplified version of Algorithm 30 where all $P_{v_i}^*$s were degenerated to be line segments.

> Thousands of experimental results *indicated* that $\kappa(\varepsilon)$ does not depend on the number k of segments but the value of ε.

We selected $\varepsilon = 10^{-15}$ and k was in between 4 and 20,000, the observed maximal value of $\kappa(\varepsilon)$ was 380,000. It shows that the smallest upper bound of $\kappa(\varepsilon) \geq \kappa(10^{-15}) \geq 380,000$. In other words, the number of iterations in the while-loop can be huge even for some small value of k. On the other hand, all these experimental results indicated that $|L_m - L_{m+1}| \leq 1.2$, when $m > 200$ and L was between 10,000 and 2,000,000. It showed that $\kappa(1.2) \leq 200$ and the relative error $|L_m - L_{m+1}|/L \leq 1.2 \times 10^{-4}$.

> In other words, these experiments showed that the algorithm already reached an approximate ESP with a very minor relative error after 200 iterations of the while loop; the remaining iterations were 'just' spent on improving a very small fraction of the length of the path.

8.4 Examples: Three NP-Complete or NP-Hard Problems

We apply Algorithms 31 and 32 for approximate solutions of hard problems, characterised below (by appropriate references) as being NP-complete or NP-hard. Let $p, q \in \Pi$ such that $p.z < q.z$. Let $V_{pq} = \{v : p.z < v.z < q.z \wedge v \in V\}$, where V is the set of all vertices of Π. For doing so, we are allowing for input polyhedrons different from the bounded type-2 polyhedrons so far, but only input polyhedrons which allow us to use those algorithms without any further modification.

We consider unbounded polyhedrons (which also satisfy the type-2 constraint), and, thus, generalised critical polygons.

Example 8.1 Let Π be a simply-connected polyhedron such that each critical polygon is the complement of an axis-aligned rectangle. Following Sect. 8.3, the Euclidean shortest path between p and q inside of Π can be approximately computed

Fig. 8.11 A path from p to q which does not intersect any of the shown *rectangles* at an inner point

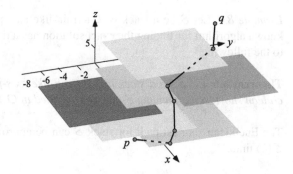

in $\kappa(\varepsilon) \cdot \mathcal{O}(|V_{pq}|)$ time. Therefore, the 3D ESP problem can be approximately solved efficiently in such a special case. Finding the exact solution is NP-complete because of the following[3]

Theorem 8.3 *It is NP-complete to decide whether there exists an obstacle-avoiding path of Euclidean length at most L among a set of stacked axis-aligned rectangles. The problem is (already) NP-complete for the special case that the axis-aligned rectangles are all q-rectangles of types 1 or 3.*

For stacked axis-aligned rectangles, see Fig. 8.11.   □

Example 8.2 Let Π be a simply-connected polyhedron such that each critical polygon is the complement of a triangle. Finding the exact solution is NP-hard because of the following[4]

Theorem 8.4 *It is NP-hard to decide whether there exists an obstacle-avoiding path of Euclidean length at most L among a set of stacked triangles.*

Following Sect. 8.3, the Euclidean shortest path between p and q inside of Π can be approximately computed in $\kappa(\varepsilon) \cdot \mathcal{O}(|V_{pq}|)$ time.   □

Example 8.3 Let S be a stack of k horizontal or vertical strips. Finding the exact solution is NP-complete because of the following[5]

Theorem 8.5 *It is NP-complete to decide whether there exists an obstacle-avoiding path of Euclidean length at most L among a finite number of stacked horizontal and vertical strips.*

The Euclidean shortest path for stack S can be approximately computed in $\kappa(\varepsilon) \cdot \mathcal{O}(k)$ time.   □

---

[3] See [20, Theorem 4].

[4] See [7].

[5] See [20, Theorem 5].

Example 8.4 Let $\mathcal{S}$ be a stack of k terrain-like axis-parallel rectangles. The best known algorithm for finding the exact solution has a time complexity in $\mathcal{O}(k^4)$ due to the following[6]

Theorem 8.6 *Let $\mathcal{S}$ be a stack of k terrain-like axis-parallel rectangles. The Euclidean shortest path among $\mathcal{S}$ can be computed in $\mathcal{O}(k^4)$ time.*

The Euclidean shortest path for stack $\mathcal{S}$ can be approximately computed in $\kappa(\varepsilon) \cdot \mathcal{O}(k)$ time. $\square$

8.5 Conclusions for the General 3D ESP Problem

This chapter described an algorithm for solving the 3D ESP problem when the domain Π is a type-2 simply connected polyhedron; the algorithm has a time complexity in $\kappa(\varepsilon) \cdot \mathcal{O}(M|V|) + \mathcal{O}(M|E| + |\mathcal{F}| + |V|\log|V|)$ (where $\mathcal{O}(M|E| + |\mathcal{F}| + |V|\log|V|)$ is the time for preprocessing). It was also shown that the algorithm approximately solves three NP-complete or NP-hard problems in time $\kappa(\varepsilon) \cdot \mathcal{O}(k)$, where k is the number of layers in the given stack of polygons.

Our algorithm has straightforward applications on ESP problems in 3D imaging (where proposed solutions depend on geodesics), or when 'flying' over a polyhedral terrain. The best result so far for the latter problem was an

$$\mathcal{O}\big((n/\varepsilon)(\log n)(\log\log n)\big)$$

algorithm for computing a $(2^{(p-1)/p} + \varepsilon)$-approximation to the L_p-shortest path above a polyhedral terrain.

As there does not exist an algorithm for finding exact solutions to the general 3D ESP problem,[7] the presented method defines a new opportunity to find approximate (and efficient!) solutions to the discussed classical, fundamental, hard, and general problems.

8.6 Problems

Problem 8.1 Prove Lemma 8.1.

Problem 8.2 Provide a high level description for Procedure 16.

---

[6]See [20, Theorem 6].
[7]See [3, Theorem 9].

Problem 8.3 Algorithm 30 is also modified from Algorithm 7. Underline those modifications of Algorithm 30 in Fig. 8.8.

Problem 8.4 Consider the proof of Theorem 8.1. Why does Algorithm 30 define a continuous function?

Problem 8.5 Provide an example of a type-2 polyhedron is of such that a rotation of this polyhedron is not anymore of type-2.

Problem 8.6 Can Theorem 8.1 also be true if Π is a type-2 polyhedron?

Problem 8.7 (Programming exercise) Implement and test Algorithm 32 for inputs of varying complexity.

Problem 8.8 (Programming exercise) Implement the generic RBA (i.e., Algorithm 11) for the 3D case, as described in Fig. 3.24, and compare its performance with that of Algorithm 32.

Problem 8.9 (Programming exercise) Implement Papadimitriou's algorithm as described in Fig. 2.2, and compare its performance with that of the generic RBA for the 3D case.

Problem 8.10 (Research problem) Prove the correctness of Theorem 8.1 based on convex analysis (if possible at all).

Problem 8.11 (Research problem) Generalise Algorithm 32 such that input Π might be a 'tree shaped' polyhedron that is homeomorphic to a sphere.

8.7 Notes

Section 8.1 follows [20].

Since the pioneering work in [10] on finding contours as minimal paths between two end points, minimal paths in volume images have raised interest in computer vision and image analysis; see, for example, [5, 11, 13]. In medical image analysis, minimal paths were extracted in 3D images and applied to virtual endoscopy [11]. However, so far, minimal path computation is typically based on the *Fast Marching Method* which only considers grid points as the possible vertices of the minimal paths; but there exist Euclidean shortest paths such that *none* of its vertices is a grid point; see, e.g., the example in Sect. 4 of [17]. Thus, paths detected by the Fast Marching Method cannot be always the exact Euclidean shortest paths.

There already exist several approximation algorithms for 3D ESP calculations, and we briefly recall those. Pioneering the field, [22] presents an

$$\mathcal{O}\big(n^4\big(m + \log(n/\varepsilon)\big)^2/\varepsilon^2\big)$$

algorithm for the general 3D ESP problem, where n is the descriptional complexity of polyhedral scene elements (that is, vertices, edges, and faces of the polyhedron), ε the accuracy of the algorithm, and m the maximum number of bits for representing a single integer coordinate of elements of the polyhedral scene. This was followed by [9], which presents an approximation algorithm for computing an $(1 + \varepsilon)$-shortest path from p to q in time

$$\mathcal{O}\big(n^2\lambda(n)\log(n/\varepsilon)/\varepsilon^4 + n^2\log nr\log(n\log r)\big)$$

where r is the ratio of the Euclidean distance $d_e(p, q)$ to the length of the longest edge of any given obstacle, and

$$\lambda(n) = \alpha(n)^{\mathcal{O}(\alpha(n)^{\mathcal{O}(1)})}$$

where $\alpha(n) = A^{-1}(n, n)$ is the inverse Ackermann function (see, e.g., [18]), which grows very slowly.

Assume a finite set of polyhedral obstacles in $\mathbb{R}^3$. Let p, q be two points outside of the union of all obstacles, and $0 < \varepsilon < 1$. Reference [12] gives an $\mathcal{O}(\log(n/\varepsilon))$ algorithm to compute an $(1 + \varepsilon)$-shortest path from p to q such that it does not intersect the interior of any obstacle. However, this algorithm requires a subdivision of $\mathbb{R}^3$ which may be computed in $\mathcal{O}(n^4/\varepsilon^6)$.

Given a convex partition of the free space, [2] presents an

$$\mathcal{O}\big((n/\varepsilon^3)(\log 1/\varepsilon)(\log n)\big)$$

algorithm for the general 3D ESP problem. More recently, [1] proposes algorithms for calculating approximate ESPs amid a set of convex obstacles. For results related to surface ESPs, see [4].

Altogether, the task of finding efficient and easy-to-implement solutions in this field is certainly challenging; see, for example, [19] saying on page 666 the following, when addressing mainly exact solutions: "*The problem is difficult even in the most basic Euclidean shortest-path problem... in a three-dimensional polyhedral domain Π, and even if the obstacles are convex, or the domain Π is simply connected.*"

RBAs are characterised by sets of steps, defining possible locations of vertices of Euclidean shortest paths, a local optimisation strategy, and a termination criterion [6, 14–17, 21]. The given RBA solves approximately three NP-complete or NP-hard 3D ESP problems in time $\kappa(\varepsilon) \cdot \mathcal{O}(k)$, where k is the number of layers in a stack, which was introduced as the problem environment. The presented RBA has straightforward applications for ESP problems when analysing polyhedral objects (e.g., in 3D imaging; for the extensive work using geodesics, we just cite [25] as one example), or for 'flying' over a polyhedral terrain. The best known result for the latter problem is due to [27] by proposing an $\mathcal{O}((n/\varepsilon)(\log n)(\log\log n))$ algorithm for computing a $(2^{(p-1)/p} + \varepsilon)$-approximation of an L_p-shortest path above a polyhedral terrain.

References

1. Agarwal, P.K., Sharathkumar, R., Yu, H.: Approximate Euclidean shortest paths amid convex obstacles. In: Proc. ACM-SIAM Sympos. Discrete Algorithms, pp. 283–292 (2009)
2. Aleksandrov, L., Maheshwari, A., Sack, J.-R.: Approximation algorithms for geometric shortest path problems. In: Proc. ACM Sympos. Theory Comput., pp. 286–295 (2000)
3. Bajaj, C.: The algebraic complexity of shortest paths in polyhedral spaces. In: Proc. Allerton Conf. Commun. Control Comput., pp. 510–517 (1985)
4. Balasubramanian, M., Polimeni, J.R., Schwartz, E.L.: Exact geodesics and shortest paths on polyhedral surfaces. IEEE Trans. Pattern Anal. Mach. Intell. **31**, 1006–1016 (2009)
5. Benmansour, F., Cohen, L.D.: Fast object segmentation by growing minimal paths from a single point on 2D or 3D images. J. Math. Imaging Vis. **33**, 209–221 (2009)
6. Buelow, T., Klette, R.: Rubber band algorithm for estimating the length of digitized space-curves. In: Proc. ICPR, vol. III, pp. 551–555. IEEE Comput. Soc., Los Alamitos (2000)
7. Canny, J., Reif, J.H.: New lower bound techniques for robot motion planning problems. In: Proc. IEEE Conf. Foundations Computer Science, pp. 49–60 (1987)
8. Choi, J., Sellen, J., Yap, C.-K.: Precision-sensitive Euclidean shortest path in 3-space. In: Proc. ACM Sympos. Computational Geometry, pp. 350–359 (1995)
9. Clarkson, K.L.: Approximation algorithms for shortest path motion planning. In: Proc. ACM Sympos. Theory Comput., pp. 56–65 (1987)
10. Cohen, L.D., Kimmel, R.: Global minimum for active contour models: a minimal path approach. Int. J. Comput. Vis. **24**, 57–78 (1997)
11. Deschamps, T., Cohen, L.D.: Fast extraction of minimal paths in 3D images and applications to virtual endoscopy. Med. Image Anal. **5**, 281–299 (2001)
12. Har-Peled, S.: Constructing approximate shortest path maps in three dimensions. In: Proc. ACM Sympos. Computational Geometry, pp. 125–130 (1998)
13. Klette, R., Rosenfeld, A.: Digital Geometry. Morgan Kaufmann, San Francisco (2004)
14. Li, F., Klette, R.: The class of simple cube-curves whose MLPs cannot have vertices at grid points. In: Proc. Discrete Geometry Computational Imaging. LNCS, vol. 3429, pp. 183–194. Springer, Berlin (2005)
15. Li, F., Klette, R.: Exact and approximate algorithms for the calculation of shortest paths. Report 2141, IMA, Minneapolis. www.ima.umn.edu/preprints/oct2006 (2006)
16. Li, F., Klette, R.: Rubberband algorithms for solving various 2D or 3D shortest path problems. In: Proc. Computing: Theory and Applications, The Indian Statistical Institute, Kolkata, pp. 9–18. IEEE Comput. Soc., Los Alamitos (2007)
17. Li, F., Klette, R.: Analysis of the rubberband algorithm. Image Vis. Comput. **25**, 1588–1598 (2007)
18. Liu, Y.A., Stoller, S.D.: Optimizing Ackermann's function by incrementalization. In: Proc. ACM SIGPLAN Sympos. Partial Evaluation Semantics-Based Program Manipulation, pp. 85–91 (2003)
19. Mitchell, J.S.B.: Geometric shortest paths and network optimization. In: Sack, J.-R., Urrutia, J. (eds.) Handbook of Computational Geometry, pp. 633–701. Elsevier, Amsterdam (2000)
20. Mitchell, J.S.B., Sharir, M.: New results on shortest paths in three dimensions. In: Proc. ACM Sympos. Computational Geometry, pp. 124–133 (2004)
21. Pan, X., Li, F., Klette, R.: Approximate shortest path algorithms for sequences of pairwise disjoint simple polygons. In: Proc. Canadian Conf. Computational Geometry, pp. 1–4. Winnipeg, Canada (2010)
22. Papadimitriou, C.H.: An algorithm for shortest path motion in three dimensions. Inf. Process. Lett. **20**, 259–263 (1985)
23. Sharir, M., Schorr, A.: On shortest paths in polyhedral spaces. SIAM J. Comput. **15**, 193–215 (1986)
24. Wachsmuth, B.G.: Interactive real analysis. http://web01.shu.edu/projects/reals/topo/index.html (2009). Accessed July 2011

25. Wang, Y., Peterson, B.S., Staib, L.H.: 3D brain surface matching based on geodesics and local geometry. Comput. Vis. Image Underst. **89**, 252–271 (2003)
26. Yap, C.-K.: Towards exact geometric computation. Comput. Geom. **7**, 3–23 (1997)
27. Zadeh, H.Z.: Flying over a polyhedral terrain. Inf. Process. Lett. **105**, 103–107 (2008)

Chapter 9
Paths in Cube-Curves

When I was a Boy Scout, we played a game when new Scouts joined the troop. We lined up chairs in a pattern, creating an obstacle course through which the new Scouts, blindfolded, were supposed to manoeuvre. The Scoutmaster gave them a few moments to study the pattern before our adventure began. But as soon as the victims were blindfolded, the rest of us quietly removed the chairs.—I think life is like this game.

Pierce Vincent Eckhart

This chapter discusses a problem defined in a 3-dimensional regular grid. Such a grid is commonly used in 3D image analysis. We may also assume that a general 3D space (e.g., for a robot) is regularly subdivided into cubes of uniform size. The chapter considers shortest paths in such a *cuboidal world*.

9.1 The Cuboidal World

A grid point $(i, j, k) \in \mathbb{Z}^3$ is assumed to be the centre point of a *grid cube*, with *faces* parallel to the coordinate planes, with *edges* of length 1, and *vertices* at its corners. *Cells* are either cubes, faces, edges, or vertices. The intersection of two cells is either empty or a joint *side* of both cells.

A *cube-curve* g is a loop of face-connected grid cubes in the 3D orthogonal grid; the union **g** of those cubes defines the *tube* of g.

This chapter discusses ESPs in such tubes which are also known as *minimum-length polygonal curves* (MLP).

See Fig. 9.1 for a polygonal curve defined by midpoints of cubes that can be used as an initial curve for a rubberband algorithm for calculating an approximate MLP.

F. Li, R. Klette, *Euclidean Shortest Paths*,
DOI 10.1007/978-1-4471-2256-2_9, © Springer-Verlag London Limited 2011

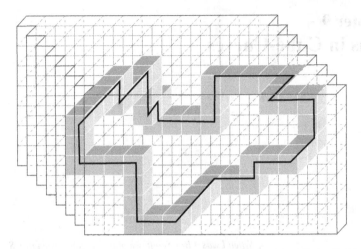

Fig. 9.1 A cuboidal world: the *bold curve* is an initial guess for a 3D walk (or flight) through the given loop of shaded cubes, being incident with the midpoints of those cubes. A 3D walk of minimum length defines the *minimum-length polygon* (MLP) of this loop of shaded cubes

A *cuboidal world* is defined by cubes in $\mathbb{Z}^3$ as specified above, where objects are sets of cubes.

A cuboidal world is a discrete subdivision of real spaces. The unit 1 in $\mathbb{Z}^3$ corresponds to a defined physical length in the real world.

This chapter provides three general and κ-linear approximate shortest path algorithms, where the problem size equals the number of *critical edges* in the given cube-curve, defined as follows:

Definition 9.1 A *critical edge* of a simple cube-curve g is a grid edge in $\mathbb{Z}^3$ which is incident with exactly three different cubes contained in g.

Linear runs of cubes in a given cube-curve do not have an essential impact on the computational complexity of the presented ESP (or MLP) algorithms.

Find the critical edges in Fig. 9.1! Note that it is already not so easy to count the critical edges for this relatively short cube-curve, especially in the upper left part. Non-critical edges are incident with two co-planar adjacent faces, or incident with two faces that form a right angle.

Figure 9.2 shows all the critical edges (in red colour) of a cube-curve. Later we will characterise this curve as being an example for a 'first-class simple cube-curve that has both middle and also end-angles'. A 'simple cube-curve' is, informally speaking, a cube-curve that is not 'crossing' or 'touching' itself. A formal definition is as follows:

Fig. 9.2 Example of a simple cube-curve. The critical edges are numbered from 0 to 21

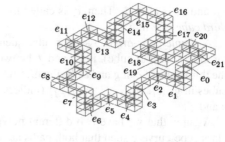

Fig. 9.3 *Left*: a first-class simple cube-curve. *Right*: a non-first-class simple cube-curve

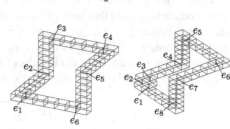

Definition 9.2 A cube-curve is *simple* iff $n \geq 4$ and for any two cubes $c_i, c_k \in g$ with $|i - k| \geq 2(\mathrm{mod}\, n + 1)$, if $c_i \cap c_k \neq \phi$ then either $|i - k| = 2(\mathrm{mod}\, n + 1)$ and $c_i \cap c_k$ is an edge, or $|i - k| \geq 3(\mathrm{mod}\, n + 1)$ and $c_i \cap c_k$ is a face.

Curve ρ is *complete* with respect to tube **g** if ρ intersects every cube in g. A shortest Euclidean curve ρ, that is contained and complete in a simple cube-curve g, is always a polygonal curve (i.e., the MLP). The MLP is uniquely defined as long as the cube-curve is not only contained in a single layer of cubes of the 3D grid. If it is contained in such a single layer, then the MLP is uniquely defined up to a translation orthogonal to that layer. However, we speak about *the* MLP of a simple cube-curve, thus ignoring this extreme case of a single layer for language simplicity. We just cite the following theorem[1] (without giving a proof here):

Theorem 9.1 *Let g be a simple cube-curve. Critical edges are the only possible locations of vertices of the MLP of g.*

Definition 9.3 A simple cube-curve g is called *first-class* iff each critical edge of g contains exactly one vertex of the MLP of g.

Figure 9.3 shows a first-class simple cube-curve (left) and a non-first-class simple cube-curve (right). For the latter one, note that the vertices of the MLP must be in e_1, e_3, e_4, e_5, e_6, e_7, and e_8, but the critical edge e_2 does not contain any vertex of the MLP of this simple cube-curve.

Let $c_0, c_1, \ldots, c_{n-1}$ be the sequence of all consecutive cubes of cube-curve g. Cube c_j is said to be *after* c_i if $j = i + 1(\mathrm{mod}\, n)$. Let e be a critical edge of g. Let c_i, c_j, and c_k be three consecutive cubes of g such that $e \in c_i \cap c_j \cap c_k$, c_k is after

---

[1]Published by *Reinhard Klette* and *Thomas Bülow* in 2000; see [32].

c_j, and c_j is after c_i. Then, c_i is called *the first cube* of e in g and c_k is called *the third cube* of e in g.

A *simple cube-arc* is a proper subsequence of a simple cube-curve (i.e., not returning to its start cube). Let e and f be two different critical edges of g. Let c_i be the third cube of e in g and c_j be the first cube of f in g. Let n be the total number of cubes in g. $\langle c_i, c_{i+1}, \ldots, c_{j-1}, c_j \rangle$ (indices $\bmod n$) is *the arc* (in g) *between edges e and f*.

Assume that we have two different polygonal paths ρ and ρ' in the same first-class cube-curve g such that both paths are complete and contained in tube **g**. Both paths have vertices $p_i = (x_i, y_i, z_i)$ and $p_i' = (x_i', y_i', z_i')$ on critical edges e_i, respectively, assuming n critical edges $e_1, \ldots, e_n$ in a cyclic order in g. The following theorem allows us to compare the lengths of both paths.

Let $\max\{|x_i' - x_i|, |y_i' - y_i|, |y_i' - y_i|\} < \delta$, for $i = 1, 2, \ldots, n$. Furthermore, let

$$d_e(i) = \sqrt{(x_{i+1} - x_i)^2 + (y_{i+1} - y_i)^2 + (z_{i+1} - z_i)^2},$$

$$d_e'(i) = \sqrt{\left(x_{i+1}' - x_i'\right)^2 + \left(y_{i+1}' - y_i'\right)^2 + \left(z_{i+1}' - z_i'\right)^2}$$

for $i = 1, 2, \ldots, n (\bmod n)$. Then, $d = \sum_{i=1}^{n} d_e(i)$ or $d' = \sum_{i=1}^{n} d_e'(i)$ is the length of path ρ or ρ', respectively.

Theorem 9.2 *Let $|d_e(i)| \geq M_2$, for $i = 1, 2, \ldots, n$. We can define a constant $M_1 > 0$ such that*

$$|d - d'| < 12n M_1 \delta / M_2.$$

Proof Let

$$\delta x_{i+1} = x_{i+1} - x_{i+1}', \qquad \delta x_i = x_i' - x_i,$$

$$\delta y_{i+1} = y_{i+1} - y_{i+1}', \qquad \delta y_i = y_i' - y_i,$$

$$\delta z_{i+1} = z_{i+1} - z_{i+1}', \qquad \delta z_i = z_i' - z_i,$$

$$\delta x = \delta x_{i+1} + \delta x_i, \qquad \delta y = \delta y_{i+1} + \delta y_i, \quad \text{and} \quad \delta z = \delta z_{i+1} + \delta z_i.$$

Then we have that

$$|\delta x| \leq 2\delta, \qquad |\delta y| \leq 2\delta, \quad \text{and} \quad |\delta z| \leq 2\delta.$$

Since

$$(x_{i+1} - x_i)^2 + (y_{i+1} - y_i)^2 + (z_{i+1} - z_i)^2$$

$$= \left[(x_{i+1}' - x_i') + (x_{i+1} - x_{i+1}') + (x_i' - x_i)\right]^2$$

$$+ \left[(y_{i+1}' - y_i') + (y_{i+1} - y_{i+1}') + (y_i' - y_i)\right]^2$$

$$+ \left[(z'_{i+1} - z'_i) + (z_{i+1} - z'_{i+1}) + (z'_i - z_i) \right]^2$$
$$= \left[(x'_{i+1} - x'_i) + \delta_x \right]^2 + \left[(y'_{i+1} - y'_i) + \delta_y \right]^2 + \left[(z'_{i+1} - z'_i) + \delta_z \right]^2$$

and Lemma 9.2 below, we have that

$$\left| d_e(i) - d'_e(i) \right| < 6M_1 \times 2\delta/M_2 = 12M_1\delta/M_2.$$

Thus, $|d - d'| = |\sum_{i=1}^{n}(d_e(i) - d'_e(i))| \leq \sum_{i=1}^{n} |d_e(i) - d'_e(i)| < 12nM_1\delta/M_2.$ $\quad\square$

Lemma 9.2 will be shown via Lemma 9.1. Let g_l, g_r, g_f, g_b, g_d, and g_u be six numbers in $\mathbb{R}$ which define a hypercube $[g_l, g_r] \times [g_f, g_b] \times [g_d, g_u]$, with $g_l \leq g_r$, $g_f \leq g_b$, and $g_d \leq g_u$. Let $(x, y, z) \in \mathbb{R}^3$ and

$$f(x, y, z) = \sqrt{x^2 + y^2 + z^2}.$$

Let $M > 0$ be such that $|f(x_1, y_1, z_1)| \geq M$ for any point (x_1, y_1, z_1) in this hypercube.

Lemma 9.1 *There is a real $\delta > 0$ such that* $\max\{|x_2 - x_1|, |y_2 - y_1|, |z_2 - z_1|\} < \delta$, *for any (x_2, y_2, z_2) in this hypercube. It also follows that $|f(x_2, y_2, z_2)| > M/2$.*

Proof Since the hypercube $[g_l, g_r] \times [g_f, g_b] \times [g_d, g_u]$ is a compact set, the function $f(x, y, z)$ is uniformly continuous at any point of this hypercube. For $\varepsilon_0 = M/2 > 0$, it follows that there is a real $\delta > 0$ such that

$$\left| f(x_2, y_2, z_2) - f(x_1, y_1, z_1) \right| < M/2$$

for any two points $(x_1, y_1, z_1), (x_2, y_2, z_2) \in [g_l, g_r] \times [g_f, g_b] \times [g_d, g_u]$ which satisfy

$$\max\{|x_2 - x_1|, |y_2 - y_1|, |z_2 - z_1|\} < \delta.$$

That is,

$$\left| f(x_1, y_1, z_1) \right| - M/2 < f(x_2, y_2, z_2) < \left| f(x_1, y_1, z_1) \right| + M/2.$$

With $|f(x_1, y_1, z_1)| \geq M$ it follows that $|f(x_2, y_2, z_2)| > M/2$. $\quad\square$

Now let $M_1 = \max\{|g_l|, |g_r|, |g_f|, |g_b|, |g_d|, |g_u|\} > 0$, $|f(x_1, y_1, z_1)| \geq M_2$, and consider two points (x_1, y_1, z_1) and (x_2, y_2, z_2) in the hypercube which $\max\{|x_2 - x_1|, |y_2 - y_1|, |z_2 - z_1|\} < \delta$. Then we have the following:

Lemma 9.2 $|f(x_2, y_2, z_2) - f(x_1, y_1, z_1)| < 6M_1\delta/M_2.$

Proof Without loss of generality, assume that $x_1 < x_2$, $y_1 < y_2$ and $z_1 < z_2$. By the mean-value theorem,

$$\left| f(x_2, y_2, z_2) - f(x_1, y_1, z_1) \right|$$

$$= \left| \left[f(x_2, y_2, z_2) - f(x_1, y_2, z_2) \right] + \left[f(x_1, y_2, z_2) - f(x_1, y_1, z_2) \right] \right.$$

$$\left. + \left[f(x_1, y_1, z_2) - f(x_1, y_1, z_1) \right] \right|$$

$$= \left| f_x(\xi_x, y_2, z_2)(x_2 - x_1) + f_y(x_1, \eta_y, z_2)(y_2 - y_1) + f_z(x_1, y_1, \zeta_z)(z_2 - z_1) \right|$$

$$= \left| \frac{\xi_x}{\sqrt{\xi_x^2 + y_2^2 + z_2^2}}(x_2 - x_1) + \frac{\eta_y}{\sqrt{x_1^2 + \eta_y^2 + z_2^2}}(y_2 - y_1) \right.$$

$$\left. + \frac{\zeta_z}{\sqrt{x_1^2 + y_1^2 + \zeta_z^2}}(z_2 - z_1) \right|$$

where $\xi_x \in (x_1, x_2)$, $\eta_y \in (y_1, y_2)$, and $\zeta_z \in (z_1, z_2)$. By Lemma 9.1, we have

$$\left| f(x_2, y_2, z_2) - f(x_1, y_1, z_1) \right| < 2M_1\delta/M_2 + 2M_1\delta/M_2 + 2M_1\delta/M_2$$

$$= 6M_1\delta/M_2.$$

This proves our lemma.   □

We make use of Theorem 9.2 further below.

9.2 Original and Revised RBA for Cube-Curves

This section reports about the historical origin of RBAs. The first rubberband algorithm was[2] designed for calculating an MLP in a simple cube-curve. We show later that it actually requires two repairs to make it deliver convergent solutions in the general case. However, it is correctly converging to accurate solutions for the case of first-class simple cube-curves, and with this property it also proves to be useful as a subprocess in another algorithm.

Let $\rho = (p_0, p_1, \ldots, p_m)$ be a polygonal curve contained in tube **g** of a cube-curve g. A polygonal curve γ is a *g-transform* of ρ iff γ is obtained from ρ by a finite number of operations, where each operation is a replacement of a triple a, b, c of vertices by a polygonal sequence $a, b_1, \ldots, b_k, c$ such that the polygonal sequence $a, b_1, \ldots, b_k, c$ is still contained in the same set of cubes of cube-curve g as the polygonal sequence a, b, c was before.

Consider a polygonal curve $\rho = (p_0, p_1, \ldots, p_m)$ and vertex indices $i - 1, i$, and $i + 1$ in this curve. Let ρ be contained and complete in tube **g**. Three different *options* for vertices p_{i-1}, p_i, and p_{i+1} result in corresponding g-transforms, and those are the local optimisations in the *original rubberband algorithm* applied in repeated iterations, starting with an initial polygonal path that is complete and contained in tube **g**:

---

[2]Published between 2000 and 2002 in papers by *Thomas Bülow* and *Reinhard Klette*; see [11].

Fig. 9.4 Illustration for the (original) Option 2

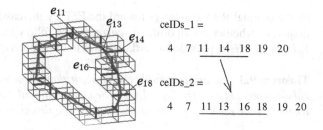

$$\text{ceIDs\_1} =$$
$$4 \quad 7 \quad 11 \quad 14 \quad 18 \quad 19 \quad 20$$

$$\text{ceIDs\_2} =$$
$$4 \quad 7 \quad 11 \quad 13 \quad 16 \quad 18 \quad 19 \quad 20$$

(O_1) Point p_i is deleted if $p_{i-1}p_{i+1}$ is a line segment within the tube. Then the subsequence (p_{i-1}, p_i, p_{i+1}) is replaced in the curve by (p_{i-1}, p_{i+1}). In this case, the algorithm continues with vertices $p_{i-1}, p_{i+1}, p_{i+2}$.

(O_2) The closed triangular region $\triangle(p_{i-1}, p_i, p_{i+1})$ intersects more than just three critical edges of p_{i-1}, p_i, and p_{i+1} (i.e., a simple deletion of p_i would not be sufficient anymore). This situation is solved by calculating a convex arc and by replacing point p_i by a sequence of vertices $q_1, \ldots, q_k$ on this convex arc between p_{i-1} and p_{i+1} such that the sequence of line segments $p_{i-1}q_1, \ldots, q_k p_{i+1}$ lies in the triangular region and within the tube. In this case, the algorithm continues with a triple of vertices starting with the calculated new vertex q_k.

If (O_1) and (O_2) do not lead to any change, the third option may lead to an improvement (i.e., a shorter polygonal curve which is still contained and complete in the given tube):

(O_3) Point p_i may be moved on its critical edge to obtain an optimum position p_{new} minimising the total length of both line segments $p_{i-1}p_{\text{new}}$ and $p_{\text{new}}p_{i+1}$. First, find $p_{\text{opt}} \in l_e$ such that $|p_{\text{opt}} - p_{i-1}| + |p_{\text{opt}} - p_{i+1}| = \min_{p \in l_e} L(p)$ with $L(p) = |p - p_{i-1}| + |p - p_{i+1}|$, where l_e is the straight line containing the critical edge e. Then, if p_{opt} lies on the closed critical edge e, let $p_{\text{new}} = p_{\text{opt}}$. Otherwise, let p_{new} be that vertex bounding e and lying closest to p_{opt}.

See Fig. 9.4 for (O_2). Here, vertices on critical edges e_{11}, e_{14}, and e_{18} are replaced by a convex arc with vertices on critical edges e_{11}, e_{13}, e_{16}, and e_{18}, and (in general) it may be e_{11}, e_{14}, and e_{18} again within a subsequent loop—of course, for a reduced length of the calculated path at this stage.

Option 3 of this *original rubberband algorithm* is not asking for testing inclusion of the generated new segments within tube **g**. The authors of this book realised that there are (rather complex) cube-curves where this test needs to be added, and that's the first (easy) repair for this *Bülow–Klette algorithm* for applying it to the general case of simple cube-curves.

The situation with the original RBA was in 2002 as follows: Even for very small values of $\varepsilon > 0$, the measured time complexity indicated $\mathcal{O}(n)$, where n is the number of cubes in g. However, there was no proof for the asymptotic time complexity of the original RBA. For a small number of test examples, calculated paths seemed (!) to converge toward the ESP. However, no implemented algorithm for calculating a correct ESP was available, and, more generally, no proof whether the path provided

by the original RBA converges toward the ESP by increasing the number of iterations (i.e., whether the algorithm is *correct*). The following theorem from 2005 by
F. Li and *R. Klette* provided a partial answer to those open questions:

Theorem 9.3 *The original rubberband algorithm is correct for first-class simple
cube-curves. Let* $\kappa(\varepsilon) = (L_0 - L)/\varepsilon$. *The computational complexity is* $\kappa(\varepsilon) \cdot \mathcal{O}(n)$,
where n is the number of critical edges of the simple cube-curve.

L_0 is the length of the initial path, L is the length of the MLP (i.e., of the ideal
solution), and ε is the usual numerical accuracy constant of an RBA.

9.3 An Algorithm with Guaranteed Error Limit

Based on Theorems 9.1, 9.2, and 9.3, we are now in a position to formulate Algorithm 33 in Fig. 9.5, which combines an application of graph theory (Dijkstra's
algorithm) with the original rubberband algorithm.

In Line 3 of Algorithm 33, p_{i_0} and p_{i_m} are the end points of e_i. In Line 5, G is
a visibility graph as commonly used in computational geometry. In Line 7, $N(v)$ is
the set of all neighbours of vertex v. In Line 13, $l(\rho_v)$ is the length of the cycle ρ_v.
Line 21 is motivated by Theorem 9.3.

Example 9.1 We illustrate the algorithm by means of an input example; see Fig. 9.6.
Table 9.1 shows the data of the critical edges of a (short) simple cube-curve g, shown
in Fig. 9.6, upper left. Figure 9.6, upper left, also shows the subdivision points of
Line 1, where $m = 3$.

Figure 9.6, upper right, shows the graph, denoted by $G = [V, E]$, constructed
in Line 5. Note that each edge of G is fully contained in tube **g**. To compute the
weight of an edge, say, of $p_2^3 p_3^1$ of G, see Table 9.1: the coordinates of p_2^3 and p_3^1
are $(0.5, 1, 0.5)$ and

$$(0.5, 1, 1.5) + \frac{1}{m} \times \left[(0.5, 2, 1.5) - (0.5, 1, 1.5) \right] = (0.5, 1.3333, 1.5),$$

respectively. It follows that the weight of segment $p_2^3 p_3^1$ equals

$$d_e\left(p_2^3, p_3^1\right) = \sqrt{(0.5 - 0.5)^2 + (1.3333 - 1)^2 + (1.5 - 0.5)^2} = 1.0541.$$

We illustrate Lines 6–21 by the (small) directed weighted graph shown in
Fig. 9.6, lower left. For Line 6, we have $V = \{v_1, v_2, v_3, v_4\}$, with $N(v_1) = \{v_3, v_4\}$,
$N(v_2) = \{v_1\}$, $N(v_3) = \{v_2\}$, and $N(v_4) = \{v_3\}$.

Consider vertex v_1 and Line 9; for $v_3 \in N(v_1)$, the directed shortest path from v_1
to v_3 equals (v_1, v_2, v_3) with weight 3; the directed shortest path from v_1 to v_4 equals
(v_1, v_2, v_3, v_4) with weight 6. By Line 10, the local minimum-weight directed cycle,
which contains $v_3 v_1$, equals (v_1, v_2, v_3, v_1) with weight 8; $(v_1, v_2, v_3, v_4, v_1)$ (with
weight 10) is the local minimum-weight directed cycle which contains $v_4 v_1$.

Algorithm 33 (Algorithm with guaranteed error limit)
Input: The set S of all critical edges of a simple cube-curve g.
Output: An approximate MLP of g.

1: Initialise an integer $m \geq 2$ and a minimum length $L = +\infty$.
2: **for** each critical edge $e_i \in S$ **do**
3:     Let $p_{i_0}, p_{i_1}, \ldots, p_{i_m}$ be subdivision points on e_i such that the Euclidean distance between p_{i_j} and $p_{i_{j+1}}$ equals $\frac{1}{m}$, for $j = 0, 1, 2, \ldots, m - 1$.
4: **end for**
5: Construct the weighted directed graph $G = [V, E]$, where

$$V = \{p_{i_j} : p_{i_j} \in e_i \wedge i = 1, 2, \ldots, n \wedge j = 1, 2, \ldots, m\},$$

$$E = \{p_{i_j} p_{k_l} : p_{i_j} p_{k_l} \text{ completely contained in the cube-arc}$$

$$\text{between } e_i \text{ and } e_k \text{ in } g\}.$$

The weight of straight line segment $p_{i_j} p_{k_l}$ is defined to be the Euclidean distance between p_{i_j} and p_{k_l}, where $i = 1, 2, \ldots, n$ and $j = 1, 2, \ldots, m$.
6: **for** each $v \in V$ **do**
7:     Let $N(v) = \{u : \overrightarrow{uv} \in E\}$.
8:     **for** each $u \in N(v)$ **do**
9:         Apply Dijkstra's algorithm to compute the directed shortest path from v to u.
10:        Find the local minimum-weight directed cycle which contains directed line segment $\overrightarrow{uv}$, denoted by ρ_v.
11:        Let S_v be the set of critical edges such that each of the vertices of the local minimum-weight directed cycle is on one of the edges in S_v.
12:    **end for**
13:    **if** $l(\rho_v) < L$ **then**
14:        Let $L = l(\rho_v)$ and an approximate MLP: AMLP $= \rho_v$.
15:    **end if**
16: **end for**
17: Let $S' = \emptyset$.
18: **for** each vertex v of AMLP **do**
19:    Find a critical edge e such that $v \in e$; let $S' = S' \cup \{e\}$.
20: **end for**
21: Apply the original rubberband algorithm on S' and AMLP, and compute and output the minimum-weight directed cycle.

Fig. 9.5 An algorithm with guaranteed error limit for computing an approximate MLP of a simple cube-curve

Now consider vertex v_2 and Line 9; for $v_1 \in N(v_2)$, the directed shortest paths from v_2 to v_1 are (v_2, v_3, v_4, v_1) with weight 9, and (v_2, v_3, v_1) with weight 7. By Line 10, the local minimum-weight directed cycle, which contains $v_1 v_2$, equals (v_2, v_3, v_1, v_2) with weight 8.

Fig. 9.6 Input example for Algorithm 33; see text for details

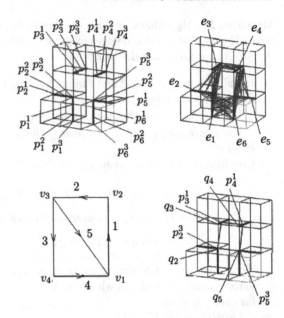

Next consider vertex v_3 and Line 9; for $v_2 \in N(v_3)$, the directed shortest paths from v_3 to v_2 are (v_3, v_4, v_1, v_2) with weight 8, and (v_3, v_1, v_2) with weight 6. By Line 10, the local minimum-weight directed cycle, which contains $v_2 v_3$, equals (v_3, v_1, v_2, v_3) with weight 8.

Finally, consider vertex v_4 and Line 9; for $v_3 \in N(v_4)$, the directed shortest path from v_4 to v_3 equals (v_4, v_1, v_2, v_3) with weight 7. By Line 10, the local minimum-weight directed cycle, which contains $v_3 v_4$, equals $(v_4, v_1, v_2, v_3, v_4)$ with weight 10.

For Lines 13–15, compare all the local minimum-weight directed cycles obtained in Line 10; the global minimum-weight directed cycle equals (v_1, v_2, v_3, v_1) with weight 8.

The global minimum-weight directed cycle of G equals $(p_2^3, p_3^1, p_4^1, p_5^3, p_2^3)$, which is denoted as AMLP; see Fig. 9.6, lower right. This AMLP has the weight 4.1082. The "associated" set of critical edges S' equals $\{e_2, e_3, e_4, e_5\}$.

Table 9.1 Coordinates of endpoints of critical edges in Fig. 9.6

Critical edge	x_{i1}	y_{i1}	z_{i1}	x_{i2}	y_{i2}	z_{i2}
e_1	0.5	1	0.5	0.5	1	−0.5
e_2	−0.5	1	0.5	0.5	1	0.5
e_3	0.5	1	1.5	0.5	2	1.5
e_4	1.5	1	1.5	1.5	2	1.5
e_5	2.5	1	0.5	1.5	1	0.5
e_6	1.5	1	0.5	1.5	1	−0.5

For Line 21, apply the original rubberband algorithm on S' and AMLP to compute the minimum-weight directed cycle $\rho = (q_2, q_3, q_4, q_5, q_2)$; see Fig. 9.6, lower right. Cycle ρ has weight 4. (Actually, it is the true MLP of this simple cube-curve.)   □

For Algorithm 33, Lines 1–5 can be preprocessed. The main computation occurs in Line 9, which is done in $\mathcal{O}(m^2n^2)$ time where n is the number of critical edges and m the number of subdivision points on each critical edge.[3] In Line 7, $|N(v)| \leq mn$. Thus, we have that the time complexity of Lines 7–15 is $\mathcal{O}(m^3n^3)$. Since $|V| \leq mn$, it follows that the time complexity of Lines 6–16 is $\mathcal{O}(m^4n^4)$. Since $|V(\text{AMLP})| \leq n$, Lines 18–20 can be computed in $\mathcal{O}(n^2)$.

By Theorem 9.3, Line 21 runs in time $\kappa(\varepsilon) \cdot \mathcal{O}(n)$, where $\kappa(\varepsilon)$ is as in Theorem 9.3. Therefore, the time complexity of Algorithm 33 is $\mathcal{O}(m^4n^4 + \kappa(\varepsilon) \cdot n)$.

Let L^* be the length of the approximate MLP computed by Algorithm 33. By Theorem 9.2, we have that

$$0 < L^* - L(P) < 12nM_1\delta/M_2.$$

This implies that

$$0 < L^* < L(P) + 12nM_1\delta/M_2 = \left(1 + \frac{12M_1}{M_2L(P)}n\delta\right)L(P)$$

where M_1, M_2, n, and δ are as in Theorem 9.2. In other words, Algorithm 33 is a δ'-approximation, where

$$\delta' = 1 + \frac{12M_1}{M_2L(P)}n\delta.$$

Obviously, we may select m to be sufficiently large, say, $m > n^2$. It follows that $\delta < 1/n^2$, and we have

$$\delta' < 1 + \frac{12M_1}{M_2L(P)}n \cdot \frac{1}{n^2} = 1 + \frac{12M_1}{M_2L(P)} \cdot \frac{1}{n}.$$

Therefore, mathematically, we can take a sufficiently large value of m to obtain the approximate MLP of g with good accuracy.

Example 9.2 However, in practise, our experiments for $N \leq 106$ and $n \leq 31$ showed that $m = 5$ is "quite sufficient", with a running time of about 173.1 seconds; see Table 9.2 for a Pentium 4 PC using Matlab 7.04.   □

Following Lemma 9.2, if it were possible to find better bounds than M_1 and M_2, then this would result in a smaller value of m.

---

[3]See, for example, [17], pages 595–601.

Table 9.2 Time complexity for Algorithm 33, where N is the number of cubes, n the number of critical edges, and m the number of subdivision points on each critical edge; the time is in seconds (on a Pentium 4 PC using Matlab 7.04)

N	n	m	Time	AMLP	MLP
20	9	5	8.6	11.30	11.24
20	9	5	9.5	11.28	11.24
20	10	5	14.7	10.20	10.15
32	8	5	5.8	22.57	22.51
38	12	5	16.4	26.40	26.39
48	15	5	32.3	34.73	34.71
48	15	5	29.4	33.61	33.55
74	20	5	51.0	55.66	55.57
76	20	5	55.8	56.46	56.38
70	21	5	59.0	50.26	50.15
88	27	5	110.4	63.94	63.89
84	27	5	116.1	61.22	61.14
90	31	5	173.1	65.47	65.44
106	30	5	140.3	79.95	79.85

In summary, we presented a δ'-approximation algorithm for computing the MLP of a general simple cube-curve, where

$$\delta' = 1 + \frac{12M_1}{M_2 L(P)} n\delta$$

M_1, M_2, n, and δ are as in Theorem 9.2. It runs in time $\mathcal{O}(m^4 n^4 + \kappa(\varepsilon) \cdot n)$, where $\kappa(\varepsilon)$ is as in Theorem 9.3, m is the chosen number of possible subdivision points on any of the critical edges, and n is the number of critical edges. We recall also a theorem of algorithmic graph theory:

Theorem 9.4 *Assume as input a directed graph G. The all-pairs shortest paths problem can be solved in $\mathcal{O}(n^3 (\log\log n / \log n)^{5/7})$ time, where n is the number of vertices of G.*

Since the graph G constructed in the algorithm has mn vertices, by Theorem 9.4, the time complexity can be reduced to

$$\mathcal{O}\left(m^3 n^3 \left[\frac{\log\log(mn)}{\log(mn)}\right]^{5/7} + \kappa(\varepsilon) \cdot n\right),$$

but certainly not to κ-linear.

This (slow) graph-theoretical algorithm allowed in 2005 for the first time to evaluate results obtained by the original RBA in comparison to provably approximate MLPs.

9.4 MLPs of Decomposable Simple Cube-Curves

This section develops a provably correct MLP algorithm for a special type of first-class simple cube-curves which can be decomposed (in a particular way) into a finite number of simple cube-arcs. (The next section will also show that the formulas derived for these simple cube-arcs are important for a correctness proof for the original rubberband algorithm, considering the same special type of cube-curves.)

In this section, we analyse the "geometric structure" of simple cube-curves. By introducing the concept of an "end-angle" (see Definition 9.4 below), we will be able to decompose a special first-class simple cube-curve into finite simple cube-arcs in a unique way. (In other words, the tube of such a special first-class simple cube-curve is the union of the tubes of these simple cube-arcs.) We focus on such a special subset of the family of all first-class simple cube-curves[4] which have at least one end-angle (e.g., as the cube-curve shown in Fig. 9.2). We also study a non-trivial example of such a curve and show that the path calculated by the original rubberband algorithm is converging toward the MLP. We also present an algorithm for the computation of approximate MLPs for the special class (as defined in this section) of simple cube-curves.

Definition 9.4 Assume a simple cube-curve g and a triple of consecutive critical edges e_1, e_2, and e_3 such that $e_i \perp e_j$, for $i, j = 1, 2, 3$ and $i \neq j$. Let σ be a metavariable for x, y, or z. If e_2 is parallel to the σ-axis and the σ-coordinates of e_1 and e_3 are identical (for $\sigma = x$, $\sigma = y$, or $\sigma = z$), then we say that e_1, e_2, and e_3 form an *end-angle*.

Cube-curve g *has an end-angle*, denoted by $\sphericalangle(e_1, e_2, e_3)$, in such a case. Otherwise we say that e_1, e_2, and e_3 form a *middle-angle* $\sphericalangle(e_1, e_2, e_3)$, and g *has a middle-angle* in this case. In particular, if e_2 is parallel to the σ-axis and the σ-coordinates of e_1 and e_3 are not identical (for $\sigma = x$, $\sigma = y$, or $\sigma = z$), then we say that e_1, e_2, and e_3 form an *inner-angle*.

Example 9.3 Figure 9.2 shows a simple cube-curve which has five end-angles $\sphericalangle(e_{21}, e_0, e_1)$, $\sphericalangle(e_4, e_5, e_6)$, $\sphericalangle(e_6, e_7, e_8)$, $\sphericalangle(e_{14}, e_{15}, e_{16})$, and $\sphericalangle(e_{15}, e_{16}, e_{17})$, and 17 middle-angles, such as $\sphericalangle(e_0, e_1, e_2)$, $\sphericalangle(e_1, e_2, e_3)$, or $\sphericalangle(e_2, e_3, e_4)$, and so on, where 11 of them are inner-angles: $\sphericalangle(e_0, e_1, e_2)$, $\sphericalangle(e_1, e_2, e_3)$, $\sphericalangle(e_2, e_3, e_4)$,

---

[4]Note that we can classify a simple cube-curve in linear time to be first-class or not, by using the original rubberband algorithm: the curve is first-class iff option (O_2) does not occur at all.

Fig. 9.7 Point $p_i(t_i)$ on the
critical line γ_i which is
incident with critical edge e_i.
Point (x_i, y_i, z_i) is an end
point of e_i

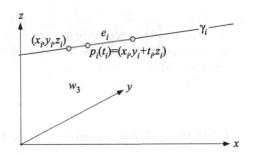

$\vartriangleleft(e_3, e_4, e_5)$, $\vartriangleleft(e_5, e_6, e_7)$, $\vartriangleleft(e_7, e_8, e_9)$, $\vartriangleleft(e_8, e_9, e_{10})$, $\vartriangleleft(e_9, e_{10}, e_{11})$, $\vartriangleleft(e_{12}, e_{13}, e_{14})$, $\vartriangleleft(e_{13}, e_{14}, e_{15})$, and $\vartriangleleft(e_{16}, e_{17}, e_{18})$, and the other six of them are non-inner-angles. $\qquad\Box$

Definition 9.5 A straight line l which is incident with a critical edge e of g (i.e., $e \subset l$) is called a *critical line* (of e in g).

A critical line l and a critical edge e are *corresponding* if they are incident (i.e., $e \subset l$).

Definition 9.6 Let $S \subseteq \mathbb{R}^3$. The set $\{(x, y, 0) : \exists z \ (z \in \mathbb{R} \wedge (x, y, z) \in S)\}$ is the *xy-projection* of S, or *projection* of S for short. Analogously, we define the *yz-* or *xz-projection* of S.

The next lemma is a basic result from computational geometry, and we already cited it (in different form) in Chap. 3:

Lemma 9.3 *There is a uniquely defined shortest path which passes through subsequent critical edges $e_1, e_2, \ldots$, and e_k in this order.*

Let e_1, e_2, and e_3 be three (not necessarily consecutive) critical edges in a simple cube-curve, and let l_1, l_2, and l_3 be the corresponding three critical lines. As a general approach in the remainder of this book, we consider points on critical lines in parameterised form, allowing to study changes in their positions by means of derivatives.

For example, point $p_2(t_2) = (x_2 + k_{x_2}t_2, y_2 + k_{y_2}t_2, z_2 + k_{z_2}t_2)$ on l_2 is parameterised by $t_2 \in \mathbb{R}$. Analogously, $p_1(t_1)$ or $p_3(t_3)$ on l_1 or l_3 are parameterised by t_1 or t_3, respectively. See Fig. 9.7.

Lemma 9.4 *Let $d_2(t_1, t_2, t_3) = d_e(p_1(t_1), p_2(t_2)) + d_e(p_2(t_2), p_3(t_3))$, for arbitrary reals t_1, t_2, and t_3. It follows that $\partial^2 d_2/\partial t_2^2 > 0$.*

Proof Let the coordinates of $p_i(t_i)$ be $(x_i + k_{x_i}t_i, y_i + k_{y_i}t_i, z_i + k_{z_i}t_i)$, where i equals 1 or 3.

Assume that $p_i \in e_i \subset l_i$ (for $i = 1, 2, 3$), where e_i is an edge of the regular 3D orthogonal grid. It follows that only one of the values k_{x_i}, k_{y_i}, or k_{z_i} can be equal to 1; the other two must be equal to zero.

Let us look at one of these cases where the coordinates of p_1 are $(x_1 + t_1, y_1, z_1)$, the coordinates of p_2 are $(x_2, y_2 + t_2, z_2)$, and the coordinates of p_3 are $(x_3, y_3, z_3 + t_3)$. Then we have that $d_2(t_1, t_2, t_3) = d_e(p_1(t_1), p_2(t_2)) + d_e(p_2(t_2), p_3(t_3))$ equals

$$\sqrt{(t_2 - (y_1 - y_2))^2 + (x_1 + t_1 - x_2)^2 + (z_1 - z_2)^2}$$
$$+ \sqrt{(t_2 - (y_3 - y_2))^2 + (x_3 - x_2)^2 + (z_3 + t_3 - z_2)^2}.$$

This can be briefly written as $d_2 = \sqrt{(t_2 - a_1)^2 + b_1^2} + \sqrt{(t_2 - a_2)^2 + b_2^2}$, where b_1 and b_2 are functions of t_1 and t_3, respectively. Then we have that

$$\frac{\partial d_2}{\partial t_2} = \frac{t_2 - a_1}{\sqrt{(t_2 - a_1)^2 + b_1^2}} + \frac{t_2 - a_2}{\sqrt{(t_2 - a_2)^2 + b_2^2}} \tag{9.1}$$

and

$$\frac{\partial^2 d_2}{\partial t_2^2} = \frac{1}{\sqrt{(t_2 - a_1)^2 + b_1^2}} - \frac{(t_2 - a_1)^2}{[(t_2 - a_1)^2 + b_1^2]^{3/2}}$$
$$+ \frac{1}{\sqrt{(t_2 - a_2)^2 + b_2^2}} - \frac{(t_2 - a_2)^2}{[(t_2 - a_2)^2 + b_2^2]^{3/2}}.$$

This simplifies to

$$\frac{\partial^2 d_2}{\partial t_2^2} = \frac{b_1^2}{[(t_2 - a_1)^2 + b_1^2]^{3/2}} + \frac{b_2^2}{[(t_2 - a_2)^2 + b_2^2]^{3/2}} > 0. \tag{9.2}$$

The other cases of positions of p_i lead, analogously, also to such a positive second derivative.   □

With this lemma we introduce the very useful (at least, for this book) concept that points $p_i(t_i)$ are studied with respect to their derivatives. This approach is followed repeatedly in the remainder of this book.

Note that we do not use $p_i(t)$ because different critical edges may have different values of t representing varying speed of movements along these edges (e.g., when adjusting the length of a curve which is passing through critical edges).

Lemma 9.4 is used in the proof of the following lemma. Let e_1, e_2, and e_3 be three critical edges (again: not necessarily consecutive critical edges), and let l_1, l_2, and l_3

Fig. 9.8 Illustration of
Case 1 in the proof of
Lemma 9.6

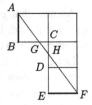

be their corresponding critical lines, respectively. Let p_1, p_2, and p_3 be three points such that p_i belongs to l_i, for $i = 1, 2, 3$. Let $p_2 = (x_2 + k_{x_2}t_2, y_2 + k_{y_2}t_2, z_2 + k_{z_2}t_2)$. Let $d_2 = d_e(p_1, p_2) + d_e(p_2, p_3)$.

Lemma 9.5 *The function* $f(t_2) = \frac{\partial d_2}{\partial t_2}$ *has a unique real root.*

Proof We refer to the proof of Lemma 9.4. Without loss of generality, we can assume that $a_1 \leq a_2$. Then, by Eq. (9.1), we have that $f(a_1) \leq 0$ as well as $f(a_2) \geq 0$. The lemma follows with Eq. (9.2). □

We use this lemma for describing our next algorithm further below.

To continue our theoretical preparations, now let e_1, e_2, and e_3 be three consecutive (!) critical edges of a simple cube-curve g.

Definition 9.7 Let $D(e_1, e_2, e_3)$ be the dimension of the linear space generated by edges e_1, e_2, and e_3.

Let l_{13} be a line segment with one end point (somewhere) on e_1, and the other end point (somewhere) on e_3. Let $d_{e_i e_j}$ be the Hausdorff distance between e_i and e_j (i.e., the minimum of all Euclidean distances between a point p on e_i, and a point q on e_j), where $i, j = 1, 2, 3$.

Lemma 9.6 *The line segment* l_{13} *(for any choice of endpoints) is not completely contained in the tube* **g** *if* $D(e_1, e_2, e_3) = 3$, $\min\{d_{e_1 e_2}, d_{e_2 e_3}\} \geq 1$, *and* $\max\{d_{e_1 e_2}, d_{e_2 e_3}\} \geq 2$, *or if* $D(e_1, e_2, e_3) \leq 2$ *and* $\min\{d_{e_1 e_2}, d_{e_2 e_3}\} \geq 2$.

Proof Case 1. Let $D(e_1, e_2, e_3) = 3$, $\min\{d_{e_1 e_2}, d_{e_2 e_3}\} \geq 1$ and $\max\{d_{e_1 e_2}, d_{e_2 e_3}\} \geq 2$. We only need to prove that the conclusion is true when $\min\{d_{e_1 e_2}, d_{e_2 e_3}\} = 1$ and $\max\{d_{e_1 e_2}, d_{e_2 e_3}\} = 2$. In this case, the parallel projection [denoted by $g'(e_1, e_2, e_3)$] of all of g's cubes contained between e_1 and e_3 is illustrated in Fig. 9.8, where AB is the projective image of e_1, EF that of e_3, and C that of one of the end points of e_2. Note that line segment AF must intercept grid edge BC at a point G, and intercept grid edge CD at a point H. Also note that line segment GH is not completely contained in $g'(e_1, e_2, e_3)$. Therefore, if l_{13} is a line segment with one end point on e_1 and the other one on e_3, then l_{13} is not completely contained in **g**.

Case 2. Let $D(e_1, e_2, e_3) = 2$ and $\min\{d_{e_1 e_2}, d_{e_2 e_3}\} \geq 2$. Without loss of generality, we can assume that $e_1 \parallel e_2$.

Fig. 9.9 Illustration of Case 2.1 in the proof of Lemma 9.6

Fig. 9.10 Illustration of Case 2.2 in the proof of Lemma 9.6

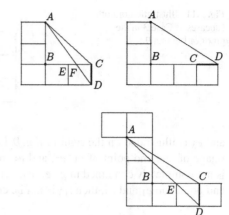

Case 2.1. e_1 and e_2 are on the same grid line; we only need to prove that the conclusion is true when $d_{e_1e_2} = 2$ and $d_{e_2e_3} = 2$. In this case, the projective image [denoted by $g'(e_1, e_2, e_3)$] of all of g's cubes contained between e_1 and e_3 is illustrated in Fig. 9.9.

Case 2.1.1. $g'(e_1, e_2, e_3)$ is as on the left in Fig. 9.9, where A and B are the projective images of either one end point of e_1 or e_2, respectively, and CD that of e_3. Note that line segment AD must intercept grid edge EC at a point F. Also note that line segments AD and AC are not completely contained in $g'(e_1, e_2, e_3)$. Therefore, if l_{13} is a line segment where one end point is on e_1, and the other on e_3, then l_{13} is not completely contained in g. Similarly, we can show that the conclusion is also true for Case 2.1.2, with $g'(e_1, e_2, e_3)$ as illustrated on the right in Fig. 9.9.

Case 2.2. Assume that e_1 and e_2 are on different grid lines. We only need to prove that the conclusion is true when $d_{e_1e_2} = \sqrt{5}$ and $d_{e_2e_3} = 2$. In this case, the projective image [denoted by $g'(e_1, e_2, e_3)$] of all of g's cubes contained between e_1 and e_3 is illustrated in Fig. 9.10, where A (B) is the projective image of one end point of e_1 (e_2), and CD that of e_3. Note that line segment AD must intercept grid edge EC at a point E. Also note that line segments AD and AC are not completely contained in $g'(e_1, e_2, e_3)$. Therefore, if l_{13} is a line segment with one end point on e_1, and one on e_3, then l_{13} is not completely contained in g.

Case 3. Let $D(e_1, e_2, e_3) = 1$ and $\min\{d_{e_1e_2}, d_{e_2e_3}\} \geq 2$. Without loss of generality, we can assume that $e_1 \parallel e_2$.

Case 3.1. e_1 and e_2 are on the same grid line. We only need to prove that the conclusion is true when $d_{e_1e_2} = 2$ and $d_{e_2e_3} = 2$. In this case, the projective image [denoted by $g'(e_1, e_2, e_3)$] of all of g's cubes contained between e_1 and e_3 is illustrated on the left of Fig. 9.11, where A, B, and C are projective images of one end point of e_1, e_2, and e_3, respectively. Note that line segment AC is not completely contained in $g'(e_1, e_2, e_3)$. Therefore, if l_{13} is a line segment with an end point on e_1 and another one on e_3, then l_{13} is not completely contained in g.

Case 3.2. Now assume that e_1 and e_2 are on different grid lines. We only need to prove that the conclusion is true when $d_{e_1e_2} = \sqrt{5}$ and $d_{e_2e_3} = 2$. In this case, the projective image [denoted by $g'(e_1, e_2, e_3)$] of all of g's cubes contained between e_1

Fig. 9.11 Illustration of both
subcases of Case 3 in the
proof of Lemma 9.6

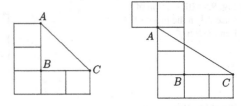

and e_3 is illustrated on the right in Fig. 9.11, where A, B, and C are the projective
image of one end point of e_1, e_2, and e_3, respectively. Note that line segment AC
is not completely contained in $g'(e_1, e_2, e_3)$. Therefore, if l_{13} is a line segment with
end points on e_1 and e_3, then l_{13} is not be completely contained in **g**.   □

This lemma and the following Lemmas 9.7 and 9.8 are used to prove the following Theorem 9.5 which states a sufficient condition for the first-class simple
cube-curve.

Let g be a simple cube-curve such that any three consecutive critical edges
e_1, e_2, and e_3 do indeed satisfy either $D(e_1, e_2, e_3) = 3$, $\min\{d_{e_1e_2}, d_{e_2e_3}\} \geq 1$,
and $\max\{d_{e_1e_2}, d_{e_2e_3}\} \geq 2$, or $D(e_1, e_2, e_3) \leq 2$ and $\min\{d_{e_1e_2}, d_{e_2e_3}\} \geq 2$. By
Lemma 9.6, we immediately obtain:

Lemma 9.7 *Every critical edge of g contains at least one vertex of g's MLP.*

Let g be a simple cube-curve, and assume that every critical edge of g contains
at least one vertex of the MLP. Then we also have the following:

Lemma 9.8 *Every critical edge of g contains at most one vertex of g's MLP.*

Proof Assume that there exists a critical edge e such that e contains at least two
vertices v and w of the MLP P of g. Without loss of generality, we can assume
that v and w are the first (in the order on P) two vertices which are on e. Let u be a
vertex of P, which is on the previous critical edge of P. Then line segments uv and
uw are completely contained in **g**.

By replacing $\{uv, uw\}$ by vw we obtain a polygon of length shorter than P,
which is in contradiction to the fact that P is an MLP of g.   □

Let g be a simple cube-curve such that any three consecutive critical edges
e_1, e_2, and e_3 do indeed satisfy either $D(e_1, e_2, e_3) = 3$, $\min\{d_{e_1e_2}, d_{e_2e_3}\} \geq 1$
and $\max\{d_{e_1e_2}, d_{e_2e_3}\} \geq 2$, or $D(e_1, e_2, e_3) \leq 2$ and $\min\{d_{e_1e_2}, d_{e_2e_3}\} \geq 2$. By Lemmas 9.7 and 9.8, we immediately obtain:

Theorem 9.5 *The specified simple cube-curve g is first-class.*

Let e_1, e_2, and e_3 be three consecutive critical edges of a simple cube-curve g. Let
p_1, p_2, and p_3 be three points such that $p_i \in e_i$, for $i = 1, 2, 3$. Let the coordinates

of p_i be $(x_i + k_{x_i}t_i, y_2 + k_{y_i}t_i, z_i + k_{z_i}t_i)$, where $k_{x_i}, k_{y_i}, k_{z_i}$ are either 0 or 1, and $0 \le t_i \le 1$, for $i = 1, 2, 3$. Let $d_2 = d_e(p_1, p_2) + d_e(p_2, p_3)$.

Theorem 9.6 $\frac{\partial d_2}{\partial t_2} = 0$ *implies that we have one of the following representations for* t_3: *we can have*

$$t_3 = \frac{-c_2 t_1 + (c_1 + c_2)t_2}{c_1} \tag{9.3}$$

if $c_1 > 0$; *we can also have*

$$t_3 = 1 - \sqrt{\frac{c_1^2(t_2 - a_2)^2}{(t_2 - t_1)^2} - c_2^2} \quad or \tag{9.4}$$

$$t_3 = \sqrt{\frac{c_1^2(t_2 - a_2)^2}{(t_2 - t_1)^2} - c_2^2} \tag{9.5}$$

if a_2 *equals either 0 or 1, and* c_1 *and* c_2 *are positive; and we can also have*

$$t_3 = 1 - \sqrt{\frac{(t_2 - a_2)^2[(t_1 - a_1)^2 + c_1^2]}{(t_2 - b_1)^2} - c_2^2} \quad or \tag{9.6}$$

$$t_3 = \sqrt{\frac{(t_2 - a_2)^2[(t_1 - a_1)^2 + c_1^2]}{(t_2 - b_1)^2} - c_2^2} \tag{9.7}$$

if $a_1, a_2,$ *and* b_1 *are either equal to 0 or 1, and* c_1 *and* c_2 *are positive reals.*

Proof We have that $p_i = (x_i + k_{x_i}t_i, y_i + k_{y_i}t_i, z_i + k_{z_i}t_i)$, with $k_{x_i}, k_{y_i}, k_{z_i}$ equals 0 or 1, and $0 \le t_i \le 1$, for $i = 1, 2, 3$. Note that only one of the values $k_{x_i}, k_{y_i}, k_{z_i}$ can be 1, and the other two must be equal to 0. It follows that for every $i, j \in \{1, 2, 3\}$, $d_e(p_i, p_j) = \sqrt{(t_j - t_i)^2 + c^2}$ or $\sqrt{(t_i - a)^2 + (t_j - b)^2 + c^2}$, where a, b are equal to 0 or 1, and $c > 0$. We have $c \neq 0$ because otherwise e_1 and e_2 would be on the same line, and that is impossible. Let $d_2 = d_e(p_1, p_2) + d_e(p_2, p_3)$. The following three cases are possible:

Case 1. $d_2 = \sqrt{(t_2 - t_1)^2 + c_1^2} + \sqrt{(t_2 - t_3)^2 + c_2^2}$, with $c_i > 0$, for $i = 1, 2$. Then we have

$$\frac{\partial d_2}{\partial t_2} = \frac{t_2 - t_1}{\sqrt{(t_2 - t_1)^2 + c_1^2}} + \frac{t_2 - t_3}{\sqrt{(t_2 - t_3)^2 + c_2^2}},$$

and equation $\frac{\partial d_2}{\partial t_2} = 0$ implies the form of Eq. (9.3).

Case 2. $d_2 = \sqrt{(t_2 - t_1)^2 + c_1^2} + \sqrt{(t_2 - a_2)^2 + (t_3 - b_2)^2 + c_2^2}$, with a_2, b_2 equals 0 or 1, and $c_i > 0$, for $i = 1, 2$. Then we have

$$\frac{\partial d_2}{\partial t_2} = \frac{t_2 - t_1}{\sqrt{(t_2 - t_1)^2 + c_1^2}} + \frac{t_2 - a_2}{\sqrt{(t_2 - a_2)^2 + (t_3 - b_2)^2 + c_2^2}},$$

and equation $\frac{\partial d_2}{\partial t_2} = 0$ implies the form of Eqs. (9.4) or (9.5).

Case 3. $d_2 = \sqrt{(t_2 - a_1)^2 + (t_1 - b_1)^2 + c_1^2} + \sqrt{(t_2 - a_2)^2 + (t_3 - b_2)^2 + c_2^2}$, with a_i, b_i equals 0 or 1, and $c_i > 0$, for $i = 1, 2$. Then we have

$$\frac{\partial d_2}{\partial t_2} = \frac{t_2 - a_1}{\sqrt{(t_2 - a_1)^2 + (t_1 - b_1)^2 + c_1^2}} + \frac{t_2 - a_2}{\sqrt{(t_2 - a_2)^2 + (t_3 - b_2)^2 + c_2^2}},$$

and equation $\frac{\partial d_2}{\partial t_2} = 0$ implies the form of Eqs. (9.6) or (9.7). $\square$

The proof of Case 3 of Theorem 9.6, and Lemma 9.6 also show the following:

Lemma 9.9 *Let g be a first-class simple cube-curve. If $e_1, e_2,$ and e_3 form a middle-angle of g then the vertex of the MLP of g on e_2 cannot be an endpoint (i.e., a grid point) on e_2.*

This lemma provides interesting information about the relationship between locations of vertices of an MLP and the geometric structure of the given simple cube-curve.

Lemma 9.10 *Let $f(x)$ be a continuous function defined on an interval $[a, b]$, and assume $f(\xi) = 0$ for some $\xi \in (a, b)$. Then, for every $\varepsilon > 0$, there exist a' and b' such that, for every $x \in [a', b']$, we have $|f(x)| < \varepsilon$.*

Proof Since $f(x)$ is continuous at $\xi \in (a, b)$, so $\lim_{n \to \xi} f(x) = f(\xi) = 0$. Then, for every $\varepsilon > 0$, there exists a $\delta > 0$ such that for every $x \in (\xi - \delta, \xi + \delta)$ we have that $|f(x)| < \varepsilon$. Let $a' = \xi - \frac{\delta}{2}$ and $b' = \xi + \frac{\delta}{2}$. Then, for every $x \in [a', b']$ we have that $|f(x)| < \varepsilon$. $\square$

We apply this auxiliary lemma to prove the following lemma:

Lemma 9.11 *Let $f(x)$ be a continuous function on an interval $[a, b]$, with $f(\xi) = 0$ at $\xi \in (a, b)$. Then, for every $\varepsilon > 0$, there are two integers $n > 0$ and $k > 0$ such that, for every $x \in [\frac{(k-1)(b-a)}{n}, \frac{k(b-a)}{n}]$, we have that $|f(x)| < \varepsilon$.*

Proof By Lemma 9.10, it follows that for every $\varepsilon > 0$, there exist a' and b' such that for every $x \in [a', b']$ we have that $|f(x)| < \varepsilon$. Select an integer $n \geq \frac{2(b-a)}{b'-a'}$. Then, $\frac{b-a}{n} \leq \frac{b'-a'}{2} \leq b' - a'$. Thus, there is an integer j (where $j = 1, 2, \ldots, n - 1$)

such that $a' \leq \frac{j(b-a)}{n} \leq b'$. If $\frac{j(b-a)}{n} \leq \frac{b'-a'}{2}$, then $a' \leq \frac{j(b-a)}{n} \leq \frac{(j+1)(b-a)}{n} \leq b'$. If $\frac{j(b-a)}{n} \geq \frac{b'-a'}{2}$, then $a' \leq \frac{(j-1)(b-a)}{n} \leq \frac{j(b-a)}{n} \leq b'$. $\qquad\square$

In the following, we discuss main ideas and operations of a numerical algorithm for computing the MLP of a first-class simple cube-curve which has at least one end-angle.

Subscripts are taken modulo $n + 1$. Let p_i be a point on e_i, where $i = 0, 1, 2, \ldots, n$. Let the coordinates of p_i be

$$(x_i + k_{x_i} t_i, y_2 + k_{y_i} t_i, z_i + k_{z_i} t_i)$$

where $i = 0, 1, \ldots, n$. Then the length of the polygon $p_0 p_1 \ldots p_n$ equals

$$d = d(t_0, t_1, \ldots, t_n) = \sum_{i=0}^{n} d_e(p_i, p_{i+1}).$$

If the polygon $p_0 p_1 \ldots p_n$ is the MLP of g, then (by Lemma 9.3) we have that $\frac{\partial d}{\partial t_i} = 0$, where $i = 0, 1, \ldots, n$.

Assume that e_i, e_{i+1}, and e_{i+2} form an end-angle, and also e_j, e_{j+1}, and e_{j+2}, and that no other three consecutive critical edges between e_{i+2} and e_j form an end-angle, where $i \leq j$ and $i, j = 0, 1, 2, \ldots, n$. By Theorem 9.6, we have

$$t_{i+3} = f_{i+3}(t_{i+1}, t_{i+2}),$$

$$t_{i+4} = f_{i+4}(t_{i+2}, t_{i+3}),$$

$$t_{i+5} = f_{i+5}(t_{i+3}, t_{i+4}),$$

$$\ldots$$

$$t_j = f_j(t_{j-2}, t_{j-1}),$$

and

$$t_{j+1} = f_{j+1}(t_{j-1}, t_j).$$

This shows that $t_{i+3}, t_{i+4}, t_{i+5}, \ldots, t_j$, and t_{j+1} can be represented by t_{i+1}, and t_{i+2}. In particular, we obtain an equation $t_{j+1} = f(t_{i+1}, t_{i+2})$, or

$$g(t_{j+1}, t_{i+1}, t_{i+2}) = 0, \tag{9.8}$$

where t_{j+1}, and t_{i+1} are already known, or

$$g_1(t_{i+2}) = 0. \tag{9.9}$$

Since e_i, e_{i+1}, and e_{i+2} form an end-angle it follows that $e_{i+1} \perp e_{i+2}$. By the proof of Theorem 9.6, we can express $\frac{\partial d_2}{\partial t_{i+2}}$ in the form

$$\frac{t_{i+2} - b_1}{\sqrt{(t_{i+1} - a_1)^2 + (t_{i+2} - b_1)^2 + c_1^2}} + \frac{t_{i+2} - a_2}{\sqrt{(t_{i+2} - a_2)^2 + (t_{i+3} - b_2)^2 + c_2^2}}. \quad (9.10)$$

Then Eq. (9.10) has a unique real root between a_2 and b_1. In other words, there are two real numbers a and b such that Eq. (9.10) has a unique root in between a and b. If $g_1(a)g_1(b) < 0$, then we can use a bisection method[5] to find an approximate root of Eq. (9.10). Otherwise, by Lemma 9.11, we can also find an approximate root of Eq. (9.10). Therefore, we can always find an approximate root for $\frac{\partial d}{\partial t_k} = 0$, where $k = i + 2, i + 3, \ldots$, and j, and an exact root for $\frac{\partial d}{\partial t_k} = 0$, where $k = i + 1$ and $j + 1$.

In this way, we find an approximate or exact root t_{k_0} for $\frac{\partial d}{\partial t_k} = 0$, where $k = 0, 1, 2, \ldots, n$. Let $t'_{k_0} = 0$ if $t_{k_0} < 0$ and $t'_{k_0} = 1$ if $t_{k_0} > 1$, where $k = 0, 1, 2, \ldots, n$. Then (by Theorem 9.5) we obtain an approximation of the MLP [its length equals $d(t'_{0_0}, t'_{1_0}, t'_{2_0}, \ldots, t'_{i_0}, \ldots, t'_{n_0})$] of the given first-class simple cube-curve.

For the numerical algorithm in Fig. 9.12, the input is a first-class simple cube-curve g with at least one end-angle. The output is an approximation of the MLP and a calculated length value. The operations of this Algorithm 34 are presented in this figure at a sufficient level of detail for proper understanding.

We give an estimate of the time complexity of Algorithm 34 depending on the number of end-angles m and accuracy parameter ε.

Let the accuracy of approximation be upper-limited by $\varepsilon = \frac{1}{2^k}$. The applied bisection method[6] needs to know initial endpoints a and b of the search interval $[a, b]$.

At best, if we can set $a = 0$ and $b = 1$ to solve all the forms of Eq. (9.9) by the bisection method. Then the algorithm completes each run in $O(mk)$ time.

In the worst case, if we have to find out the values of a and b for every of the forms of Eq. (9.9) by the bisection method. Then (by Lemma 9.11, and assuming that we need $f(k)$ operations to find out the values of a and b), the algorithm completes each run in $O(mk \cdot f(k))$ time.

Example 9.4 We provide an example where we compare the results obtained with Algorithm 34 with those obtained with the Bülow–Klette algorithm. We approximate the MLP of the first-class simple cube-curve shown in Fig. 9.2 by following the defined lines of operations in Algorithm 34:

Line 1. See Table 9.3 which lists the coordinates of the critical edges $e_0, e_1, \ldots$, e_{21} of g. Let p_i be a point on the critical line of e_i, where $i = 0, 1, \ldots, 21$.

Line 2. We calculate the coordinates of p_i, where $i = 0, 1, \ldots 21$, as follows: $(1 + t_0, 4, 7), (2, 4 + t_1, 5), (4, 5, 4 + t_2), (4 + t_3, 7, 4), (5, 7 + t_4, 2), (7, 09, 1 + t_5)$, $\ldots, (2, 2, 7 + t_{21})$.

---

[5]See, for example, [12, page 49].

[6]See [12, page 49].

Algorithm 34 (A numerical MLP approximation for cube-curves with an end-angle)
Input: A first-class simple cube-curve g with at least one end-angle.
Output: An approximation of the MLP and a calculated length value.

1: Represent g by the coordinates of the endpoints of its critical edges e_i, where
   $i = 0, 1, 2, \ldots, n$. Let p_i be a point on e_i, where $i = 0, 1, 2, \ldots, n$. Then, the
   coordinates of p_i are equal to $(x_i + k_{x_i} t_i, y_i + k_{y_i} t_i, z_i + k_{z_i} t_i)$, where only one
   of the parameters k_{x_i}, k_{y_i} and k_{z_i} can be equal to 1, and the other two are equal
   to 0, for $i = 0, 1, \ldots, n$.
2: Find all end-angles $\sphericalangle(e_j, e_{j+1}, e_{j+2}), \sphericalangle(e_k, e_{k+1}, e_{k+2}), \ldots$ of g. For every $i \in$
   $\{0, 1, 2, \ldots, n\}$, let $d_{i+1} = d_e(p_i, p_{i+1}) + d_e(p_{i+1}, p_{i+2})$. By Lemma 9.5, we
   can find a unique root $t_{(i+1)_0}$ of equation $\frac{\partial d_{i+1}}{\partial t_{i+1}} = 0$ if e_i, e_{i+1} and e_{i+2} form an
   end-angle.
3: For every pair of two consecutive end-angles

$$\sphericalangle(e_i, e_{i+1}, e_{i+2}) \quad \text{and} \quad \sphericalangle(e_j, e_{j+1}, e_{j+2})$$

   of g, apply the ideas as described in Lemma 9.10 to find the root of equation
   $\frac{\partial d_k}{\partial t_k} = 0$, where $k = i+1, i+2, \ldots$, and $j+1$.
4: Repeat Line 3 until we find an approximate or exact root t_{k_0} for $\frac{\partial d}{\partial t_k} = 0$, where
   $d = d(t_0, t_1, \ldots, t_n) = \sum_{i=1}^{n-1} d_i$, for $k = 0, 1, 2, \ldots, n$. Let $t'_{k_0} = 0$ if $t_{k_0} < 0$,
   and $t'_{k_0} = 1$ if $t_{k_0} > 1$, for $k = 0, 1, 2, \ldots, n$.
5: The output is a polygonal curve $p_0(t'_{0_0}) p_1(t'_{1_0}) \ldots p_n(t'_{n_0})$ of total length

$$d\left(t'_{0_0}, t'_{1_0}, \ldots, t'_{i_0}, \ldots, t'_{n_0}\right).$$

This curve approximates the MLP of g.

Fig. 9.12 Algorithm 34: A numerical MLP approximation for first-class simple cube-curves with
an end-angle

Line 3. Now let $d = d(t_0, t_1, \ldots, t_{21}) = \sum_{i=0}^{21} d_e(p_i, p_{i+1 \pmod{22}})$. Then we obtain

$$\frac{\partial d}{\partial t_0} = \frac{t_0 - 1}{\sqrt{(t_0 - 1)^2 + t_{21}^2 + 4}} + \frac{t_0 - 1}{\sqrt{(t_0 - 1)^2 + t_1^2 + 4}}, \tag{9.11}$$

$$\frac{\partial d}{\partial t_1} = \frac{t_1}{\sqrt{(t_0 - 1)^2 + t_1^2 + 4}} + \frac{t_1 - 1}{\sqrt{(t_1 - 1)^2 + (t_2 - 1)^2 + 4}}, \tag{9.12}$$

$$\frac{\partial d}{\partial t_2} = \frac{t_2 - 1}{\sqrt{(t_1 - 1)^2 + (t_2 - 1)^2 + 4}} + \frac{t_2}{\sqrt{t_2^2 + t_3^2 + 4}}, \tag{9.13}$$

Table 9.3 Coordinates of endpoints of critical edges in Fig. 9.2

Critical edge	x_{i1}	y_{i1}	z_{i1}	x_{i2}	y_{i2}	z_{i2}
e_0	1	4	7	2	4	7
e_1	2	4	5	2	5	5
e_2	4	5	4	4	5	5
e_3	4	7	4	5	7	4
e_4	5	7	2	5	8	2
e_5	7	8	1	7	8	2
e_6	7	10	2	8	10	2
e_7	8	10	4	8	11	4
e_8	10	10	4	10	10	5
e_9	10	8	5	11	8	5
e_{10}	11	7	7	11	8	7
e_{11}	12	7	7	12	7	8
e_{12}	12	5	7	12	5	8
e_{13}	10	4	8	10	5	8
e_{14}	9	4	10	10	4	10
e_{15}	9	2	10	9	2	11
e_{16}	7	1	10	7	2	10
e_{17}	6	2	8	7	2	8
e_{18}	6	4	7	6	4	8
e_{19}	4	4	7	4	4	8
e_{20}	3	2	7	3	2	8
e_{21}	2	2	7	2	2	8

$$\frac{\partial d}{\partial t_3} = \frac{t_3}{\sqrt{t_2^2 + t_3^2 + 4}} + \frac{t_3 - 1}{\sqrt{(t_3 - 1)^2 + t_4^2 + 4}}, \qquad (9.14)$$

$$\frac{\partial d}{\partial t_4} = \frac{t_4}{\sqrt{(t_3 - 1)^2 + t_4^2 + 4}} + \frac{t_4 - 1}{\sqrt{(t_4 - 1)^2 + (t_5 - 1)^2 + 4}}, \qquad (9.15)$$

and

$$\frac{\partial d}{\partial t_5} = \frac{t_5 - 1}{\sqrt{(t_4 - 1)^2 + (t_5 - 1)^2 + 4}} + \frac{t_5 - 1}{\sqrt{(t_5 - 1)^2 + t_6^2 + 4}}. \qquad (9.16)$$

By Eqs. (9.11) and (9.16), we obtain that $t_0 = t_5 = 1$.

Similarly, we have $t_7 = t_{15} = 0$, and $t_{16} = 1$. Therefore, we find all end-angles as follows: $\sphericalangle(e_{21}, e_0, e_1)$, $\sphericalangle(e_4, e_5, e_6)$, $\sphericalangle(e_6, e_7, e_8)$, $\sphericalangle(e_{14}, e_{15}, e_{16})$, and $\sphericalangle(e_{15}, e_{16}, e_{17})$.

By Theorem 9.6 and Eqs. (9.12–9.14), it follows that

$$t_2 = 1 - \sqrt{\frac{(t_1 - 1)^2[(t_0 - 1)^2 + 4]}{t_1^2} - 4}, \qquad (9.17)$$

$$t_3 = \sqrt{\frac{t_2^2[(t_1 - 1)^2 + 4]}{(t_2 - 1)^2} - 4}, \qquad (9.18)$$

and

$$t_4 = \sqrt{\frac{(t_3 - 1)^2[t_2^2 + 4]}{t_3^2} - 4}. \qquad (9.19)$$

By Eq. (9.15), we have

$$t_4^2\big[(t_5 - 1)^2 + 4\big] = (t_4 - 1)^2\big[(t_3 - 1)^2 + 4\big].$$

Let

$$g_1(t_1) = t_4^2\big[(t_5 - 1)^2 + 4\big] - (t_4 - 1)^2\big[(t_3 - 1)^2 + 4\big]. \qquad (9.20)$$

By Eq. (9.17), we have that $t_1 \in (0, 0.5)$, $g_1(0.4924) = 3.72978 > 0$, and also $g_1(0.4999) = -51.2303 < 0$. By Lemmas 9.3, 9.5, 9.9, and the bisection method, we obtain the following unique roots of Eqs. (9.17–9.20):

$$t_1 = 0.492416, \qquad t_2 = 0.499769, \qquad t_3 = 0.499769, \quad \text{and} \quad t_4 = 0.507584,$$

with error $g_1(t_1) = 4.59444 \times 10^{-9}$. These roots correspond to the two consecutive end-angles $\sphericalangle(e_{21}, e_0, e_1)$ and $\sphericalangle(e_4, e_5, e_6)$ of g.

Line 4. Similarly, we find the unique roots of equation $\frac{\partial d}{\partial t_i} = 0$, where $i = 6, 7, \ldots, 21$. At first we have $t_6 = 0.5$, which corresponds to the two consecutive end-angles $\sphericalangle(e_4, e_5, e_6)$ and $\sphericalangle(e_6, e_7, e_8)$; then we also obtain $t_8 = 0.492582$, $t_9 = 0.494543$, $t_{10} = 0.331074$, $t_{11} = 0.205970$, $t_{12} = 0.597034$, $t_{13} = 0.502831$, $t_{14} = 0.492339$, which correspond to the two consecutive end-angles $\sphericalangle(e_6, e_7, e_8)$ and $\sphericalangle(e_{14}, e_{15}, e_{16})$; followed by $t_{15} = 0$, $t_{16} = 1$, which correspond to the two consecutive end-angles $\sphericalangle(e_{14}, e_{15}, e_{16})$ and $\sphericalangle(e_{15}, e_{16}, e_{17})$; and finally, $t_{17} = 0.501527$, $t_{18} = 0.77824$, $t_{19} = 0.56314$, $t_{20} = 0.32265$, and $t_{21} = 0.2151$, which correspond to the two consecutive end-angles $\sphericalangle(e_{15}, e_{16}, e_{17})$ and $\sphericalangle(e_{21}, e_0, e_1)$.

Line 5. In summary, we obtain the values shown in the first two columns of Table 9.4. The calculated approximation of the MLP of g is

$$p_0(t'_{0_0}) p_1(t'_{1_0}) \cdots p_n(t'_{n_0}),$$

and its length is $d(t'_{0_0}, t'_{1_0}, \ldots, t'_{i_0}, \ldots, t'_{n_0}) = 43.767726$, where $t'_{i_0} = t_{i_0}$ for i limited to the set $\{0, 1, 2, \ldots, 21\}$.

The original RBA calculated the roots of Eqs. (9.11) through (9.16) as shown in the third column of Table 9.4. Note that there is only a finite number of iterations until the algorithm terminates (i.e., no threshold needs to be specified for the chosen

Table 9.4 Comparison of
results of both algorithms

Critical points	t_{i_0} (Algorithm 34)	t_{i_0} (Original RBA)
p_0	1	1
p_1	0.492416	0.4924
p_2	0.499769	0.4998
p_3	0.499769	0.4998
p_4	0.507584	0.5076
p_5	1	1
p_6	0.5	0.5
p_7	0	0
p_8	0.492582	0.4926
p_9	0.494543	0.4945
p_{10}	0.331074	0.3311
p_{11}	0.205970	0.2060
p_{12}	0.597034	0.5970
p_{13}	0.502831	0.5028
p_{14}	0.492339	0.4923
p_{15}	0	0
p_{16}	1	1
p_{17}	0.501527	0.5015
p_{18}	0.77824	0.7789
p_{19}	0.56314	0.5641
p_{20}	0.32265	0.3235
p_{21}	0.2151	0.2157

input curve). From Table 9.4 we can see that both algorithms converge (within some
minor numerical deviations) to 'basically' identical values.   □

In summary, this section reported a provably correct algorithm for the approximate calculation of an MLP for a special class of simple cube-curves
(namely, first-class simple cube-curves with at least one end-angle).

Mathematically, the problem is equivalent to solving equations having one variable each. Applying methods of numerical analysis, we can compute their roots with
sufficient accuracy. We illustrated the algorithm by one non-trivial example, illustrating this way that our algorithm found the same approximation of an MLP as the
original RBA did.

9.5 Analysis of the Original RBA

This section still focuses on first-class simple cube-curves; general simple cube-curves are discussed in Sect. 9.6. This section proves that the Bülow–Klette algorithm has κ-linear time complexity $\kappa(\varepsilon) \cdot \mathcal{O}(m)$ for such cube-curves, where $\kappa(\varepsilon)$ is as defined in Theorem 9.3, and m is the number of critical edges of a given simple cube-curve.

This answered in 2007 a question for this algorithm, which had been open until then. However, in the course of analysing this time complexity, counterexamples were found to Option 2 of the original RBA (for non-first-class simple cube-curves) which led to a correction of Option 3 (by adding one missing test) of the original RBA.

These counterexamples were found by studying the question:[7] *Is there a simple cube-curve such that none of the nodes of its MLP is a grid vertex?* This section constructs an example of such a simple cube-curve, and we also characterise the class of all of such cube-curves.

The basic importance of this class of cube-curves without any end-angle is that it shows the need of further algorithmic studies: for such cube-curves we cannot use the MLP algorithm proposed in Sect. 9.4 which is provably correct.

We start with two rather technical definitions used in this section:

Definition 9.8 Let e be a critical edge of g. Let P_1 and P_2 be the two end points of e. If one of the coordinates of P_1 is less than the corresponding coordinate of P_2, then P_1 is called the *first end point of e*, otherwise P_1 is called the *second end point of e*.

Let $e_1, e_2, \ldots, e_m$ be a subsequence of the sequence of all consecutive critical edges $\ldots, e_0, e_1, \ldots, e_m, e_{m+1}, \ldots$ of a cube-curve g. Let $m \geq 2$.

Definition 9.9 If $e_0 \perp e_1$, $e_m \perp e_{m+1}$, and $e_i \parallel e_{i+1}$, where $i = 1, 2, \ldots, m-1$, then $e_1, e_2, \ldots, e_m$ is a *maximal run of parallel critical edges* of g, and critical edges e_0 or e_{m+1} are called *adjacent to* this run.

Example 9.5 Figure 9.2 shows a simple cube-curve which has two maximal runs of parallel critical edges: e_{11}, e_{12} and $e_{18}, e_{19}, e_{20}, e_{21}$. The two adjacent critical edges of run e_{11}, e_{12} are e_{10} and e_{13}; they are on two different grid planes. The two adjacent critical edges of run $e_{18}, e_{19}, e_{20}, e_{21}$ are e_{17} and e_0; they are also on two different grid planes.   □

---

[7]Formulated as an open problem on [33, page 406].

Fig. 9.13 Illustration for the
proof of Lemma 9.12

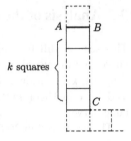

Let $e_0, e_1, e_2, \ldots, e_m$, and e_{m+1} be $m+2$ consecutive critical edges in a simple
cube-curve, and let $l_0, l_1, l_2, \ldots, l_m$, and l_{m+1} be the corresponding critical lines.
We express a point $p_i(t_i) = (x_i + k_{x_i}t_i, y_i + k_{y_i}t_i, z_i + k_{z_i}t_i)$ on l_i in general form,
with $t_i \in \mathbb{R}$, where i equals $0, 1, \ldots,$ or $m+1$.

In the following, $p_i(t_i)$ will be denoted by p_i for short, where i equals $0, 1, \ldots,$
or $m+1$.

Lemma 9.12 *If $e_1 \perp e_2$, then $\partial d_e(p_1, p_2)/\partial t_2$ can be written as $(t_2 - \alpha)\beta$, where
$\beta > 0$ is a function of t_1 and t_2, and $\alpha = 0$ if e_1 and the first end point of e_2 are on
the same grid plane, or $\alpha = 1$ otherwise.*

Proof Without loss of generality, we can assume that e_2 is parallel to the z-axis. In
this case, the parallel projection [denoted by $g'(e_1, e_2)$] of all of g's cubes, contained
between e_1 and e_2, is illustrated in Fig. 9.13, where AB is the projective image of e_1,
and C is that of one of the end points of e_2.

Case 1. e_1 and the first end point of e_2 are on the same grid plane. Let the two
end points of e_2 be (a, b, c) and $(a, b, c+1)$. Then the two end points of e_1 are
$(a-1, b+k, c)$ and $(a, b+k, c)$. Then the coordinates of p_1 and p_2 are $(a-1+
t_1, b+k, c)$ and $(a, b, c+t_2)$, respectively, and $d_e(p_1, p_2) = \sqrt{(t_1 - 1)^2 + k^2 + t_2^2}$.
Therefore,

$$\frac{\partial d_e(p_1, p_2)}{\partial t_2} = \frac{t_2}{\sqrt{(t_1 - 1)^2 + k^2 + t_2^2}}.$$

Let $\alpha = 0$ and

$$\beta = \frac{1}{\sqrt{(t_1 - 1)^2 + k^2 + t_2^2}}.$$

This proves the lemma for Case 1.

Case 2. Now assume that e_1 and the first end point of e_2 are on different grid
planes (i.e., e_1 and the second end point of e_2 are on the same grid plane). Let
the two end points of e_2 be (a, b, c) and $(a, b, c+1)$. Then the two end points of
e_1 are $(a-1, b+k, c+1)$ and $(a, b+k, c+1)$. Then the coordinates of p_1 and
p_2 are $(a-1+t_1, b+k, c+1)$ and $(a, b, c+t_2)$, respectively, and $d_e(p_1, p_2) =$

Fig. 9.14 Illustration for the proof of Lemma 9.13. *Left*: Case 1. *Right*: Case 2

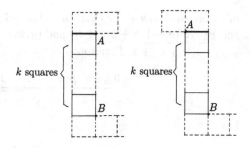

$\sqrt{(t_1 - 1)^2 + k^2 + (t_2 - 1)^2}$. Therefore,

$$\frac{\partial d_e(p_1, p_2)}{\partial t_2} = \frac{t_2 - 1}{\sqrt{(t_1 - 1)^2 + k^2 + (t_2 - 1)^2}}.$$

Let $\alpha = 1$ and

$$\beta = \frac{1}{\sqrt{(t_1 - 1)^2 + k^2 + (t_2 - 1)^2}}.$$

This proves the lemma for Case 2.   □

Lemma 9.13 *If $e_1 \parallel e_2$, then*

$$\frac{\partial d_e(p_1, p_2)}{\partial t_2} = (t_2 - t_1)\beta$$

for some $\beta > 0$, where β is a function in t_1 and t_2.

Proof Without loss of generality, we can assume that e_2 is parallel to the z-axis. In this case, the parallel projection [denoted by $g'(e_1, e_2)$] of all of g's cubes contained between e_1 and e_2 is illustrated in Fig. 9.14, where A is the projective image of one of the end points of e_1, and B is that of one of the end points of e_2.

Case 1. Edges e_1 and e_2 are on the same grid plane. Let the two end points of e_2 be (a, b, c) and $(a, b, c + 1)$. Then the two end points of e_1 are $(a, b + k, c)$ and $(a, b + k, c + 1)$. Then the coordinates of p_1 and p_2 are $(a, b + k, c + t_1)$ and $(a, b, c + t_2)$, respectively, and $d_e(p_1, p_2) = \sqrt{(t_2 - t_1)^2 + k^2}$. Therefore,

$$\frac{\partial d_e(p_1, p_2)}{\partial t_2} = \frac{t_2 - t_1}{\sqrt{(t_2 - t_1)^2 + k^2}}.$$

Let

$$\beta = \frac{1}{\sqrt{(t_2 - t_1)^2 + k^2}}.$$

This proves the lemma for Case 1.

Case 2. Now assume that edges e_1 and e_2 are on different grid planes. Let the two end points of e_2 be (a, b, c) and $(a, b, c + 1)$. Then the two end points of e_1 are $(a-1, b+k, c)$ and $(a-1, b+k, c+1)$. Then the coordinates of p_1 and p_2 are $(a-1, b+k, c+t_1)$ and $(a, b, c+t_2)$, respectively, and $d_e(p_1, p_2) = \sqrt{(t_2-t_1)^2 + k^2 + 1}$. Therefore,

$$\frac{\partial d_e(p_1, p_2)}{\partial t_2} = \frac{t_2 - t_1}{\sqrt{(t_2-t_1)^2 + k^2 + 1}}.$$

Let

$$\beta = \frac{1}{\sqrt{(t_2-t_1)^2 + k^2 + 1}}.$$

This proves the lemma for Case 2.   □

This lemma will be used later for the proof of Lemma 9.17.

Let $d_i = d_e(p_{i-1}, p_i) + d_e(p_i, p_{i+1})$, for $i = 1, 2, \ldots, m$.

Theorem 9.7 *If $e_i \perp e_j$, where $i, j = 1, 2, 3$ and $i \neq j$, then $e_1, e_2,$ and e_3 form an end-angle iff the equation*

$$\frac{\partial(d_e(p_1, p_2) + d_e(p_2, p_3))}{\partial t_2} = 0$$

has a unique root 0 or 1.

Proof Without loss of generality, we can assume that e_2 is parallel to the z-axis.

(A) If $e_1, e_2,$ and e_3 form an end-angle, then by Definition 9.4, the z-coordinates of two end points of e_1 and e_3 are equal.

Case A1. Edges $e_1, e_3,$ and the first end point of e_2 are on the same grid plane. By Lemma 9.12,

$$\frac{\partial d_e(p_1, p_2)}{\partial t_2} = (t_2 - \alpha_1)\beta_1$$

where $\alpha_1 = 0$ and $\beta_1 > 0$, and

$$\frac{\partial d_e(p_2, p_3)}{\partial t_2} = (t_2 - \alpha_2)\beta_2$$

where $\alpha_2 = 0$ and $\beta_2 > 0$. Thus, we have

$$\frac{\partial(d_e(p_1, p_2) + d_e(p_2, p_3))}{\partial t_2} = t_2(\beta_1 + \beta_2).$$

Therefore, the equation of the theorem has the unique root $t_2 = 0$.

Case A2. Edges $e_1, e_3,$ and the second end point of e_2 are on the same grid plane. By Lemma 9.12,

$$\frac{\partial d_e(p_1, p_2)}{\partial t_2} = (t_2 - \alpha_1)\beta_1$$

where $\alpha_1 = 1$ and $\beta_1 > 0$, and

$$\frac{\partial d_e(p_2, p_3)}{\partial t_2} = (t_2 - \alpha_2)\beta_2$$

where $\alpha_2 = 1$ and $\beta_2 > 0$. Thus, we have

$$\frac{\partial(d_e(p_1, p_2) + d_e(p_2, p_3))}{\partial t_2} = (t_2 - 1)(\beta_1 + \beta_2).$$

Therefore, the equation of the theorem has the unique root $t_2 = 1$.

(B) Conversely, if the equation of the theorem has a unique root 0 or 1, then e_1, e_2, and e_3 form an end-angle. Otherwise, e_1, e_2, and e_3 form a middle angle. By Definition 9.4, the z-coordinates of two end points of e_1 are not equal to z-coordinates of two end points of e_3. (Note: Without loss of generality, we can assume that $e_2 \parallel z$-axis.) So e_1 and e_3 are not on the same grid plane.

Case B1. Edge e_1 and the first end point of e_2 are on the same grid plane, while e_3 and the second end point of e_2 are on the same grid plane. By Lemma 9.12,

$$\frac{\partial d_e(p_1, p_2)}{\partial t_2} = (t_2 - \alpha_1)\beta_1$$

where $\alpha_1 = 0$ and $\beta_1 > 0$, while

$$\frac{\partial d_e(p_2, p_3)}{\partial t_2} = (t_2 - \alpha_2)\beta_2$$

where $\alpha_2 = 1$ and $\beta_2 > 0$. Thus, we have

$$\frac{\partial(d_e(p_1, p_2) + d_e(p_2, p_3))}{\partial t_2} = t_2\beta_1 + (t_2 - 1)\beta_2.$$

Therefore, $t_2 = 0$ or 1 is not a root of the equation of the theorem. This is a contradiction.

Case B2. Edge e_1 and the second end point of e_2 are on the same grid plane, while e_3 and the first end point of e_2 are on the same grid plane. By Lemma 9.12,

$$\frac{\partial d_e(p_1, p_2)}{\partial t_2} = (t_2 - \alpha_1)\beta_1$$

where $\alpha_1 = 1$ and $\beta_1 > 0$, while

$$\frac{\partial d_e(p_2, p_3)}{\partial t_2} = (t_2 - \alpha_2)\beta_2$$

where $\alpha_2 = 0$ and $\beta_2 > 0$. Thus, we have

$$\frac{\partial(d_e(p_1, p_2) + d_e(p_2, p_3))}{\partial t_2} = (t_2 - 1)\beta_1 + t_2\beta_2.$$

Therefore, $t_2 = 0$ or 1 is not a root of the equation of the theorem. This is a contradiction as well. $\qquad\square$

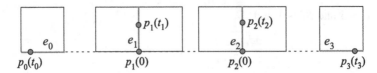

Fig. 9.15 Illustration of the proof of Lemma 9.14

Theorem 9.8 *If $e_i \perp e_j$, where $i, j = 1, 2, 3$ and $i \neq j$, then e_1, e_2, and e_3 form an inner-angle iff the equation*

$$\frac{\partial (d_e(p_1, p_2) + d_e(p_2, p_3))}{\partial t_2} = 0$$

has a root t_{2_0} such that $0 < t_{2_0} < 1$.

Proof If edges e_1, e_2, and e_3 form an inner-angle, then by Definition 9.4, e_1, e_2, and e_3 do not form an end-angle. By Theorem 9.7, 0 or 1 is not a root of the equation given in the theorem. By Lemma 9.12, we have

$$\frac{\partial (d_e(p_1, p_2) + d_e(p_2, p_3))}{\partial t_2} = (t_2 - \alpha_1)\beta_1 + (t_2 - \alpha_2)\beta_2$$

where α_1, α_2 are 0 or 1, $\beta_1 > 0$ is a function of t_1 and t_2, and $\beta_2 > 0$ is a function of t_2 and t_3. Thus, $\alpha_1 \neq \alpha_2$ (i.e., $\alpha_1 = 0$ and $\alpha_2 = 1$ or $\alpha_1 = 1$ and $\alpha_2 = 0$). Therefore, the equation of the theorem has a root t_{2_0} such that $0 < t_{2_0} < 1$.

Conversely, if the equation of the theorem has a root t_{2_0} such that $0 < t_{2_0} < 1$ then, by Theorem 9.7, critical edges e_1, e_2, and e_3 do not form an end-angle. By Definition 9.4, e_1, e_2, and e_3 do form an inner-angle. □

Assume that $e_0 \perp e_1$, $e_2 \perp e_3$, and $e_1 \parallel e_2$. Assume that $p(t_{i_0})$ is a vertex of the MLP of g, where $i = 1$ or $i = 2$. Then we have the following:

Lemma 9.14 *If e_0, e_3, and the first end point of e_1 are on the same grid plane, and t_{i_0} is a root of*

$$\frac{\partial d_i}{\partial t_i} = 0,$$

then $t_{i_0} = 0$, where $i = 1$ or $i = 2$.

Proof From $p_0(t_0) p_1(0) \perp e_1$ it follows that

$$d_e(p_0(t_0) p_1(0)) = \min\{d_e(p_0(t_0), p_1(t_1)) : t_1 \in [0, 1]\}.$$

See Fig. 9.15. Analogously, we have

$$d_e(p_2(0) p_3(t_3)) = \min\{d_e(p_2(t_2), p_3(t_3)) : t_2 \in [0, 1]\}$$

and

$$d_e\big(p_1(0)p_2(0)\big) = \min\{d_e\big(p_1(t_1), p_2(t_2)\big) : t_1, t_2 \in [0, 1]\}.$$

Therefore, we have

$$\min\{d_e\big(p_0(t_0), p_1(t_1)\big) + d_e\big(p_1(t_1), p_2(t_2)\big) + d_e\big(p_2(t_2), p_3(t_3)\big) : t_1, t_2 \in [0, 1]\}$$
$$\geq d_e\big(p_0(t_0), p_1(0)\big) + d_e\big(p_1(0), p_2(0)\big) + d_e\big(p_2(0), p_3(t_3)\big).$$

This proves the lemma. $\qquad\square$

Assume that we have $e_0 \perp e_1$, $e_m \perp e_{m+1}$, and $e_i \parallel e_{i+1}$ (i.e., the set $\{e_1, e_2, \ldots, e_m\}$ is a maximal run of parallel critical edges of g, and e_0 or e_{m+1} is the adjacent critical edge of this set). Furthermore, let $p(t_{i_0})$ be a vertex of the MLP of g, where $i = 1, 2, \ldots, m$. Analogously to the previous lemma, we also have the following two lemmas:

Lemma 9.15 *If e_0, e_{m+1}, and the first point of e_1 are on the same grid plane, and t_{i_0} is a root of*

$$\frac{\partial d_i}{\partial t_i} = 0,$$

then $t_{i_0} = 0$, where $i = 1, 2, \ldots, m$.

Lemma 9.16 *If e_0, e_{m+1}, and the second end point of e_1 are on the same grid plane, and t_{i_0} is a root of*

$$\frac{\partial d_i}{\partial t_i} = 0,$$

then $t_{i_0} = 1$, where $i = 1, 2, \ldots, m$.

Now we study the case that critical edges are on different grid planes. (Note that even two parallel edges can be on different grid planes.)

Lemma 9.17 *If e_0 and e_{m+1} are on different grid planes, and t_{i_0} is a root of*

$$\frac{\partial d_i}{\partial t_i} = 0,$$

where $i = 1, 2, \ldots, m$, then $0 < t_1 < t_2 < \cdots < t_m < 1$.

Proof Assume that e_0 and the first end point of e_1 are on the same grid plane, and e_{m+1} and the second end point of e_1 are on the same grid plane. Then (by Lemmas 9.12 and 9.13), the derivatives $\frac{\partial d_i}{\partial t_i}$, where $i = 1, 2, \ldots, m$, have the following forms:

$$\frac{\partial d_1}{\partial t_1} = t_1 b_{1_1} + (t_1 - t_2)b_{1_2},$$

$$\frac{\partial d_2}{\partial t_2} = (t_2 - t_1)b_{2_1} + (t_2 - t_3)b_{2_2},$$

$$\frac{\partial d_3}{\partial t_3} = (t_3 - t_2)b_{3_1} + (t_3 - t_4)b_{3_2},$$

$$\cdots \tag{9.21}$$

$$\frac{\partial d_{m-1}}{\partial t_{m-1}} = (t_{m-1} - t_{m-2})b_{m-1_1} + (t_{m-1} - t_m)b_{m-1_2}, \quad \text{and}$$

$$\frac{\partial d_m}{\partial t_m} = (t_m - t_{m-1})b_{m_1} + (t_m - 1)b_{m_2},$$

where $b_{i_1} > 0$, b_{i_1} is a function of t_i and t_{i-1}, $b_{i_2} > 0$, and b_{i_2} is a function of t_i and t_{i+1}, for $i = 1, 2, \ldots, m$.

If $t_{1_0} < 0$, then (due to $\frac{\partial d_1}{\partial t_1} = 0$) we have that $t_{1_0} b_{1_1} + (t_{1_0} - t_{2_0})b_{1_2} = 0$. Since $b_{1_1} > 0$ and $b_{1_2} > 0$, we also have $t_{1_0} - t_{2_0} > 0$ (i.e., $t_{1_0} > t_{2_0}$).

Analogously, because of $\frac{\partial d_2}{\partial t_2} = 0$ we have $(t_{2_0} - t_{1_0})b_{2_1} + (t_{2_0} - t_{3_0})b_{2_2} = 0$. This means that we also have $t_{2_0} > t_{3_0}$.

Analogously we can also verify that $t_{3_0} > t_{4_0}, \ldots,$ and $t_{m-1_0} > t_{m_0}$. Therefore, by Eq. (9.21), we have $t_{m_0} - 1 > 0$. Altogether we have $0 > t_{1_0} > t_{2_0} > t_{3_0} > \cdots > t_{m_0} > 1$. This is an obvious contradiction.

If $t_{1_0} = 0$, then (since $\frac{\partial d_1}{\partial t_1} = 0$) we have that $t_{2_0} = 0$. Analogously, $\frac{\partial d_2}{\partial t_2} = 0$ implies $t_{3_0} = 0$, and we also have $t_{4_0} = 0, \ldots, t_{m_0} = 0$ due to the same argument. But, by Eq. (9.21), we have

$$\frac{\partial d_m}{\partial t_m} = (t_m - 1)b_{m_2} = -b_{m_2} < 0.$$

This contradicts $\frac{\partial d_m}{\partial t_m} = 0$.

If $t_{1_0} \geq 1$, then (due to $\frac{\partial d_1}{\partial t_1} = 0$) we have $t_{1_0} b_{1_1} + (t_{1_0} - t_{2_0})b_{1_2} = 0$. Due to $b_{1_1} > 0$ and $b_{1_2} > 0$ we have $t_{1_0} - t_{2_0} < 0$ (i.e., $t_{1_0} < t_{2_0}$). Analogously, from $\frac{\partial d_2}{\partial t_2} = 0$ it follows that $(t_{2_0} - t_{1_0})b_{2_1} + (t_{2_0} - t_{3_0})b_{2_2} = 0$. Then we have $t_{2_0} < t_{3_0}$, and we also have $t_{3_0} < t_{4_0}, \ldots, t_{m-1_0} < t_{m_0}$. Therefore, by Eq. (9.21), we have $t_{m_0} - 1 < 0$. Altogether we have $1 \leq t_{1_0} < t_{2_0} < t_{3_0} < \cdots < t_{m_0} < 1$, which is again an obvious contradiction. □

Let t_{i_0} be a root of $\frac{\partial d_i}{\partial t_i} = 0$, where $i = 1, 2, \ldots, m$. We apply Lemmas 9.15, 9.16, and 9.17 and obtain

Theorem 9.9 *Edges* e_0 *and* e_{m+1} *are on different grid planes iff* $0 < t_{1_0} < t_{2_0} < \cdots < t_{m_0} < 1$.

Fig. 9.16 Illustration for
Lemma 9.19

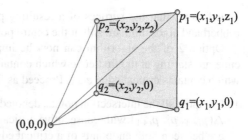

The previous three theorems characterise end-angles, inner-angles, and non-coplanar critical edges. This knowledge will be useful for verifying a necessary correction of Option 3 of the original RBA while analysing its time complexity in the following.

The initial path of the original RBA is defined by midpoints of critical edges.

Let p_1, p_2 be points on a critical edge e_i of curve g, and p a point on a critical edge e_j of g.

Lemma 9.18 *If the line segments pp_1, pp_2 are contained and complete in tube $\mathbf{g}$, then the triangular region $\triangle(p_1, p_2, p)$ is also contained and complete in $\mathbf{g}$.*

Proof Without loss of generality, we can assume that $i < j$. Let $a(e_i, e_j)$ be the arc from the first cube which contains the critical edge e_i to the last cube which contains the critical edge e_j. (Note that a set of consecutive critical edges will uniquely define a cube-curve.) If line segments pp_1, pp_2 are contained and complete in $\mathbf{g}$, then the xy- (yz- and xz-) projection of $\triangle(p_1, p_2, p)$ is contained and complete in the xy- (yz- and xz-) projection of $a(e_i, e_j)$. Therefore, the triangular region $\triangle(p_1, p_2, p)$ is contained and complete in the tube of $a(e_i, e_j)$.   □

Let $O = (0, 0, 0)$, and let $q_i(x_i, y_i, 0)$ be the projection of $p_i(x_i, y_i, z_i)$ onto the xy-plane, where $i = 1, 2$; see Fig. 9.16.

Lemma 9.19 *If q_2 is on the left of Oq_1 then p_2 is on the left of Op_1.*

Proof This follows because (see Fig. 9.16) $\triangle Op_1p_2$ can be obtained by continuously moving q_i toward p_i, where $i = 1, 2$.   □

Lemma 9.20 *Option 2 of the original rubberband algorithm (see Sect. 9.2) can be computed in $\mathcal{O}(m)$ time, where m is the number of critical edges intersected by the polygonal path between p_{i-1} and p_{i+1}.*

Proof We start with vertices of the initial polygon at the centre points of all critical edges of the given cube-curve (as defined for the initial path of the original RBA).

It follows that the vertices of a resulting polygon, using only Option 1 of the rubberband algorithm, are still at the centre points of critical edges.

Option 2 of the algorithm can now be implemented as follows: Let $\mathcal{A}$ be the cube-arc starting at the first cube which contains critical edge e_{i-1}, to the last cube which contains critical edge e_{i+1}. Proceed as follows:

1. Compute all the intersection points, denoted by S_I, of the closed triangular region $\triangle(p_{i-1}, p_i, p_{i+1})$ with consecutive critical edges from e_{i-1} to e_{i+1} (note: they are between both endpoints of a critical edge). This can be computed in $\mathcal{O}(m_1)$ time, where $m_1 = |S_I| \leq$ is the number of critical edges in $\mathcal{A}$.
2. Let S_P be the set of three planes: xy-plane, yz-plane, and zx-plane. Select a plane $\alpha \in S_P$, such that α is not perpendicular to $\triangle(p_{i-1}, p_i, p_{i+1})$. This can be computed in $\mathcal{O}(1)$ time,
3. Project S_I onto α. Let the resulting set be S_I'.
4. Apply the Melkman algorithm (i.e., Algorithm 15 which has linear time complexity, see [58]) to find the convex arc, denoted by $\mathcal{A}'$ in α.
5. By Lemma 9.19 (the projection of the convex hull of S_I onto α is the convex hull of S_I'), compute a convex arc, denoted by $\mathcal{A}''$, in $\triangle(p_{i-1}, p_i, p_{i+1})$ such that $\mathcal{A}'$ is the projection of $\mathcal{A}''$ onto α.
6. If each edge uw of $\mathcal{A}''$ is fully contained in the tube g, then $\mathcal{A}''$ is the required shortest convex arc from p_{i-1} to p_{i+1}. Otherwise, do not replace the arc from p_{i-1} to p_{i+1}.

Altogether, it follows that the convex arc can be computed in $\mathcal{O}(m)$ time, where m is the number of critical edges intersecting the arc between p_{i-1} and p_{i+1}. □

Let $\kappa(\varepsilon)$ be as in Theorem 9.3, and m is the number of critical edges of the tube **g**. Due to the analysis in Sect. 3.5, we also know the following:

Lemma 9.21 *Option 3 of the original RBA can be computed in $\kappa(\varepsilon) \cdot \mathcal{O}(m)$ time.*

Together with Lemma 9.20 (which implies that all operations in Option 2 of the original RBA can be computed in $\mathcal{O}(m)$ time), we have

Theorem 9.10 *The original RBA is κ-linear.*

In other words, the time complexity is $\kappa(\varepsilon) \cdot \mathcal{O}(m)$, where $\kappa(\varepsilon)$ is as in Theorem 9.3, and m is the number of critical edges of the given simple cube-curve.

Example 9.6 We provide an example to show (in generalisation of the example) that there are simple cube-curves such that none of the vertices of their MLPs is a grid vertex.[8] See Fig. 9.17 and Table 9.5 for an example of such a cube-curve, which lists

---

[8]This leads to two new open problems (smallest simple cube-curve without end-angle, and smallest simple cube-curve where none of the MLP vertices is a grid point). See Problem 9.7 at the end of

Fig. 9.17 A simple cube-curve such that none of the vertices of its MLP is a grid vertex

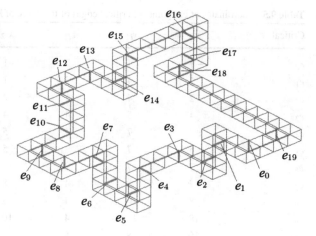

the coordinates of the critical edges $e_0, e_1, \ldots, e_{19}$ of g. Let $v(t_0), v(t_1), \ldots, v(t_{19})$ be the vertices of the MLP of g such that $v(t_i)$ is on e_i and t_i is in $[0, 1]$, where $i = 0, 1, 2, \ldots, 19$.

See the Appendix for a complete list of all $\frac{\partial d_i}{\partial t_i}$ (for $i = 0, 1, \ldots, 19$) for this cube-curve g. It follows that there is no end-angle in g, but we have six inner-angles, namely:

$$\sphericalangle(e_2, e_3, e_4), \qquad \sphericalangle(e_3, e_4, e_5), \qquad \sphericalangle(e_6, e_7, e_8),$$

$$\sphericalangle(e_9, e_{10}, e_{11}), \qquad \sphericalangle(e_{10}, e_{11}, e_{12}), \quad \text{and} \quad \sphericalangle(e_{13}, e_{14}, e_{15}).$$

By Theorem 9.8 we have that $t_3, t_4, t_7, t_{10}, t_{11}$, and t_{14} are all in the open interval $(0, 1)$. Figure 9.17 shows that $e_1 \parallel e_2$, and e_0 and e_3 are on different grid planes. By Theorem 9.9, it follows that t_1 and t_2 are in $(0, 1)$, too. Analogously we have that t_5 and t_6 are in $(0, 1)$, t_8 and t_9 are in $(0, 1)$, t_{12} and t_{13} are in $(0, 1)$, t_{15}, t_{16}, and t_{17} are in $(0, 1)$, and t_{18}, t_{19}, and t_0 are in $(0, 1)$. Therefore, each t_i is in the open interval $(0, 1)$, where $i = 0, 1, \ldots, 19$, which proves that g is a simple cube-curve such that none of the vertices of its MLP is a grid vertex.   □

Example 9.7 We discuss a first counterexample (a simple cube-curve) to Option 2 of the original RBA; see Fig. 9.18. Critical edges and centres of the cubes of this simple cube-curve, denoted by g, are shown in Tables 9.6 and 9.7, respectively. Table 9.8 shows the vertices of the final polygon obtained by original rubberband algorithm.

We discuss why the vertices of the final polygon are not those of the MLP, for curve g as shown in Fig. 9.19:

The cyan polyline $q_3 q_5 q_7 q_{11}$ is shorter than the red polyline $p_3 p_7 p_{11}$. However, according to Option 2, $p_3 p_7 p_{11}$ cannot be improved. In other words, limited by (the

---

this chapter. We consider that the second problem (i.e., all MLP vertices not at a grid point) is more difficult to solve.

Table 9.5 Coordinates of endpoints of critical edges of the curve of Fig. 9.17

Critical edge	x_{i1}	y_{i1}	z_{i1}	x_{i2}	y_{i2}	z_{i2}
e_0	−1	4	7	−1	4	8
e_1	1	4	7	1	5	7
e_2	2	4	5	2	5	5
e_3	4	5	4	4	5	5
e_4	4	7	4	5	7	4
e_5	5	7	2	5	8	2
e_6	7	7	2	7	8	2
e_7	7	8	4	8	8	4
e_8	8	10	4	8	10	5
e_9	10	10	4	10	10	5
e_{10}	10	8	5	11	8	5
e_{11}	11	7	7	11	8	7
e_{12}	12	7	7	12	7	8
e_{13}	12	5	7	12	5	8
e_{14}	10	4	8	10	5	8
e_{15}	9	4	10	10	4	10
e_{16}	9	0	10	10	0	10
e_{17}	9	0	8	10	0	8
e_{18}	9	1	7	9	1	8
e_{19}	−1	2	7	−1	2	8

Fig. 9.18 All critical edges of a simple cube-curve used for testing Option 2 of the original RBA; see Example 9.7

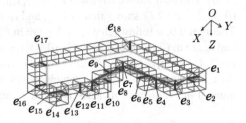

original) Option 2, we cannot use the shorter polyline $q_3q_5q_7q_{11}$ instead of polyline $p_3p_7p_{11}$. This is because the vertices q_3 and q_7 are not in the set of the intersection points between any critical edge in the set $\{e_i : i = 4, 5, 6, 7, 8, 9, 10\}$ and the closed triangular region $\triangle(p_3, p_7, p_{11})$.   □

Example 9.8 We also discuss a second counterexample (see Fig. 9.20) to Option 2 of the original RBA. The critical edges and centres of the cubes of this simple cube-curve, again denoted by g, are shown in Tables 9.9 and 9.10, respectively. Table 9.11 shows the vertices of the final polygon obtained when applying the original rubber-band algorithm.

Table 9.6 Coordinates of endpoints of critical edges of the curve of Fig. 9.18

Critical edge	x_{i1}	y_{i1}	z_{i1}	x_{i2}	y_{i2}	z_{i2}
e_1	−0.5	1	0.5	0.5	1	0.5
e_2	−0.5	1	0.5	−0.5	2	0.5
e_3	−1.5	1	0.5	−1.5	1	1.5
e_4	−2.5	0	0.5	−2.5	0	1.5
e_5	−2.5	−1	0.5	−2.5	0	0.5
e_6	−3.5	−1	0.5	−2.5	−1	0.5
e_7	−3.5	−3	−0.5	−3.5	−3	0.5
e_8	−3.5	−4	0.5	−3.5	−3	0.5
e_9	−4.5	−4	0.5	−4.5	−3	0.5
e_{10}	−4.5	−4	1.5	−4.5	−3	1.5
e_{11}	−5.5	−4	1.5	−5.5	−3	1.5
e_{12}	−5.5	−4	1.5	−5.5	−4	2.5
e_{13}	−6.5	−4	1.5	−6.5	−4	2.5
e_{14}	−7.5	−5	2.5	−7.5	−4	2.5
e_{15}	−8.5	−5	2.5	−7.5	−5	2.5
e_{16}	−8.5	−7	2.5	−7.5	−7	2.5
e_{17}	−7.5	−8	0.5	−7.5	−7	0.5
e_{18}	−0.5	−7	−0.5	−0.5	−7	0.5

Table 9.7 Centres of the cubes in the simple cube-curve shown in Fig. 9.18

i	(x_i, y_i, z_i)	i	(x_i, y_i, z_i)	i	(x_i, y_i, z_i)	i	(x_i, y_i, z_i)
1	$(0, -7.5, 0)$	12	$(-1, 1.5, 1)$	23	$(-5, -3.5, 1)$	34	$(-8, -7.5, 1)$
2	$(0, -6.5, 0)$	13	$(-2, 1.5, 1)$	24	$(-5, -3.5, 2)$	35	$(-8, -7.5, 0)$
3	$(0, -5.5, 0)$	14	$(-2, 0.5, 1)$	25	$(-6, -3.5, 2)$	36	$(-7, -7.5, 0)$
4	$(0, -4.5, 0)$	15	$(-2, -0.5, 1)$	26	$(-6, -4.5, 2)$	37	$(-6, -7.5, 0)$
5	$(0, -3.5, 0)$	16	$(-3, -0.5, 1)$	27	$(-7, -4.5, 2)$	38	$(-5, -7.5, 0)$
6	$(0, -2.5, 0)$	17	$(-3, -0.5, 0)$	28	$(-8, -4.5, 2)$	39	$(-4, -7.5, 0)$
7	$(0, -1.5, 0)$	18	$(-3, -1.5, 0)$	29	$(-8, -4.5, 3)$	40	$(-3, -7.5, 0)$
8	$(0, -0.5, 0)$	19	$(-3, -2.5, 0)$	30	$(-8, -5.5, 3)$	41	$(-2, -7.5, 0)$
9	$(0, 0.5, 0)$	20	$(-3, -3.5, 0)$	31	$(-8, -6.5, 3)$	42	$(-1, -7.5, 0)$
10	$(0, 1.5, 0)$	21	$(-4, -3.5, 0)$	32	$(-8, -7.5, 3)$	43	$(0, -7.5, 0)$
11	$(0, 1.5, 1)$	22	$(-4, -3.5, 1)$	33	$(-8, -7.5, 2)$	44	$(0, -6.5, 0)$

We discuss why the vertices of the final polygon are not those of the MLP of curve g shown in Fig. 9.20:

The cyan polyline $q_9 q_{10} q_{12} q_{14}$ is shorter than the red polyline $p_9 p_{11} p_{14}$. However, according to Option 2, $p_9 p_{11} p_{14}$ cannot be improved. In other words, by Op-

Table 9.8 An example	i	$p_i(x_i, y_i, z_i)$
output of the original RBA.		
i is the index of the critical		
edge e_i. Point p_i is a vertex	1	$(-0.5, 1, 0.5)$
(on e_i) of a polygon	3	$(-1.5, 1, 1)$
contained in the simple	7	$(-3.5, -3, 0)$
cube-curve shown in Fig. 9.18	11	$(-5.5, -4, 1.5)$
	15	$(-7.5, -5, 2.5)$
	16	$(-7.5, -7, 2.5)$
	17	$(-7.5, -7, 0.5)$
	18	$(-0.5, -7, 0.5)$

Fig. 9.19 Three critical
edges of the simple
cube-curve used for testing
Option 2 of the original RBA;
see Example 9.7

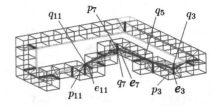

tion 2 we cannot identify the actual shorter polyline $q_9 q_{10} q_{12} q_{14}$ when considering the polyline $p_9 p_{11} p_{14}$. This is because the vertex q_{10} is not in the set of the intersection points between any critical edge in the set $\{e_i : i = 10, 11, 12, 13\}$ and the closed triangular region $\triangle(p_9, p_{11}, p_{14})$.   □

Example 9.9 Figure 9.21 shows a non-first-class simple cube-curve. The figure shows resulting polygons when applying the original RBA, or the revision of the original RBA (called *the revised RBA* for short; to be detailed below), respectively. At the top of the figure, edge $p(t_{9_0}) p(t_{13_0})$ is not contained in the tube **g** while $p(\bar{t}_{9_0}) p(\bar{t}_{13_0})$ is contained in it. The bottom of the figure shows the same polygons as at the top, but with all the cubes removed.

See Table 9.12 for the data of this curve shown in Fig. 9.21 and for the final t values obtained via the original or the revised RBA. $p(t_{9_0}) p(t_{13_0})$ is not contained in tube **g** (see also Fig. 9.21). $p(\bar{t}_{9_0}) p(\bar{t}_{13_0})$ is contained in the curve.

We start with the polygonal curve L_1. After applying Option 1, we obtain the curve L_2. Then we apply Option 2 and obtain the curve L_3. Finally, we apply Option 3 as given in the original RBA, and we obtain curve L_4 as the final result.

For the resulting polygon L_4, note that edge $p(t_{9_0}) p(t_{13_0})$ is not contained in the tube **g**. This means that the final polygon is not contained in the tube g! This is because Option 3 of the original RBA did not check whether $p_{i-1} p_{new}$ and $p_{new} p_{i+1}$ are both contained in the tube **g**. A minor but essential correction is required to fix this problem.

The figure also shows the corrected polygon L_5. Note that edge $p(\bar{t}_{9_0}) p(\bar{t}_{13_0})$ is now contained in the tube **g**.   □

Fig. 9.20 Three critical edges of a simple cube-curve also used for testing Option 2 of the original RBA; see Example 9.8

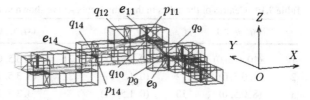

Figure 9.22 shows that there are cases where none of the two endpoints of an edge of the polygonal curve (resulting from Option 2) is allowed to do any move along a critical edge. This leads to a further modification of the original Option 3.

We consider cubes c_1 and c_2, and two different critical edges e_1 and e_2. Line $p_1 p_2$ is contained and complete in the arc from the cube which contains e_1 to the cube which contains e_2. $p_1 p_2$ intersects with c_1 and c_2 only at a single point each. If p_1 moves to the left along e_1, then $p_1 p_2$ will not intersect with c_2 anymore. If p_1 moves to the right along e_1, then $p_1 p_2$ will not intersect with c_1 anymore. If p_2 moves up along e_2, then $p_1 p_2$ will not intersect with c_2 anymore. If p_2 moves down along e_2, then $p_1 p_2$ will not intersect with c_1 anymore.

Table 9.9 Coordinates of endpoints of critical edges of the curve of Fig. 9.20

Critical edge	x_{i1}	y_{i1}	z_{i1}	x_{i2}	y_{i2}	z_{i2}
e_1	0.5	1	−0.5	0.5	1	−0.5
e_2	−0.5	1	−0.5	0.5	1	−0.5
e_3	−0.5	2	−0.5	0.5	2	−0.5
e_4	0.5	4	−1.5	0.5	4	−0.5
e_5	0.5	4	−0.5	0.5	5	−0.5
e_6	1.5	4	−0.5	1.5	5	−0.5
e_7	1.5	4	−0.5	1.5	4	0.5
e_8	2.5	4	−0.5	2.5	4	0.5
e_9	2.5	3	0.5	2.5	4	0.5
e_{10}	2.5	3	1.5	3.5	3	1.5
e_{11}	3.5	3	1.5	3.5	3	2.5
e_{12}	4.5	2	1.5	4.5	3	1.5
e_{13}	5.5	2	1.5	5.5	3	1.5
e_{14}	6.5	2	0.5	6.5	3	0.5
e_{15}	7.5	2	−0.5	7.5	3	−0.5
e_{16}	9.5	2	−0.5	9.5	3	−0.5
e_{17}	9.5	2	−0.5	10.5	2	−0.5
e_{18}	9.5	1	−0.5	9.5	1	0.5

Table 9.10 Centres of the cubes in the simple cube-curve shown in Fig. 9.20

i	(x_i, y_i, z_i)	i	(x_i, y_i, z_i)	i	(x_i, y_i, z_i)	i	(x_i, y_i, z_i)
1	$(10, 0.5, 0)$	11	$(0, 0.5, 0)$	21	$(3, 3.5, 0)$	31	$(7, 2.5, -1)$
2	$(9, 0.5, 0)$	12	$(0, 1.5, 0)$	22	$(3, 3.5, 1)$	32	$(8, 2.5, -1)$
3	$(8, 0.5, 0)$	13	$(0, 1.5, -1)$	23	$(3, 3.5, 2)$	33	$(9, 2.5, -1)$
4	$(7, 0.5, 0)$	14	$(0, 2.5, -1)$	24	$(3, 2.5, 2)$	34	$(10, 2.5, -1)$
5	$(6, 0.5, 0)$	15	$(0, 3.5, -1)$	25	$(4, 2.5, 2)$	35	$(10, 2.5, 0)$
6	$(5, 0.5, 0)$	16	$(0, 4.5, -1)$	26	$(5, 2.5, 2)$	36	$(10, 1.5, 0)$
7	$(4, 0.5, 0)$	17	$(1, 4.5, -1)$	27	$(5, 2.5, 1)$	37	$(10, 0.5, 0)$
8	$(3, 0.5, 0)$	18	$(1, 4.5, 0)$	28	$(6, 2.5, 1)$	38	$(9, 0.5, 0)$
9	$(2, 0.5, 0)$	19	$(2, 4.5, 0)$	29	$(7, 2.5, 1)$		
10	$(1, 0.5, 0)$	20	$(2, 3.5, 0)$	30	$(7, 2.5, 0)$		

Table 9.11 An example output of the original RBA. i is the index of the critical edge e_i. Point p_i is a vertex (on e_i) of a polygon contained in the simple cube-curve shown in Fig. 9.20

i	$p_i(x_i, y_i, z_i)$
1	$(0.5, 1, -0.5)$
5	$(0.5, 4.3626, -0.5)$
9	$(2.5, 3.6374, 0.5)$
11	$(3.5, 3, 2)$
14	$(6.5, 2.5044, 0.5)$
15	$(7.5, 2.2955, -0.5)$
16	$(9.5, 2, -0.5)$
18	$(9.5, 1, -0.5)$

The following (revised) Option 3 ensures that the final polygon is always contained and complete in the tube **g**.

We use Fig. 9.23 for an illustration of the revised Option 3.

Let $p_i = p_i(t_i)$ and $p_{new} = p_i(t_{i_0})$. By Option 3 in the original RBA, $t_i, t_{i_0} \in [0, 1]$. Let ε be a sufficiently small positive real number.

(*Case 1.* $t_i < t_{i_0}$.) See Fig. 9.23 on the left.

(*Case 1.1.* Both $p_{i-1}p(t_i + \varepsilon)$ and $p_{i+1}p(t_i + \varepsilon)$ are inside the arc from p_{i-1} to p_{i+1}.) If both $p_{i-1}p_{new}$ and $p_{i+1}p_{new}$ are inside the arc from e_{i-1} to e_{i+1}, then $\bar{p}_{new} = p_{new}$. Otherwise, by Lemmas 9.18 and 9.4, use binary search to find a value $\bar{t}_{i_0} \in (t_i, t_{i_0})$, and then let $\bar{p}_{new} = p(\bar{t}_{i_0})$.

(*Case 1.2.* Either $p_{i-1}p(t_i + \varepsilon)$ or $p_{i+1}p(t_i + \varepsilon)$ is outside the arc from e_{i-1} to e_{i+1}.) Then let $\bar{p}_{new} = p_i(t_i) = p_i$.

(*Case 2.* $t_{i_0} < t_i$.) See Fig. 9.23 on the right.

(*Case 2.1.* Both $p_{i-1}p(t_i - \varepsilon)$ and $p_{i+1}p(t_i - \varepsilon)$ are inside the arc from p_{i-1} to p_{i+1}.) If both $p_{i-1}p_{new}$ and $p_{i+1}p_{new}$ are inside the arc from e_{i-1} to e_{i+1}, then

Fig. 9.21 An example of a non-first-class simple cube-curve. See text for details

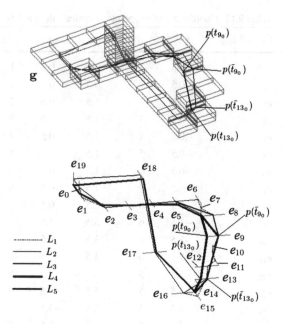

$\bar{p}_{\text{new}} = p_{\text{new}}$. Otherwise, (again by Lemmas 9.18 and 9.4) use binary search to find a value $\bar{t}_{i_0} \in (t_{i_0}, t_i)$, and then let $\bar{p}_{\text{new}} = p(\bar{t}_{i_0})$.

(*Case 2.2.* Either $p_{i-1}p(t_i - \varepsilon)$ or $p_{i+1}p(t_i - \varepsilon)$ is outside the arc from e_{i-1} to e_{i+1}.) Then let $\bar{p}_{\text{new}} = p_i(t_i) = p_i$.

This *revised Option 3* now contains the test of inclusion (which was missing in the original algorithm), and it details the operations for minimising the length of the calculated polygonal curve, providing a more specific description of Option 3 compared to the original RBA.

The revised Option 3 defines the *revised RBA*, which is short for 'revision of the original RBA'.

To summarise this section, we constructed a non-trivial simple cube-curve such that none of the vertices of its MLP is a grid vertex. Indeed, Theorems 9.7 and 9.9, and Lemmas 9.16 and 9.17 allow for the following conclusion:

Corollary 9.1 *Given a simple first-class cube-curve* g. *None of the vertices of its MLP is at a grid point position iff* g *has no end angle and, for every maximal run of parallel edges of* g, *its two adjacent critical edges are not on the same grid plane.*

It follows that the (provably correct) MLP algorithm proposed in Sect. 9.4 cannot be applied to such a cube-curve because this algorithm requires at least one end-angle for decomposing a given cube-curve into arcs. Of course, the original or revised RBA is applicable, and will produce a result (i.e., a polygonal curve).

We also proved that the original or revised RBA has κ-linear time complexity $\kappa(\varepsilon) \times \mathcal{O}(m)$, where m is the number of critical edges of a given simple cube-curve, and $\varepsilon > 0$ the accuracy parameter.

Table 9.12 Coordinates of endpoints of critical edges in Fig. 9.21. See text for details

Critical edge	x_{i1}	y_{i1}	z_{i1}	x_{i2}	y_{i2}	z_{i2}	t_{i0}	$\bar{t}_{i0}$
e_0	0.5	1	−0.5	0.5	1	0.5	1	1
e_1	−0.5	1	0.5	0.5	1	0.5	−	−
e_2	−0.5	2	1.5	0.5	2	1.5	0.7574	0.7561
e_3	−0.5	3	1.5	0.5	3	1.5	0.5858	0.5837
e_4	−0.5	4	1.5	0.5	4	1.5	0.4142	0.4113
e_5	−0.5	5	1.5	0.5	5	1.5	0.2426	0.2388
e_6	−0.5	6	1.5	0.5	6	1.5	−	−
e_7	−0.5	6	1.5	−0.5	6	2.5	1	0.9581
e_8	−0.5	6	2.5	−0.5	7	2.5	−	−
e_9	−1.5	6	3.5	−1.5	7	3.5	0	0.5
e_{10}	−2.5	6	3.5	−2.5	6	4.5	−	−
e_{11}	−3.5	6	4.5	−2.5	6	4.5	−	−
e_{12}	−3.5	5	4.5	−3.5	6	4.5	−	−
e_{13}	−3.5	5	5.5	−3.5	6	5.5	0.2612	0.5
e_{14}	−4.5	5	5.5	−3.5	5	5.5	−	−
e_{15}	−4.5	5	6.5	−3.5	5	6.5	−	−
e_{16}	−3.5	4	6.5	−3.5	5	6.5	1	1
e_{17}	1.5	4	6.5	1.5	5	6.5	0.5455	0.5455
e_{18}	1.5	4	0.5	2.5	4	0.5	0	0
e_{19}	1.5	1	−0.5	1.5	1	0.5	1	1

Fig. 9.22 An example where any move of one of the two end points of a line segment along critical edges is impossible

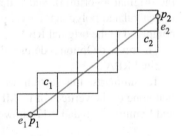

9.6 RBAs for MLP Calculation in Any Simple Cube-Curve

In this section, we finally present two provably correct RBAs for the MLP calculation in any simple cube-curve. At first, we summarise the revised RBA under the name *edge-based RBA*; it uses Option 2 of the original RBA and revised Option 3. The step set of this algorithm is given by critical edges. Furthermore, we also propose a *face-based RBA* where the step set is defined by cube faces incident with critical edges. The face-based RBA moves vertices of constructed curves within faces of cubes, rather than along critical edges; it does not use Option 2 at all but it is conceptually somehow more complicated than the edge-based algorithm.

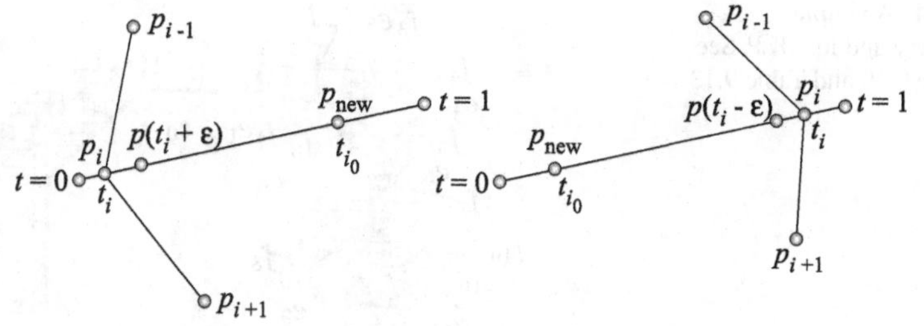

Fig. 9.23 Illustration for revised Option 3. *Left*: Case 1. *Right*: Case 2

With respect to asymptotic time, both the edge-based and the face-based RBA are κ-linear in the number of critical edges of a given simple cube-curve. We start with defining a simple but very useful notion:

Definition 9.10 If f is a face of a cube in g and one of f's edges is a critical edge e in g then f is called a *critical face of e in g*, or simply a *critical face*. In this case, we say that e is in critical face f.

The basic computational task of the graph-theoretical algorithm in Sect. 9.3 (addressed by Lines 1 through 20) consists in selecting this set, but this graph-theoretical algorithm is not time-efficient for large inputs. Option 2 of the original RBA was also designed having this goal in mind, but (we discussed that) it was flawed.

We need to undertake a very close observation of the geometric structure of simple cube-curves, and introduce for this purpose a few rather technical definitions:

Definition 9.11 Let e be a critical edge of a simple cube-curve g and f_1, f_2 be two critical faces of e in g. Let c_1, c_2 be the centres of f_1, f_2, respectively. Then a polygonal curve can go in the direction from c_1 to c_2, or from c_2 to c_1, to visit all cubes in g so that each cube is visited exactly once. If e is to the left of line segment $c_1 c_2$, then the orientation from c_1 to c_2 is called *counter-clockwise orientation* of g. f_1 is called *the first* critical face of e in g. If e is to the right of line segment $c_1 c_2$, then the direction from c_1 to c_2 is called *clockwise orientation* of g.

Definition 9.12 A *minimum-length pseudo-polygon* of a simple cube-curve g, denoted by MLPP, is a shortest curve ρ which is contained and complete in tube **g** such that each vertex of ρ is on the first critical face of a critical edge in g.

The number of vertices of an MLPP is the number of all critical edges of g. Let f_{i1} and f_{i2} be two critical faces of e_i in g, for $i = 1, 2$. Let c_{i1} and c_{i2} be the centres of f_{i1} and f_{i2}, respectively, for $i = 1, 2$. Obviously, the counter-clockwise orientation of g defined by c_{11} and c_{12} is identical to the one defined by c_{21} and c_{22}.

Definition 9.13 Let $e_0, e_1, e_2, \ldots, e_m$, and e_{m+1} be all consecutive critical edges of g in counter-clockwise orientation of g. Let f_i be the first critical face of e_i in

Fig. 9.24 A simple
cube-curve and its MLP. See
Example 9.10 and Table 9.13

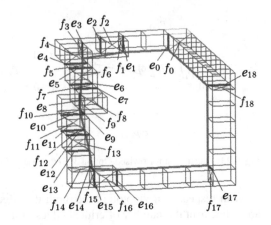

g, and p_i be a point on f_i, for $i = 0, 1, 2, \ldots, m, m + 1$. Then the polygonal curve $p_0 p_1 \ldots p_m p_{m+1}$ is called an *approximate minimum-length pseudo-polygon* of g, denoted by AMLPP.

Example 9.10 Figure 9.24 shows 19 critical faces f_i, where $i = 0, 1, \ldots, 18$. This cube-curve is not first-class because there are no vertices of the MLP on the following critical edges: $e_1, e_4, e_5, e_6, e_8, e_9, e_{10}, e_{11},$ and e_{14}. (If not yet clear at this point then we refer to experiments which are later reported in this section for the shown curve.)

Figure 9.24 shows all critical edges $(e_0, e_1, e_2, \ldots, e_{18})$ and their first critical faces $(f_0, f_1, f_2, \ldots, f_{18})$ of a simple cube-curve, denoted by g_{19}.

Curve $p_0^4 p_1^4 \ldots p_{18}^4$ (see Table 9.15) is the MLPP of g as shown in Fig. 9.24. The polygonal curve $p_0^1 p_1^1 \ldots p_{18}^1$ (see Table 9.14) is an AMLPP of g_{18} shown in Fig. 9.24. □

Definition 9.14 Let p_1, p_2, and p_3 be three consecutive vertices of an AMLPP of a simple cube-curve g. If p_1, p_2, and p_3 are collinear, then p_2 is called a *trivial* vertex of the AMLPP of g. Point p_2 is called a *non-trivial* vertex of the AMLPP of g if it is not a trivial vertex of that AMLPP of g.

We recall that a simple cube-arc is an alternating sequence

$$\rho = (f_0, c_0, f_1, c_1, \ldots, f_k, c_k, f_{k+1})$$

of faces f_i and cubes c_i with $f_{k+1} \neq f_0$, denoted by $\rho = (f_0, f_1, \ldots, f_{k+1})$, or $\rho(f_0, f_{k+1})$ for short; it is a connected part of a simple cube-curve. A *subarc* of an arc $\rho(f_0, f_{k+1})$ is an arc $\rho(f_i, f_j)$, where $0 \leq i \leq j \leq k + 1$.

Definition 9.15 Let a polygonal curve $\rho = p_0 p_1 \cdots p_m p_{m+1}$ be an AMLPP of g and $p_i \in f_i$, where f_i is a critical face of g, for $i = 0, 1, 2, \ldots, m + 1$. A cube-arc $\rho(f_i, f_j)$ is called

Table 9.13 Coordinates of endpoints of critical edges shown in Fig. 9.24: these data are used later in an experiment

Critical edge	x_{i1}	y_{i1}	z_{i1}	x_{i2}	y_{i2}	z_{i2}
e_0	−0.5	1	−0.5	−0.5	1	0.5
e_1	−0.5	2	−0.5	−0.5	2	0.5
e_2	−1.5	3	−0.5	−1.5	3	0.5
e_3	−2.5	3	−0.5	−2.5	4	−0.5
e_4	−3.5	3	−0.5	−3.5	4	−0.5
e_5	−3.5	3	−1.5	−3.5	4	−1.5
e_6	−4.5	3	−1.5	−4.5	4	−1.5
e_7	−5.5	4	−2.5	−5.5	4	−1.5
e_8	−6.5	4	−2.5	−5.5	4	−2.5
e_9	−6.5	4	−2.5	−6.5	5	−2.5
e_{10}	−6.5	4	−3.5	−6.5	5	−3.5
e_{11}	−7.5	4	−3.5	−7.5	5	−3.5
e_{12}	−7.5	4	−4.5	−7.5	5	−4.5
e_{13}	−8.5	4	−5.5	−7.5	4	−5.5
e_{14}	−8.5	4	−6.5	−8.5	4	−5.5
e_{15}	−8.5	3	−6.5	−8.5	3	−5.5
e_{16}	−9.5	−1	−5.5	−8.5	−1	−5.5
e_{17}	−8.5	−2	−0.5	−8.5	−1	−0.5
e_{18}	−0.5	−1	−0.5	−0.5	−1	0.5

- a *(2,3)-cube-arc* with respect to ρ if each vertex p_k is identical to p_{k-1} or p_{k+1}, where $k = i + 1, \ldots, j - 1$,[9]
- a *maximal (2,3)-cube-arc* with respect to ρ if it is a (2,3)-cube-arc and p_i is not identical to p_{i+1} and p_{i-1}, and p_j is not identical to p_{j-1} and p_{j+1},
- a *3-cube-arc unit* with respect to ρ if it is a (2,3)-cube-arc such that $j = i + 4 \pmod{m + 2}$ and p_{i+1}, p_{i+2}, p_{i+3} are identical,
- a *2-cube-arc* with respect to ρ if it is a (2,3)-cube-arc and no three consecutive vertices of ρ on a are identical,
- a *maximal 2-cube-arc* with respect to ρ if it is both a maximal (2,3)-cube-arc and a 2-cube-arc as well,
- a *2-cube-arc unit* with respect to ρ if it is a 2-cube-arc such that $j = i + 3 \pmod{m + 2}$ and p_{i+1} is identical to p_{i+2},
- a *regular cube-arc unit* with respect to ρ if $a = (f_i, f_{i+1}, f_j)$ such that p_i is not identical to p_{i+1} and p_j is not identical to p_{i+1},
- a *cube-arc unit* with respect to ρ if a is a regular cube-arc unit, 2-cube-arc unit or 3-cube-arc unit, or

---

[9]Note that it is impossible that four consecutive vertices of ρ are identical.

Table 9.14 Comparison of results of operations of the face-based RBA. Points $p_0^1, p_1^1, \ldots, p_{18}^1$ are the results of Line 2; points $p_0^2, p_1^2, \ldots, p_{18}^2$ are the results of Line 3

p_{1i}	x_{1i}	y_{1i}	z_{1i}	p_{2i}	x_{2i}	y_{2i}	z_{2i}
p_0^1	−0.5	1	0	p_0^2	−0.5	1	−0.21
p_1^1	−0.5	1	0	p_1^2	−0.5	1	−0.21
p_2^1	−1.5	3	−0.34	p_2^2	−1.5	3	−0.34
p_3^1	−2.5	3.29	−0.5	p_3^2	−2.5	3.23	−0.5
p_4^1	−2.5	3.29	−0.5	p_4^2	−2.5	3.23	−0.5
p_5^1	−3.5	3.5	−1.11	p_5^2	−3.5	3.45	−1.11
p_6^1	−4.15	3.64	−1.5	p_6^2	−4.15	3.64	−1.5
p_7^1	−5.5	3.94	−2.32	p_7^2	−5.5	3.94	−2.32
p_8^1	−5.8	4	−2.5	p_8^2	−5.69	4	−2.5
p_9^1	−5.8	4	−2.5	p_9^2	−5.69	4	−2.5
p_{10}^1	−6.5	4	−3.32	p_{10}^2	−6.5	4	−3.32
p_{11}^1	−6.65	4	−3.5	p_{11}^2	−6.65	4	−3.5
p_{12}^1	−7.5	4	−4.5	p_{12}^2	−7.5	4	−4.5
p_{13}^1	−7.95	4	−5.5	p_{13}^2	−8	4	−5.5
p_{14}^1	−7.95	4	−5.5	p_{14}^2	−8	4	−5.5
p_{15}^1	−8.5	3	−5.5	p_{15}^2	−8.5	3	−5.5
p_{16}^1	−8.5	−1	−5.5	p_{16}^2	−8.5	−1	−5.5
p_{17}^1	−8.5	−1	−0.5	p_{17}^2	−8.5	−1	−0.5
p_{18}^1	−0.5	−1	−0.1	p_{18}^2	−0.5	−1	−0.1

- a *regular cube-arc* with respect to ρ if no two consecutive vertices of ρ on a are identical.

Example 9.11 This example continues with Example 9.10. Tables 9.14 and 9.15 are also used further down for the face-based RBA which has not yet been defined at this point.

Let $\rho_{18}^i = p_0^i p_1^i \cdots p_{18}^i$ (see Tables 9.14 and 9.15), where $i = 1, 2, 3, 4$. Then there are four maximal 2-cube-arcs with respect to ρ_{18}^i: $(p_{18}^i, p_0^i, p_1^i, p_2^i)$, $(p_2^i, p_3^i, p_4^i, p_5^i)$, $(p_7^i, p_8^i, p_9^i, p_{10}^i)$, and $(p_{12}^i, p_{13}^i, p_{14}^i, p_{15}^i)$ in total, where $i = 1, 2, 3$. They are also maximal 2-cube-arcs and 2-cube-arc units with respect to ρ_{18}^i, where $i = 1, 2, 3$. There are no 3-cube-arc units with respect to ρ_{18}^i, where $i = 1, 2$. (p_1^i, p_2^i, p_3^i) is a regular cube-arc unit with respect to ρ_{18}^i, and $(p_4^i, p_5^i, p_6^i, p_7^i, p_8^i)$ is a regular cube-arc with respect to ρ_{18}^i, where $i = 1, 2, 3$.

There are three maximal 2-cube-arcs with respect to ρ_{18}^4: $(p_{18}^4, p_0^4, p_1^4, p_2^4)$, $(p_2^4, p_3^4, p_4^4, p_5^4)$, and $(p_{12}^4, p_{13}^4, p_{14}^4, p_{15}^4)$ in total. They are also maximal 2-cube-arcs and 2-cube-arc units with respect to ρ_{18}^4. $(p_6^4, p_7^4, p_8^4, p_9^4, p_{10}^4, p_{11}^4, p_{12}^4)$ is a (2,3)-

Table 9.15 Comparison of results of operations of the face-based RBA. Points $p_0^3, p_1^3, \ldots, p_{18}^3$ are the results of Line 4; points $p_0^4, p_1^4, \ldots, p_{18}^4$ are the results of Line 7

p_{3i}	x_{3i}	y_{3i}	z_{3i}	p_{4i}	x_{4i}	y_{4i}	z_{4i}
p_0^3	−0.5	1	−0.21	p_0^4	−0.5	1	−0.5
p_1^3	−0.5	1	−0.21	p_1^4	−0.5	1	−0.5
p_2^3	−1.5	3	−0.41	p_2^4	−1.5	3	−0.5
p_3^3	−2.5	3.23	−0.5	p_3^4	−2.5	3.22	−0.5
p_4^3	−2.5	3.23	−0.5	p_4^4	−2.5	3.22	−0.5
p_5^3	−3.5	3.47	−1.13	p_5^4	−3.5	3.48	−1.17
p_6^3	−4.09	3.62	−1.5	p_6^4	−4	3.61	−1.5
p_7^3	−5.5	3.95	−2.38	p_7^4	−5.5	4	−2.5
p_8^3	−5.69	4	−2.5	p_8^4	−5.5	4	−2.5
p_9^3	−5.69	4	−2.5	p_9^4	−5.5	4	−2.5
p_{10}^3	−6.5	4	−3.4	p_{10}^4	−6.5	4	−3.5
p_{11}^3	−6.59	4	−3.5	p_{11}^4	−6.5	4	−3.5
p_{12}^3	−7.5	4	−4.5	p_{12}^4	−7.5	4	−4.5
p_{13}^3	−8	4	−5.5	p_{13}^4	−8	4	−5.5
p_{14}^3	−8	4	−5.5	p_{14}^4	−8	4	−5.5
p_{15}^3	−8.5	3	−5.5	p_{15}^4	−8.5	3	−5.5
p_{16}^3	−8.5	−1	−5.5	p_{16}^4	−8.5	−1	−5.5
p_{17}^3	−8.5	−1	−0.5	p_{17}^4	−8.5	−1	−0.5
p_{18}^3	−0.5	−1	−0.27	p_{18}^4	−0.5	−1	−0.5

cube-arc with respect to P_{18}^4. $(p_6^4, p_7^4, p_8^4, p_9^4, p_{10}^4)$ is a unique 3-cube-arc unit with respect to ρ_{18}^4.　　□

Definition 9.16 Let $\rho = (f_i, f_{i+1}, \ldots, f_j)$ be a simple cube-arc and $p_k \in f_k$, for $k = i, j$. A *minimum-length arc* with respect to p_i and p_j of ρ, denoted by MLA(p_i, p_j), is a shortest arc (from p_i to p_j) which is contained and complete in ρ such that each vertex of MLA(p_i, p_j) is on the first critical face of a critical edge in ρ.

Let s_i and s_i' be ith side of faces f and f', respectively ($i = 1, 2, 3, 4$). If f contains f' and the Euclidean distance between s_i and s_i' is ε ($i = 1, 2, 3, 4$), then we say that f' is *obtained from* f by ε-*dilation*, or, in short, is a (first critical) *dilation face* (see Fig. 9.25).

The following lemma is used for the description of the edge-based RBA further below. Let $p_i \in f_i$, where f_i is the first critical face of e_i in g, for $i = 0, 1, 2, \ldots, m + 1$. We consider a polygonal curve $\rho = p_0 p_1 \cdots p_m p_{m+1}$.

Lemma 9.22 *Let p_i and p_{i+1} be two consecutive vertices of an AMLPP of g. If p_i is identical to p_{i+1} then p_i and p_{i+1} are on a critical edge of g.*

Fig. 9.25 Illustration of
ε-dilation. *Left*: a first critical
face. *Right*: a first critical
ε-dilation face

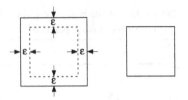

Analogously to the proof of Theorem 9.1, we also obtain (see Definition 9.14) the following:

Lemma 9.23 *If a vertex p of an AMLPP of g is on a first critical face f but not on a critical edge of it, then p is a trivial vertex of the AMLPP.*

We present below three algorithms for MLP calculation in simple cube-curves which are all κ-linear-time and provably correct (i.e., the calculated curves are converging to the MLP of a simple cube-curve).

We start with describing some useful procedures which are part of the first two of those three algorithms (in the sense of subroutines). The third algorithm is simple and does not apply any of the following procedures.

Given a critical e in g, and two points p_1 and p_3 in g such that neither p_1 nor p_3 is an endpoint of e,[10] by Procedure 17 (see Fig. 9.26),[11] we can find a unique point p_2 in e such that

$$d_{p_1 p_2} + d_{p_3 p_2} = \min\{d_{p_1 p} + d_{p_3 p} : p \in e\}.$$

For the next Procedure 18 (see Fig. 9.27), assume a critical face f of a critical edge in g, and two points p_1 and p_3 in the tube **g**; by Procedure 18 we find a point p_2 in f such that $d_{p_1 p_2} + d_{p_3 p_2} = \min\{d_{p_1 p} + d_{p_3 p} : p \in f\}$.

Procedure 17
Input: A critical edge e in g, and two points p_1 and p_3 in g such that neither p_1 nor p_3 is an endpoint of e.
Output: A unique point p_2 in e such that $d_{p_1 p_2} + d_{p_3 p_2} = \min\{d_{p_1 p} + d_{p_3 p} : p \in e\}$.

1: Let a and b be the two endpoints of e.
2: Let $p_2 = a + t \times (b - a)$, where $t = -(A_1 B_2 + A_2 B_1)/(B_2 + B_1)$; A_1, A_2, B_1, and B_2 are functions of the coordinates of p_1, p_3, a, and b.

Fig. 9.26 Procedure for computing an optimal point on a critical edge

---

[10]Otherwise, we update e by removing a sufficiently small segment(s) from its endpoint. This is another way to handle the degenerate case (see Sect. 3.4) of the used RBA.

[11]For Line 2, see Lemma 6 in [40].

Procedure 18

Input: A critical face f of a critical edge in g, and two points p_1 and p_3 in tube **g**.
Output: A point p_2 in f such that $d_{p_1 p_2} + d_{p_3 p_2} = \min\{d_{p_1 p} + d_{p_3 p} : p \in f\}$.

1: **if** $p_1 p_3$ and f are on the same plane **then**
2:     **if** $p_1 p_3 \cap f \neq \emptyset$ **then**
3:         Let p_2 be that end point of this segment which is closer to p_1.
4:     **else**
5:         Apply Procedure 17 on the four edges of f, denoted by e_1, e_2, e_3, and e_4;
         we obtain p_{2_i} such that $d_{p_1 p_{2_i}} + d_{p_3 p_{2_i}} = \min\{d_{p_1 p} + d_{p_3 p} : p \in e_i\}$, where
         $i = 1, 2, 3, 4$.
6:         Compute a point p_2 such that $d_{p_1 p_2} + d_{p_3 p_2} = \min\{d_{p_1 p_{2_i}} + d_{p_3 p_{2_i}} : i =$
         $1, 2, 3, 4\}$.
7:     **end if**
8: **else**
9:     **if** $p_1 p_3 \cap f \neq \emptyset$ **then**
10:         $p_1 p_3 \cap f$ must be a unique point. Let p_2 be this point.
11:     **else**
12:         Compute p_2 exactly the same way as Lines 4–6.
13:     **end if**
14: **end if**

Fig. 9.27 Procedure for computing an optimal point on a critical face

Fig. 9.28 Position of point
p_2

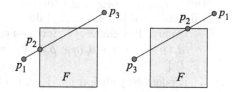

In Line 3 of Procedure 18, in this case, $p_1 p_3 \cap f$ is a line segment. See Fig. 9.28.
In Line 5, by Lemma 8.1, p_2 must be on the edges of f. By Lemma 9.4, p_2 must be
uniquely on one of the edges of f.

The next Procedure 19 (see Fig. 9.29) is used to convert an MLPP into an MLP.
Given are a polygonal curve $\rho = p_0 p_1 \cdots p_m p_{m+1}$ and three pointers addressing
vertices at positions $i - 1$, i, and $i + 1$ in this curve. Delete p_i if p_{i-1}, p_i, and
p_{i+1} are collinear. Next, the subsequence (p_{i-1}, p_i, p_{i+1}) is replaced in the curve
by (p_{i-1}, p_{i+1}). Then, continue with vertices $(p_{i-1}, p_{i+1}, p_{i+2})$ until $i + 2$ equals
$m + 1$.

Let $p_i \in l_i \subset f_i, \ldots, p_j \in l_j \subset f_j$ be a sequence of some consecutive vertices
of the AMLPP of g, where $f_i, \ldots, f_j$ are some consecutive critical faces of g, and
l_k is a line segment on f_k, $k = i, i + 1, \ldots, j$. Let $\varepsilon = 10^{-10}$ be an example for
an accuracy parameter. We can apply the method of Option 3 of the original RBA
(see Sect. 9.2), also including its correction in the previous section, for a cube-
arc $\rho(f_i, f_j)$ and for finding an approximate *minimum-length arc* MLA(p_i, p_j).

Procedure 19

Input: A polygonal curve $\rho = (p_0, p_1, \ldots, p_m, p_{m+1})$.
Output: A polygonal curve $(p_0', p_1', \ldots, p_{m'}', p_{m'+1}')$ such that any three consecutive vertices p_{i-1}', p_i', and p_{i+1}' are not collinear, where $i = 1, 2, \ldots, m'$ and $m' \leq m$.

1: Let $i = 1$.
2: **while** $i + 2 < m + 1$ **do**
3:    **while** p_{i-1}, p_i, and p_{i+1} are collinear **do**
4:       Update the curve ρ by replacing the subsequence (p_{i-1}, p_i, p_{i+1}) in the
   curve by (p_{i-1}, p_{i+1}).
5:    **end while**
6:    Let $i = i + 1$.
7: **end while**

Fig. 9.29 Procedure for removing redundant vertices from a polygonal curve

Procedure 20 (RBA for finding an approximate MLA)

Input: Let $p_i \in l_i \subset f_i, \ldots, p_j \in l_j \subset f_j$ be a sequence of some consecutive vertices of the AMLPP of g, where $f_i, \ldots, f_j$ are some consecutive critical faces of g, and l_k is a line segment on f_k, $k = i, i+1, \ldots, j$ and an accuracy parameter $\varepsilon > 0$.
Output: An approximate $MLA(p_i, p_j)$.

1: $L_{current} \leftarrow \sum_{j'=i}^{j} d_e(p_{j'}, p_{j'+1})$; and let $L_{previous} \leftarrow \infty$.
2: **while** $L_{previous} - L_{current} \geq \varepsilon$ **do**
3:    **for each** $j' \in \{i, i+1, \ldots, j\}$ **do**
4:       Apply Procedure 17 for computing a point $q_{j'} \in l_{j'}$ such that $d_e(p_{j'-1}, q_{j'}) + d_e(q_{j'}, p_{j'+1}) = \min\{d_e(p_{j'-1}, p) + d_e(p, p_{j'+1}) : p \in l_{j'}\}$.
5:       Update the path $(p_i, p_{i+1}, \ldots, p_j)$ by replacing $p_{j'}$ by $q_{j'}$.
6:    **end for**
7:    Let $L_{previous} \leftarrow L_{current}$ and $L_{current} \leftarrow \sum_{j'=i}^{j} d_e(p_{j'}, p_{j'+1})$.
8: **end while**
9: Return the path $(p_i, p_{i+1}, \ldots, p_j)$.

Fig. 9.30 RBA for finding an approximate MLA

This application is as in Procedure 20, shown in Fig. 9.30 (and is used later on in Procedure 22).

Let $e_0, e_1, e_2, \ldots, e_m$, and e_{m+1} be all the consecutive critical edges of g in the counter-clockwise orientation of g. Let f_i be the first critical face of e_i in g, and c_i be the centre of f_i, for $i = 0, 1, 2, \ldots, m+1$. All indices of points, edges and faces are taken $\mod(m+2)$. Using the following Procedure 21 (see Fig. 9.31), we can compute an AMLPP of g and its length.

Given is an m'-cube-arc unit $(f_i, \ldots, f_j)$ with respect to a polygonal curve ρ of g, where $m' = 2$ or 3. Let $p_i \in f_i$ and $p_j \in f_j$. We can calculate an $MLA(p_i, p_j)$ by applying Procedure 22 (see Fig. 9.32).

Procedure 21 (RBA for finding an approximate AMLPP)

Input: Let $e_0, e_1, e_2, \ldots, e_m$, and e_{m+1} be all consecutive critical edges of g in the counter-clockwise orientation of g. Let f_i be the first critical face of e_i in g, and c_i be the centre of f_i, for $i = 0, 1, 2, \ldots, m + 1$; and an accuracy constant $\varepsilon > 0$.

Output: An approximate AMLPP and its length.

1: Let ρ be a polygonal curve $(p_0, p_1, \ldots, p_m, p_{m+1})$, where $p_i = c_i$, the centre of the first critical face f_i of e_i in g, $i = 0, 1, 2, \ldots, m + 1$.
2: $L_{\text{current}} \leftarrow \sum_{j=0}^{m+1} d_e(p_j, p_{j+1})$; and let $L_{\text{previous}} \leftarrow \infty$.
3: **while** $L_{\text{previous}} - L_{\text{current}} \geq \varepsilon$ **do**
4:   **for** each $j \in \{0, 1, \ldots, m + 1\}$ **do**
5:     Apply Procedure 18 for computing a point $q_j \in f_j$ such that $d_e(p_{j-1}, q_j) + d_e(q_j, p_{j+1}) = \min\{d_e(p_{j-1}, p) + d_e(p, p_{j+1}) : p \in f_j\}$.
6:     Update the curve $(p_0, p_1, \ldots, p_{m+1})$ by replacing p_j by q_j.
7:   **end for**
8:   Let $L_{\text{previous}} \leftarrow L_{\text{current}}$ and $L_{\text{current}} \leftarrow \sum_{j=0}^{m+1} d_e(p_j, p_{j+1})$.
9: **end while**
10: Return the curve $(p_0, p_1, \ldots, p_{m+1})$ and its length.

Fig. 9.31 RBA for finding an approximate AMLPP

Lemma 9.24 *For each cube-arc unit $\rho(f_i, f_j)$ with respect to ρ, MLA(p_i, p_j) can be computed in $\mathcal{O}(1)$ time.*

Proof If ρ is a regular cube-arc unit, then MLA(p_i, p_j) can be found by Procedure 18 which has complexity $\mathcal{O}(1)$. Otherwise, ρ is an m'-cube-arc unit, where $m' = 2$ or 3. Then, by Lemma 9.23, MLA(p_i, p_j) can be found by Procedure 22, which can be computed in $\mathcal{O}(1)$ because $m' = 2$ or 3. $\qquad\square$

We extend now the revised RBA (see the previous section) into the following (provably correct) Algorithms 35 and 36.

The presentation of the edge-based RBA (i.e., Fig. 9.33) has been sufficiently prepared in prior discussions.

In Line 3 of Algorithm 36 (see Fig. 9.34), by Lemma 9.22, the input line segments of Procedure 20 are critical edges. As usual for RBAs, the updated AMLPP is "sufficiently accurate" if the previous length minus the current length is smaller than a defined accuracy parameter $\varepsilon > 0$.

The first difficult task for applying an RBA is to find a proper step set. Another issue when applying an RBA is to deal with the degenerative case of the RBA. The following Algorithm 37 (see Fig. 9.35) overcomes the first difficulty by simply taking all the initial critical faces as the step set. It handles the second task by ε_2-dilation, for some $\varepsilon_2 > 0$.

Procedure 22 (Exhaustive search for computing an approximate MLA)
Input: m'-cube-arc unit $(f_i, \ldots, f_j)$ with respect to a polygonal curve ρ of g, where $m' = 2$ or 3. Let $p_i \in f_i$ and $p_j \in f_j$.
Output: An approximate MLA(p_i, p_j).

1: Compute the set $E = \{e : e$ is a critical edge of $f_k \wedge k = i+1, \ldots, j-1\}$.
2: Let $I = 1$ and $L = +\infty$.
3: **while** $I < m'$ **do**
4:      Compute the set $S_E = \{S : S \subseteq E \wedge |S| = I\}$.
5:      **for** each $S \in S_E$ **do**
6:          Input $p_i, e_1, \ldots, e_l, p_j$ to Procedure 20 to compute an approximate MLA(p_i, p_j) such that it has minimal length with respect to all sets $S \in S_E$, denoted by AMLA(I, S_E), where $e_k \in S$, for $k = 1, 2, \ldots, l$ and $l = |S|$.
7:          **if** the length of AMLA$(I, S_E) < L$ **then**
8:              Let MLA$(p_i, p_j) =$ AMLA(I, S_E) and L equal the length of AMLA(I, S_E).
9:          **end if**
10:     **end for**
11:     Let $I = I + 1$.
12: **end while**

Fig. 9.32 Exhaustive search for computing an approximate MLA

Regarding the output of Algorithm 37, it follows that $\lim_{\varepsilon_1 \to 0} r(\varepsilon_1) = 0$. In this algorithm, all subscripts are taken modulo $m + 1$. Line 4 is illustrated by Fig. 9.36; on the left it shows that $p_{i-1}p_{i+1} \cap f_i = q_i$, the middle shows that $p_{i-1}p_{i+1} \cap f_i = \emptyset$ and q_i is on one side of f_i, the right shows that $p_{i-1}p_{i+1} \cap f_i = q_i q_i'$ ($q_i q_i'$ is a line segment inside of f_i).

9.7 Correctness Proof

We first prove the correctness of Algorithm 37 (i.e., the convergence of the output polygonal path toward the MLP if the accuracy parameter goes to zero), and then that of Algorithms 35 and 36. We apply basic results of convex analysis:

Theorem 9.11 *Let S_1 and S_2 be convex sets in $\mathbb{R}^m$ and $\mathbb{R}^n$, respectively. Then $S_1 \times S_2$ is a convex set in $\mathbb{R}^{m+n}$, where $m, n \in \mathbb{N}$.*

Proposition 9.1 *Each norm on $\mathbb{R}^n$ is a convex function; a nonnegative weighted sum of convex functions is a convex function.*

Proposition 9.2 *Let f be a convex function. If x is a point where f has a finite local minimum, then x is a point where f has its global minimum.*

Algorithm 35 (The edge-based RBA for cube-curves)
Input: Let $e_0, e_1, e_2, \ldots, e_m$, and e_{m+1} be all consecutive critical edges of g in counter-clockwise orientation of g. Let f_i be the first critical face of e_i in g, and c_i be the centre of f_i, for $i = 0, 1, 2, \ldots, m + 1$.
Output: An approximate MLP of g.

1: Let ρ_0 be the polygon obtained by the revised RBA.
2: Find a point $p_i \in f_i$ such that p_i is the intersection point of an edge of ρ_0 with f_i, for $i = 0, 1, 2, \ldots, m + 1$. Let ρ be a polygonal curve $(p_0, p_1, \ldots, p_m, p_{m+1})$. Let *Search* = true.
3: **while** *Search* = true **do**
4:    **for** each cube-arc unit $\rho = (f_i, f_{i+1}, \ldots, f_j)$ with respect to ρ **do**
5:       Apply Procedure 22 to update the arc $(p_i, p_{i+1}, \ldots, p_j)$.
6:    **end for**
7:    Let *Search'* = true.
8:    **for** each cube-arc unit $\rho = (f_i, f_{i+1}, \ldots, f_j)$ with respect to ρ and *Search'* = true **do**
9:       **if** the arc $(p_i, p_{i+1}, \ldots, p_j) = \text{MLA}(p_i, p_j)$ **then**
10:          Let *Search* = false.
11:       **else**
12:          Let *Search* = true and *Search'* = false.
13:       **end if**
14:    **end for**
15: **end while**
16: Apply Procedure 19 to polygonal curve $(p_0, p_1, \ldots, p_m, p_{m+1})$ to obtain the final MLP.

Fig. 9.33 The edge-based RBA for calculating an MLP in a simple cube-curve

For a start, we state for later reference:

Proposition 9.3 *Each face or dilated face is a convex set.*

By Theorem 9.11 and Propositions 9.1 and 9.3, we have the following

Corollary 9.2 L_g: $f_0 \times f_1 \times \cdots \times f_{m+1} \times f_0 \to R$ is a convex function, where L_g is a mapping from an AMLPP to its length, and f_i is defined in Algorithm 37, for $i = 0, 1, \ldots, m + 1$.

Theorem 9.12 *If the chosen accuracy value ε is sufficiently small, then Algorithm 37 outputs a $\{1 + 4(m + 1) \cdot [r(\varepsilon_1) + \sqrt{2} \times \varepsilon_2]/L\}$-approximate global MLP.*

Proof By Proposition 9.2, Algorithm 37 outputs an approximate global MLP. For each $i \in \{0, 1, 2, \ldots, m + 1\}$, the error of the difference between $d_e(p_i, p_{i+1})$ and $d_e(v_i, v_{i+1})$ is at most $4 \times r(\varepsilon_1) + \sqrt{2} \times \varepsilon_2$ because of $d_e(p_i, v_i) \le r(\varepsilon) + \sqrt{2} \times \varepsilon_2$.

Algorithm 36 (The face-based RBA for cube-curves)
Input: The same as for Algorithm 35.
Output: The same as for Algorithm 35.

1: Take a point $p_i \in f_i$, for $i = 0, 1, 2, \ldots, m + 1$.
2: Apply Procedure 21 to find an AMLPP of g, denoted by ρ.
3: Find all maximal 2-cube-arcs with respect to ρ, apply Procedure 20 to update
   the vertices of the AMLPP, which are on one of the 2-cube-arcs. Repeat this
   operation until the length of the updated AMLPP is sufficiently accurate.
4: Apply Procedure 21 to update the current AMLPP.
5: Find all maximal (2,3)-cube-arcs with respect to the current ρ; apply Procedure 20 to update those vertices of the current AMLPP which are on one of
   the (2,3)-cube-arcs. The input line segments of Procedure 20 can be found such
   that they are on the critical face and parallel or perpendicular to the critical
   edge of the face. Repeat this operation until the length of the updated AMLPP
   is sufficiently accurate.
6: Apply Procedure 21 to update the current AMLPP.
7: Apply Procedure 22 to all cube-arc units of ρ. If the arc $(p_i, p_{i+1}, \ldots, p_j)$ is
   equal to $MLA(p_i, p_j)$ for each cube-arc unit $\rho = (f_i, f_j)$ with respect to P,
   then ρ is the MLPP of g; go to Line 8. Otherwise, go to Line 3.
8: Apply Procedure 19 to obtain the final MLP.

Fig. 9.34 Main operations of the face-based RBA for simple cube-curves

Algorithm 37 (Approximate MLP algorithm)
Input: Let $f_0, f_1, f_2, \ldots, f_m$, and f_{m+1} be all the consecutive critical faces of g in
counter-clockwise orientation of g; chose two accuracy values $\varepsilon_1 > 0$ and $\varepsilon_2 > 0$.
Output: An updated closed $\{1 + 4(m + 1) \times [r(\varepsilon_1) + \sqrt{2} \times \varepsilon_2]/L\}$-approximation
path (MLP) $\rho(p_0, p_1, \ldots, p_{m+1})$, where L is the length of an optimal path, $r(\varepsilon_1)$ the
upper error bound for distances between p_i and the corresponding optimal vertex
p_i': $d_e(p_i, p_i') \leq r(\varepsilon_1)$, for $i = 0, 1, \ldots, m + 1$.

1: For each $i \in \{0, 1, \ldots, m+1\}$, update face f_i by ε_2-dilation; let p_i be the centre
   of f_i; let L_0 be $\sum_{i=0}^{m+1} d_e(p_i, p_{i+1})$; and let L_1 be ∞.
2: **while** $L_1 - L_0 \geq \varepsilon_1$ **do**
3:     **for** each $i \in \{0, 1, \ldots, m+1\}$ **do**
4:         Compute $q_i \in f_i$ such that $d_e(p_{i-1}, q_i) + d_e(q_i, p_{i+1}) = \min\{d_e(p_{i-1}, q) + d_e(q, p_{i+1}) : q \in f_i\}$;
           update ρ by replacing p_i by q_i.
5:     **end for**
6:     Let L_0 be L_1; calculate the perimeter L_1 of ρ.
7: **end while**
8: Output ρ and its length L_1.

Fig. 9.35 Simplified approximate MLP algorithm for simple cube-curves

Fig. 9.36 Illustration of
Line 4 in Algorithm 37

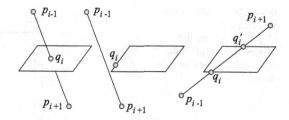

We obtain that

$$L \leq \sum_{i=0}^{m+1} d_e(p_i, p_{i+1}) \leq \sum_{i=0}^{m+1} \left[d_e(v_i, v_{i+1}) + 4 \times r(\varepsilon_1) + \sqrt{2} \times \varepsilon_2 \right]$$

$$= L + 4(m+1) \times \left[r(\varepsilon_1) + \sqrt{2} \times \varepsilon_2 \right].$$

Thus, the output path is a $\{1 + 4(m+1) \times [r(\varepsilon_1) + \sqrt{2} \times \varepsilon_2]/L\}$-approximation path. □

Note that, both (see Lines 15 and 7) the edge-based rubberband algorithm and the face-based rubberband algorithm return an AMLPP which converges to the output of Algorithm 37. By Theorem 9.12, both algorithms are thus correct as well. Thus, we obtain the following

Theorem 9.13 ρ is an MLPP of g iff for each cube-arc unit $\rho(f_i, f_j)$ with respect to ρ, the arc $(p_i, p_{i+1}, \dots, p_j)$ is equal to $\mathrm{MLA}(p_i, p_j)$.

9.8 Time Complexities and Examples

We discuss the time complexity of the edge-based and the face-base RBA. For a start, Procedures 17 and 18 can be computed in $\mathcal{O}(1)$, and Procedure 19 can be computed in $\mathcal{O}(m)$, where m is the number of critical edges of g.

By Lemma 9.21, Procedure 20 can be computed in $\kappa(\varepsilon) \cdot \mathcal{O}(n)$ time, where $\kappa(\varepsilon)$ is as in Theorem 9.3, and n is the number of vertices of the arc. Analogously, Procedure 21 can be computed in $\kappa(\varepsilon) \cdot \mathcal{O}(m)$ time, where m is the number of vertices of the polygonal curve.

By Theorem 9.10, the original RBA can be executed in $\kappa(\varepsilon) \cdot \mathcal{O}(m)$ time, where m is again the number of critical edges of g. The main additional operations of the edge-based RBA are Lines 3–15 (i.e., the while-loop) which can be computed in $\mathcal{O}(m)$, where m is the number of critical edges of g (by Lemma 9.24). It follows that the edge-based rubberband algorithm can be executed in $\kappa(\varepsilon) \cdot \mathcal{O}(m)$ time.

For the face-based RBA, Line 1 requires $\mathcal{O}(m)$ time. Lines 2, 4, and 6 have the same time complexity as Procedure 21. Again, by Lemma 9.21, Line 3 can be computed in $\kappa(\varepsilon) \cdot \mathcal{O}(m)$ time, where m is the number of vertices of the polygonal curve.

Table 9.16 Results of the edge-based RBA. $p_0^4, p_1^4, \ldots, p_{18}^4$ are the vertices of the MLP of the simple cube-curve shown in Fig. 9.24

final$_{p_i}$	x_i	y_i	z_i
p_0^4	-0.5	1	-0.5
p_2^4	-1.5	3	-0.5
p_3^4	-2.5	3.22	-0.5
p_7^4	-5.5	4	-2.5
p_{12}^4	-7.5	4	-4.5
p_{13}^4	-8	4	-5.5
p_{15}^4	-8.5	3	-5.5
p_{16}^4	-8.5	-1	-5.5
p_{17}^4	-8.5	-1	-0.5
p_{18}^4	-0.5	-1	-0.5

Analogously, Line 5 can be computed in $\kappa(\varepsilon) \cdot \mathcal{O}(m)$ time. (Note that there is a constant number of different combinations of input line segments of Procedure 20.) By Lemma 9.24, Line 7 can be computed in $\mathcal{O}(m)$ time. Therefore, the face-based rubberband algorithm can be computed in $\kappa(\varepsilon) \cdot \mathcal{O}(m)$ time.

Regarding the time complexity of Algorithm 37, we state that the main computation is in the two stacked loops. The while-loop takes $\kappa(\varepsilon_1)$ iterations; the for-loop can be computed in time $\mathcal{O}(k)$. Thus, Algorithm 37 can be computed in time $\kappa(\varepsilon_1) \cdot \mathcal{O}(k)$, where ε_1 and k are as defined in the algorithm.

Example 9.12 We approximate the MLP of the simple cube-curve g_{19} shown in Fig. 9.24. Table 9.13 lists all coordinates of critical edges of g_{19}. We take the centres of the first critical faces of g_{19} to produce an initial polygonal curve for the face-based rubberband algorithm. The updated polygonal curves are shown in Tables 9.14 and 9.15.[12] We take the centres of each critical edge of g_{19} for the initialisation of the polygonal curve of the revised RBA. The resulting polygon is shown in Table 9.16. Table 9.17 illustrates that the edge-based and face-based RBAs converge to the same MLP of g_{19}. □

We conclude the section with a few experimental results. See Fig. 9.37 for some statistics about measured run time. Half of a simple cube-curve was generated randomly, and the second half was then generated using three straight arcs to close

Table 9.17 Lengths of calculated curves at different operations of the face-based RBA, compared with the lengths calculated by the edge-based RBA (column EBRA)

Line	Initial	2	3	4	8	EBRA
Length	35.22	31.11	31.08	31.06	31.01	31.01

[12]Two digits are used only for displaying coordinates. Obviously, in the calculations it is necessary to use higher precision.

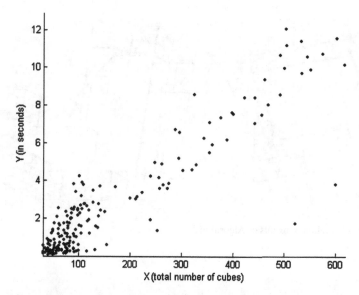

Fig. 9.37 Edge-based RBA implemented in Java, run under Matlab 7.0.4, Pentium 4, using $\varepsilon = 10^{-10}$

the curve. The number of cubes in generated curves was between 10 and 630. The break-off criterion was defined by $\varepsilon = 10^{-10}$.

Figure 9.38 shows two resulting MLPs (in red) obtained by Algorithm 37 when both chosen accuracy constants ε_1 and ε_2 are set to be 10^{-6} and 10^{-3}, respectively. The initial paths are in green. Table 9.18 shows the difference in the numbers of iterations taken in Algorithm 37 when the first accuracy constant ε_1 was set to 10^{-6} while the second accuracy constant ε_2 was set to 10^{-3} or 10^{-1}.

In this chapter, we presented an edge-based and two face-based RBAs and have shown that all three are provably correct for any simple cube-curve. We also have shown that their time complexity is $\kappa(\varepsilon) \cdot \mathcal{O}(m)$, where $\kappa(\varepsilon)$ is as in Theorem 9.3, and m is the number of critical edges of g.

> We identified one criterion (see Theorem 9.13) for testing whether a polygonal curve inside of a simple cube-curve is actually the MLP of this curve, or not.

The main idea of this test is implemented by Procedure 22.

The chapter introduced the concept of a *critical face* to deal with "combinatorial hardness".[13] It also solved the difficulty in redesigning Option 2 of the original RBA algorithm. (This option attempted to solve the "combinatorial hardness", but it was flawed in its original design.)

---

[13] This "hardness" is described in [45], page 666, or in [46].

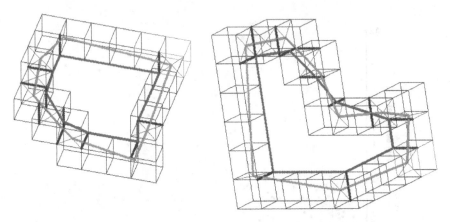

Fig. 9.38 Illustration of results of Algorithm 37

> The "most powerful" operations of the original RBA are summarised in its
> Option 3. All the RBAs discussed in this book for various applications in
> computational geometry are based on this.

Option 2 is still very useful even if it may not detect the correct subset of critical
edges which contain the vertices of the MLP of a given simple cube-curve. This
is because Option 2 is (in general) significantly speeding up and simplifying the
algorithm when comparing the edge-based with the face-based RBA.

9.9 The Non-existence of Exact Solutions

This section proves that there does not exist any exact arithmetic algorithm for solving the general MLP problem for any simple cube-curve.

Option 3 of the original RBA can be expressed as solving a system of partial
derivative equations (PDEs) involving parameters $t_i \in \mathbb{R}$ for critical edges e_i of the
step set. The result ensures that $p_i(t_i)$ is the optimum point on e_i.

Example 9.13 Considering the cube-curve illustrated in Fig. 9.39 (see also Fig. 9.2
and Table 9.3), calculating the MLP is equivalent to the problem of finding the roots
of

$$p(x) = 84x^6 - 228x^5 + 361x^4 + 20x^3 + 210x^2 + 200^x + 25$$

Table 9.18 Resulting data obtained from Algorithms 37: i and i' are the indices of experiments; m and m' the numbers of critical edges; I and I' the numbers of iterations taken; L_0 and L_0' the lengths of initial paths; L and L' the lengths of resulting paths; $\delta = L_0 - L$; and $\delta' = L_0' - L'$

i	m	I	L_0	L	δ	i'	m'	I'	L_0'	L'	δ'
1	13	37	19.35	15.85	3.49	1	12	1,559	19.73	15.59	4.14
2	19	30	29.40	24.72	4.69	2	19	1,505	33.35	26.99	6.36
3	26	27	45.02	38.97	6.04	3	25	3,832	42.94	35.04	7.90
4	33	25	54.49	46.58	7.91	4	36	1,674	43.99	35.57	8.42
5	40	34	46.25	36.53	9.72	5	40	3,610	58.00	46.84	11.16
6	48	38	69.34	57.02	12.32	6	48	5,877	75.52	64.13	11.39
7	54	92	79.30	67.67	11.63	7	59	1,831	78.29	62.95	15.34
8	58	22	103.61	87.29	16.32	8	64	2,127	106.23	88.28	17.95
9	74	48	103.57	88.49	15.08	9	69	1,777	88.33	68.27	20.06
10	78	81	95.75	78.38	17.37	10	81	2,281	116.83	94.37	22.46

Fig. 9.39 Calculation of t_1 and t_2 such that the polyline $p_0(t_0) p_1(t_1) p_2(t_2) p_3(t_3)$ is fully contained in **g**. Point p_1 is on e_1, and p_2 on e_2

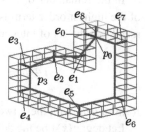

(as detailed below). In fact, this problem is not solvable by radicals over the field of rationals.[14]   □

It is well-known that there is no exact arithmetic algorithm for calculating the roots of polynomials of degree ≥ 5.[15] Our example allows a more specific corollary:

Corollary 9.3 *There is no exact arithmetic algorithm for calculating 3D ESPs.*

C. Bajaj showed this in 1985[16] for the general case of 3D ESPs based on a polynomial of degree 20. Example 9.13 shows that there is even no exact arithmetic algorithm for calculating MLPs in simple cube-curves; it defines a polynomial of degree 6 only, and for a restricted 3D ESP problem defined by simple cube-curves.

---

[14]A proof can be based on a theorem by C. Bajaj [6] and the factorisation algorithm by E.R. Berlekamp [7]. Details are given further below. *Chandrajit Bajaj* is with the University of Texas. *Elwyn R. Berlekamp* is with the University of California at Berkeley.

[15]Theorem by E. Galois; see also B.L. van der Waerden's famous example $p(x) = x^5 - x - 1$.

[16]See Theorem 9 in [5], saying that the ESP problem is in general not solvable by radicals over the field of rationals.

This fundamental non-existence of exact arithmetic algorithms is valid, no matter what magnitude of time-complexity is allowed.

As discussed earlier, the MLP is uniquely defined. This shortest path passes through subsequent line segments $e_1, e_2, \ldots, e_k$ in 3D space in this order. Obviously, vertices of a shortest path can be at real division points, and the following theorem means that they are even at points which cannot be represented by radicals over the field of rationals:

Theorem 9.14 *The MLP problem for simple cube-curves is in general not solvable by radicals over the field of rationals.*

In the remainder of this section, all polynomials are *monic* (i.e., the coefficient of the highest order term is 1) and with integer coefficients. Let ψ be a prime. We recall that the set of integers $\{0, 1, 2, \ldots, \psi - 1\}$, with operations

$$a \oplus b = a + b \bmod \psi \quad \text{and} \quad a \odot b = ab \bmod \psi$$

forms a *field* with these two operations, denoted by $\mathbb{Z}_\psi$.

Let $\deg(p(x))$ be the degree of the polynomial $p(x)$.[17] Let

$$p(x) = x^n + a_{n-1}x^{n-1} + a_{n-2}x^{n-2} + \cdots + a_1 x + a_0.$$

Then we have that $\deg(p(x)) = n$. The discriminant of $p(x)$ is defined as a $(2n - 1) \times (2n - 1)$ determinant, which equals 0 iff $p(x)$ has one or more multiple roots.

For a square matrix of order greater than 3, evaluating the determinant from its definition (as a sum of products) is hopelessly inefficient, and so its determinant should be evaluated by triangularising the matrix (with pivoting). That process preserves the determinant, except that each pivotal interchange changes the sign. Hence, if the triangularisation involves k inter-changes, then the determinant of the original matrix equals the product of the diagonal elements of the triangularised matrix, except that the sign is changed for odd k. If all potential pivots are 0 at any stage in the triangularisation, then the original matrix has zero determinant.

---

[17]The following results are well known in mathematical algebra; proofs can be found, for example, in [6, 24].

An example is shown below for the case of $n = 5$:

$$
\begin{vmatrix}
1 & a_4 & a_3 & a_2 & a_1 & a_0 & 0 & 0 & 0 \\
0 & 1 & a_4 & a_3 & a_2 & a_1 & a_0 & 0 & 0 \\
0 & 0 & 1 & a_4 & a_3 & a_2 & a_1 & a_0 & 0 \\
0 & 0 & 0 & 1 & a_4 & a_3 & a_2 & a_1 & a_0 \\
5 & 4a_4 & 3a_3 & 2a_2 & 1a_1 & 0 & 0 & 0 & 0 \\
0 & 5 & 4a_4 & 3a_3 & 2a_2 & 1a_1 & 0 & 0 & 0 \\
0 & 0 & 5 & 4a_4 & 3a_3 & 2a_2 & 1a_1 & 0 & 0 \\
0 & 0 & 0 & 5 & 4a_4 & 3a_3 & 2a_2 & 1a_1 & 0 \\
0 & 0 & 0 & 0 & 5 & 4a_4 & 3a_3 & 2a_2 & 1a_1
\end{vmatrix}
$$

A *good prime*[18] for a given polynomial $p(x)$ is a prime which does not divide the discriminant of $p(x)$. Let $\mathbb{Q}$ be the set of all rational numbers. We cite (see reference section) without proof:

Lemma 9.25 *Let $n = \deg(p(x))$. If $n > 2$ is even, then the joint occurrence of*

(i) *an $(n-1)$-cycle,*
(ii) *an n-cycle, and*
(iii) *a permutation of the type $2 + (n-3)$ on factoring the polynomial $p(x)$ modulo good primes*

implies that the Galois group of $p(x)$ over $\mathbb{Q}$ is the symmetric group S_n.

We explain the used notation. An $(n-1)$-*cycle occurs* if there exists a good prime ψ_1 such that

$$p(x) \bmod \psi_1 = f_1^1(x) f_2^1(x)$$

where $\deg(f_1^1(x)) = 1$ and $\deg(f_2^1(x)) = n - 1$, and $f_2^1(x)$ is irreducible over $\mathbb{Z}_{\psi_1}$.

An n-*cycle occurs* if there exists a good prime ψ_2 such that

$$p(x) \bmod \psi_2 = f(x)$$

where $\deg(f(x)) = n$ is irreducible over $\mathbb{Z}_{\psi_2}$.

A *permutation of the type $2 + (n-3)$ occurs* if there exists a good prime ψ_3 such that

$$p(x) \bmod \psi_3 = f_1^3(x) f_2^3(x) f_3^3(x)$$

where $\deg(f_1^3(x)) = 1$, $\deg(f_2^3(x)) = 2$, $\deg(f_3^3(x)) = n - 3$, and $f_2^3(x)$, $f_3^3(x)$ are irreducible over $\mathbb{Z}_{\psi_3}$.

We recall that the finite symmetric group S_n is the group of all permutations of n elements; it has order $n!$ and it is not Abelian (i.e., not commutative).[19] Every

---

[18] We follow [6]. However, this notion is not uniformly defined in literature; for a different use, see [62], for example.

[19] Named after the Norwegian mathematician *Niels Henrik Abel* (1802–1829).

element of S_n can be written as a product of cycles. A k-cycle is a permutation of
the given n elements such that a k-times repeated application of this permutation
maps at least one of the n elements onto itself.

A subgroup H of G is said to be a *normal subgroup* of G iff $ghg^{-1} \in H$, for
each $g \in G$ and $h \in H$. A group G is said to be *solvable* iff there exist subgroups
$G = H_0 \supset H_1 \supset H_2 \supset \cdots \supset H_r$ such that H_r is a singleton (i.e., it only contains the
identity element of G), H_i is normal in H_{i-1}, and H_{i-1}/H_i is Abelian.

Assume that $f(x) \in \mathbb{Q}[x]$; then a finite extension E of $\mathbb{Q}$ is said to be a *splitting
field* over $\mathbb{Q}$ for $f(x)$ iff $f(x)$ can be factored over E into a product of linear factors,
but not over any proper subfield of E.

The *Galois group over* $\mathbb{Q}$ of $p(x)$ is defined as a group of automorphisms of the
splitting field over $\mathbb{Q}$ for $f(x)$.[20]

Lemma 9.26 *The symmetric group S_n is not solvable for $n > 5$.*

Lemma 9.27 *If $p(x) \in \mathbb{Q}[x]$ is solvable by radicals over $\mathbb{Q}$, then the Galois group
over $\mathbb{Q}$ of $p(x)$ is a solvable group.*

The following methodology (which is used below) is a simplification of a 1967
algorithm by *Elwyn R. Berlekamp*. This algorithm was designed for the factorisation
of polynomials.[21]

Let ψ be a prime number. In this remainder of this section, all arithmetic on
polynomials is done modulo ψ. Let

$$u(x) = x^n + u_{n-1}x^{n-1} + \cdots + u_1 x + u_0$$

for $u_i \in \mathbb{Z}_\psi$ and $i = 0, 1, \ldots, n - 1$. Let

$$\alpha_0 = \begin{pmatrix} 0 \, 0 \ldots 0 \, 1 \end{pmatrix}$$

and

$$\alpha = \begin{pmatrix} a_{n-1} \, a_{n-2} \ldots a_1 \, a_0 \end{pmatrix}$$

be two n-dimensional row vectors, where $a_i \in \mathbb{Z}_\psi$ and $i = 0, 1, \ldots, n - 1$. We update α by Procedure 23 (see Fig. 9.40). We illustrate Procedure 23 by the following
example.

Example 9.14 Let $\psi = 19$. Consider the input

$$u(x) = x^6 + 12x^3 + 12x^2 + 13x + 15$$

also expressed by

$$\beta = \begin{pmatrix} 0 \, 0 \, 12 \, 12 \, 13 \, 15 \end{pmatrix}.$$

---

[20] See, for example, [24] for more details. The following two lemmas are Theorems 5.7.1 and 5.7.2
in [24].

[21] See, for example, [36].

Procedure 23 (Update a row vector α)
Input: Two n-dimensional row vectors $\alpha_0 = (0\ 0\ \ldots\ 0\ 1)$ and $\beta = (u_{n-1}\ u_{n-2}$
$\ldots\ u_1\ u_0)$, and a prime ψ.
Output: Updated row vector α.

  1: Let $k = 1$.
  2: **while** $k < \psi$ **do**
  3:    Let $\alpha = \alpha_0$.
  4:    Let $t = a_{n-1}$,

$$a_{n-1} = (a_{n-2} - tu_{n-1}) \bmod \psi,$$

$$\ldots$$

$$a_1 = (a_0 - tu_1) \bmod \psi$$

    and

$$a_0 = (-tu_0) \bmod \psi$$

  5:    Let $k = k + 1$.
  6: **end while**
  7: Output α.

Fig. 9.40 Update of a row vector α

Procedure 24 (Update of matrix Q)
Input: An $n \times n$ zero matrix Q, two n-dimensional row vectors $\alpha_0 = (0\ 0\ \ldots\ 0\ 1)$
and $\beta = (u_{n-1}\ u_{n-2}\ \ldots\ u_1\ u_0)$, and a prime ψ.
Output: Updated $n \times n$ matrix Q.

  1: Let $k = 1$.
  2: **while** $k < n$ **do**
  3:    Let α_0, β and ψ as input; apply Procedure 23 to update α.
  4:    Update Q by replacing its kth row by the reversed α.
  5:    Let $\alpha_0 = \alpha$.
  6:    Let $k = k + 1$.
  7: **end while**
  8: Output α.

Fig. 9.41 Update a matrix Q

Table 9.19 shows the updated row-vectors α_k, each corresponding to a number k of
iterations.   □

Procedure 24 (see Fig. 9.41) computes a matrix Q which is used for testing the
irreducibility of a polynomial. We provide the following example for illustrating
Procedure 24:

Table 9.19 Illustration of Procedure 23; see Example 9.14

k	$a_{k,5}$	$a_{k,4}$	$a_{k,3}$	$a_{k,2}$	$a_{k,1}$	$a_{k,0}$
0	0	0	0	0	0	1
1	0	0	0	0	1	0
2	0	0	0	1	0	0
3	0	0	1	0	0	0
4	0	1	0	0	0	0
5	1	0	0	0	0	0
6	0	0	7	7	6	4
7	0	7	7	6	4	0
8	7	7	6	4	0	0
9	7	6	15	11	4	9
10	6	15	3	15	13	9
11	15	3	0	17	7	5
12	3	0	8	17	0	3
13	0	8	0	2	2	12
14	8	0	2	2	12	0
15	0	2	1	11	10	13
16	2	1	11	10	13	0
17	1	11	5	8	12	8
18	11	5	15	0	14	4
19	5	15	1	15	13	6

Example 9.15 We start with $u(x) = x^6 + 12x^3 + 12x^2 + 13x + 15$, as obtained in the example before; and reverse the last row in Table 9.19. We have

$$\beta = (6\ 13\ 15\ 1\ 15\ 5)$$

as input for Procedure 24. Table 9.20 shows the updated vector α corresponding to the number $k = 3$ of iterations. Thus,

$$(2\ 5\ 6\ 2\ 13\ 0)$$

is now the third row of Q.   □

Let I be the $n \times n$ identity matrix and $\deg(u(x)) = n$. The following proposition by *Elwyn R. Berlekamp*[22] gives a necessary and sufficient condition for testing the irreducibility of a given polynomial; this condition is later used below:

Proposition 9.4 $u(x)$ *is irreducible iff the rank of matrix $Q - I$ equals $n - 1$.*

---

[22] See [36], page 441.

Table 9.20 Illustration of Procedure 24: calculation of the third row of Q (before reversing)

k	$a_{k,5}$	$a_{k,4}$	$a_{k,3}$	$a_{k,2}$	$a_{k,1}$	$a_{k,0}$
0	5	15	1	15	13	6
1	15	1	12	10	17	1
2	1	12	1	8	15	3
3	12	1	15	3	9	4
4	1	15	11	17	0	10
5	15	11	5	7	16	4
6	11	5	17	7	18	3
7	5	17	8	0	12	6
8	17	8	16	9	17	1
9	8	16	14	3	8	11
10	16	14	2	7	2	13
11	14	2	5	0	14	7
12	2	5	3	17	15	18
13	5	3	12	10	11	8
14	3	12	7	8	0	1
15	12	7	10	2	0	12
16	7	10	10	8	8	10
17	10	10	0	0	14	9
18	10	0	13	8	12	2
19	0	13	2	6	5	2

Example 9.16 For illustrating this theorem, we consider the simple cube-arc ρ between e_0 and e_3 in Fig. 9.2 (see also Table 9.3). The coordinates of $p_0(t_0)$ and $p_3(t_3)$ are equal to $(1, 4, 7)$ and $(4, 7, 4)$, respectively, where $t_0 = t_3 = 0$.

We want to detect $t_1, t_2 \in [0, 1]$ such that a polyline $p_0(t_0)p_1(t_1)p_2(t_2)p_3(t_3)$ is fully contained in ρ. By Eqs. (9.12) and (9.13), we have that

$$0 = \frac{t_1}{\sqrt{t_1^2 + 5}} + \frac{t_1 - 1}{\sqrt{(t_1 - 1)^2 + (t_2 - 1)^2 + 4}}, \tag{9.22}$$

$$0 = \frac{t_2 - 1}{\sqrt{(t_1 - 1)^2 + (t_2 - 1)^2 + 4}} + \frac{t_2}{\sqrt{t_2^2 + 4}}. \tag{9.23}$$

In the following, we show that the MLP problem defined by the simple cube-curve in Fig. 9.2 is reduced to finding the roots of the polynomial

$$p(x) = 84x^6 - 228x^5 + 361x^4 + 20x^3 + 210x^2 - 200x + 25.$$

Let $t_1 = x$ and $t_2 = y$ in Eqs. (9.22) and (9.23), respectively. Equations (9.24), (9.26)–(9.32) represent, step by step, the simplification of Eq. (9.22); Eqs. (9.25), (9.27)–(9.33) show the operations of simplifying Eq. (9.23):

$$0 = \frac{x}{\sqrt{x^2 + 5}} + \frac{x - 1}{\sqrt{(x - 1)^2 + (y - 1)^2 + 4}}, \tag{9.24}$$

$$0 = \frac{y - 1}{\sqrt{(x - 1)^2 + (y - 1)^2 + 4}} + \frac{y}{\sqrt{y^2 + 4}}, \tag{9.25}$$

$$\frac{x}{\sqrt{x^2 + 5}} = -\frac{x - 1}{\sqrt{(x - 1)^2 + (y - 1)^2 + 4}}, \tag{9.26}$$

$$\frac{y - 1}{\sqrt{(x - 1)^2 + (y - 1)^2 + 4}} = -\frac{y}{\sqrt{y^2 + 4}}, \tag{9.27}$$

$$x\sqrt{(x - 1)^2 + (y - 1)^2 + 4} = -(x - 1)\sqrt{x^2 + 5}, \tag{9.28}$$

$$(y - 1)\sqrt{y^2 + 4} = -y\sqrt{(x - 1)^2 + (y - 1)^2 + 4}, \tag{9.29}$$

$$x^2\left[(x - 1)^2 + (y - 1)^2 + 4\right] = (x - 1)^2(x^2 + 5), \tag{9.30}$$

$$(y - 1)^2(y^2 + 4) = y^2\left[(x - 1)^2 + (y - 1)^2 + 4\right], \tag{9.31}$$

$$x^2\left[(y - 1)^2 + 4\right] = 5(x - 1)^2, \tag{9.32}$$

$$4(y - 1)^2 = y^2\left[(x - 1)^2 + 4\right]. \tag{9.33}$$

Equations (9.34), (9.35), and (9.36) show the operations for representing y in terms of x, based on Eq. (9.32):

$$(y - 1)^2 = \frac{5(x - 1)^2}{x^2} - 4$$

$$= \frac{5(x - 1)^2 - 4x^2}{x^2}$$

$$= \frac{5(x^2 - 2x + 1) - 4x^2}{x^2}$$

$$= \frac{x^2 - 10x + 5}{x^2}, \tag{9.34}$$

$$y - 1 = -\frac{\sqrt{x^2 - 10x + 5}}{x}, \tag{9.35}$$

$$y = 1 - \frac{\sqrt{x^2 - 10x + 5}}{x}. \tag{9.36}$$

Substituting y in Eq. (9.33) by the right-hand side of Eq. (9.36), we have the following:

$$4\left(1 - \frac{\sqrt{x^2 - 10x + 5}}{x} - 1\right)^2 = \left(1 - \frac{\sqrt{x^2 - 10x + 5}}{x}\right)^2 [(x-1)^2 + 4]. \quad (9.37)$$

Equations (9.38), (9.39)–(9.43) show the operations of simplifying Eq. (9.37):

$$4\frac{x^2 - 10x + 5}{x^2} = \left(\frac{x - \sqrt{x^2 - 10x + 5}}{x}\right)^2 [(x-1)^2 + 4], \quad (9.38)$$

$$\frac{4(x^2 - 10x + 5)}{(x-1)^2 + 4} = \left(x - \sqrt{x^2 - 10x + 5}\right)^2$$

$$= x^2 + x^2 - 10x + 5 - 2x\sqrt{x^2 - 10x + 5}$$

$$= 2x^2 - 10x + 5 - 2x\sqrt{x^2 - 10x + 5}, \quad (9.39)$$

$$\frac{4(x^2 - 10x + 5)}{(x-1)^2 + 4} - (2x^2 - 10x + 5) = -2x\sqrt{x^2 - 10x + 5}, \quad (9.40)$$

$$\frac{4(x^2 - 10x + 5) - (2x^2 - 10x + 5)(x^2 - 10x + 5)}{x^2 - 2x + 5}$$

$$= -2x\sqrt{x^2 - 10x + 5}, \quad (9.41)$$

$$\frac{[4(x^2 - 10x + 5) - (2x^2 - 10x + 5)(x^2 - 10x + 5)]^2}{(x^2 - 2x + 5)^2}$$

$$= 4x^2(x^2 - 10x + 5), \quad (9.42)$$

$$[4(x^2 - 10x + 5) - (2x^2 - 10x + 5)(x^2 - 2x + 5)]^2$$

$$= 4x^2(x^2 - 10x + 5)(x^2 - 2x + 5)^2. \quad (9.43)$$

The left-hand side of Eq. (9.43) can be further simplified as follows:

$$[4(x^2 - 10x + 5) - (2x^2 - 10x + 5)(x^2 - 2x + 5)]^2,$$

$$[4x^2 - 40x + 20 - (2x^4 - 14x^3 + 35x^2 - 60x + 25)]^2,$$

$$[-2x^4 + 14x^3 - 31x^2 + 20x - 5]^2,$$

$$4x^8 - 56x^7 + 320x^6 - 948x^5 + 1541x^4 - 1380x^3 + 710x^2 - 200x + 25.$$

The right-hand side of Eq. (9.43) can be further simplified as follows:

$$4x^2(x^2 - 10x + 5)(x^2 - 2x + 5)^2,$$

$$4x^2(x^2 - 10x + 5)(x^4 - 4x^3 + 14x^2 - 20x + 25),$$
$$4x^2(x^6 - 14x^5 + 59x^4 - 180x^3 + 295x^2 - 350x + 125),$$
$$4x^8 - 56x^7 + 236x^6 - 720x^5 + 1180x^4 - 1400x^3 + 500x^2.$$

Altogether, we obtain the following polynomial

$$84x^6 - 228x^5 + 361x^4 + 20x^3 + 210x^2 - 200x + 25 = 0$$

as a simplification of Eq. (9.43). In the following, we consider this polynomial

$$p(x) = 84x^6 - 228x^5 + 361x^4 + 20x^3 + 210x^2 - 200x + 25.$$

After having the polynomial derived, we need to find three good primes.

To prove Theorem 9.14, by Lemmas 9.25 to 9.27, it suffices to find three good primes, ψ_1, ψ_2, and ψ_3 for $p(x)$ such that:

1. $p(x) \bmod \psi_1$ can be factorised as one irreducible polynomial $f_1^1(x)$ over the field $\mathbb{Z}_{\psi_1}$ such that

$$\deg(f_1^1(x)) = \deg(p(x));$$

2. $p(x) \bmod \psi_2$ can be factorised as two irreducible polynomials $f_1^2(x)$ and $f_2^2(x)$ over the field $\mathbb{Z}_{\psi_2}$ such that

$$\deg(f_1^2(x)) = 1 \quad \text{and} \quad \deg(f_2^2(x)) = \deg(p(x)) - 1;$$

3. $p(x) \bmod \psi_3$ can be factorised as three irreducible polynomials $f_1^3(x)$, $f_2^3(x)$ and $f_3^3(x)$ over the field $\mathbb{Z}_{\psi_3}$ such that

$$\deg(f_1^3(x)) = 1, \qquad \deg(f_2^3(x)) = 2 \quad \text{and} \quad \deg(f_3^3(x)) = \deg(g(x)) - 3.$$

Indeed, using mathematical software[23] we identified three good primes as desired (for further details, see the Appendix), namely:

1. $\psi_1 = 19$:

$$f_1^1(x) = Z(19)^3 * x^6 + x^3 + x^2 + Z(19)^8 * x + Z(19)^{14};$$

2. $\psi_2 = 37$:

$$f_1^2(x) = Z(37)^{24} * x + Z(37)^0,$$

---

[23] We used GAP, see [22].

$$f_2^2(x) = x^5 + Z(37)^{26} * x^4 + Z(37)^{22} * x^3$$
$$+ Z(37)^{30} * x^2 + Z(37)^9 * x + Z(37)^{10};$$

3. $\psi_3 = 13$:

$$f_1^3(x) = Z(13)^5 * x + Z(13)^{10},$$
$$f_2^3(x) = x^2 + Z(13)^5,$$
$$f_3^3(x) = x^3 + Z(13)^3 * x^2 + Z(13)^2 * x + Z(13)^3.$$

We need to show that all those polynomials are irreducible.

Case of $f_1^1(x)$: We prove that $f_1^1(x)$ is irreducible.

(1) We simplify $f_1^1(x)$ to a monic polynomial. Because 2 is a primitive element of Z_{19},[24] we have

$$f_1^1(x) = Z(19)^3 * x^6 + x^3 + x^2 + Z(19)^8 * x + Z(19)^{14}$$
$$= 2^3 * x^6 + x^3 + x^2 + 2^8 * x + 2^{14}$$
$$= 8x^6 + x^3 + x^2 + 9x + 6.$$

Because $8 \times 12 \equiv 1 \bmod 19$,

$$f_1^1(x) = 8x^6 + x^3 + x^2 + 9x + 6$$
$$= 12(8x^6 + x^3 + x^2 + 9x + 6)$$
$$= x^6 + 12x^3 + 12x^2 + 13x + 15.$$

The final row is the desired monic representation.

(2) We apply Proposition 9.4 to prove that $f_1^1(x)$ is irreducible. We summarise the computation here; for details see Tables 9.19 and 9.20, and the remaining tables in the Appendix. Note that the rows computed in these tables have to be reversed to be the rows of the following matrix Q. We obtain that

$$Q = \begin{bmatrix} 1 & 0 & 0 & 0 & 0 & 0 \\ 6 & 13 & 15 & 1 & 15 & 5 \\ 2 & 5 & 6 & 2 & 13 & 0 \\ 2 & 10 & 13 & 13 & 0 & 14 \\ 17 & 10 & 10 & 1 & 2 & 7 \\ 4 & 16 & 8 & 10 & 18 & 3 \end{bmatrix} \quad \text{and} \quad Q - I = \begin{bmatrix} 0 & 0 & 0 & 0 & 0 & 0 \\ 6 & 12 & 15 & 1 & 15 & 5 \\ 2 & 5 & 5 & 2 & 13 & 0 \\ 2 & 10 & 13 & 12 & 0 & 14 \\ 17 & 10 & 10 & 1 & 1 & 7 \\ 4 & 16 & 8 & 10 & 18 & 2 \end{bmatrix}.$$

---

[24]For each $a \in Z_{19}$, there exists a number $j \in \{0, 1, \ldots, 18\}$ such that $2^j \equiv a \bmod 19$.

Since

$$
\begin{vmatrix}
6 & 12 & 15 & 1 & 15 \\
2 & 5 & 5 & 2 & 13 \\
2 & 10 & 13 & 12 & 0 \\
17 & 10 & 10 & 1 & 1 \\
4 & 16 & 8 & 10 & 18
\end{vmatrix} = -160{,}520 \equiv 11 \bmod 19,
$$

we finally obtain that $\mathrm{rank}(Q - I) = 5$. By Proposition 9.4, we can conclude that $f_1^1(x)$ is irreducible.

Case of $f_2^2(x)$: To prove that $f_2^2(x)$ is irreducible, we proceed as follows:

(1) We simplify $f_2^2(x)$. Because 2 is a primitive element of $\mathbb{Z}_{37}$, we have the following:

$$
\begin{aligned}
f_2^2(x) &= x^5 + Z(37)^{26} \times x^4 + Z(37)^{22} \times x^3 \\
&\quad + Z(37)^{30} \times x^2 + Z(37)^9 \times x + Z(37)^{10} \\
&= x^5 + 2^{26} \times x^4 + 2^{22} \times x^3 + 2^{30} \times x^2 + 2^9 \times x + 2^{10} \\
&= x^5 + 3x^4 + 21x^3 + 11x^2 + 31x + 25.
\end{aligned}
$$

(2) Now we apply Proposition 9.4 to prove that $f_2^2(x)$ is irreducible. We summarise the computation here; for details see the tables in the Appendix. We obtain that

$$
Q = \begin{bmatrix}
1 & 0 & 0 & 0 & 0 \\
3 & 1 & 11 & 21 & 36 \\
28 & 29 & 22 & 34 & 33 \\
23 & 7 & 1 & 22 & 31 \\
32 & 35 & 24 & 35 & 28
\end{bmatrix}
\quad \text{and} \quad
Q - I = \begin{bmatrix}
0 & 0 & 0 & 0 & 0 \\
3 & 0 & 11 & 21 & 36 \\
28 & 29 & 21 & 34 & 33 \\
23 & 7 & 1 & 21 & 31 \\
32 & 35 & 24 & 35 & 27
\end{bmatrix}.
$$

Since

$$
\begin{vmatrix}
3 & 0 & 11 & 21 \\
28 & 29 & 21 & 34 \\
23 & 7 & 1 & 21 \\
32 & 35 & 24 & 35
\end{vmatrix} = 9{,}835 \equiv 30 \bmod 37,
$$

we finally have that $\mathrm{rank}(Q - I) = 4$. By Proposition 9.4, this proves that $f_2^2(x)$ is irreducible.

Case of $f_2^3(x)$: To prove that $f_2^3(x)$ is irreducible, we proceed as follows:

(1) We simplify $f_2^3(x)$. Because 2 is a primitive element of $\mathbb{Z}_{13}$, we have the following:

$$
\begin{aligned}
f_2^3(x) &= x^2 + Z(13)^5 \\
&= x^2 + 6.
\end{aligned}
$$

(2) We apply Proposition 9.4 to prove that $f_2^3(x)$ is irreducible. For details see the table in the Appendix. We obtain that

$$Q = \begin{bmatrix} 1 & 0 \\ 0 & 12 \end{bmatrix} \quad \text{and} \quad Q - I = \begin{bmatrix} 0 & 0 \\ 0 & 11 \end{bmatrix}.$$

Thus, we have that $\text{rank}(Q - I) = 1$. By Proposition 9.4, this shows that also $f_2^3(x)$ is irreducible.

Case of $f_3^3(x)$: To prove that $f_3^3(x)$ is irreducible, we proceed as follows:

(1) We simplify $f_3^3(x)$. Since 2 is a primitive element of Z_{13}, we have the following:

$$f_3^3(x) = x^3 + Z(13)^3 * x^2 + Z(13)^2 * x + Z(13)^3$$
$$= x^3 + 8x^2 + 4x + 8.$$

(2) We apply Proposition 9.4 to prove that $f_3^3(x)$ is irreducible. For details see tables in the Appendix. We obtain that

$$Q = \begin{bmatrix} 1 & 0 & 0 \\ 1 & 3 & 0 \\ 1 & 6 & 9 \end{bmatrix} \quad \text{and} \quad Q - I = \begin{bmatrix} 0 & 0 & 0 \\ 1 & 2 & 0 \\ 1 & 6 & 8 \end{bmatrix}.$$

From this we see that $\text{rank}(Q - I) = 2$. By Proposition 9.4, we know that $f_3^3(x)$ is irreducible. This concludes the example. □

We have shown that there does not exist an exact algorithm for solving the general MLP problem for simple cube-curves, thus also not for more general ESP problems.

Interestingly, this result is also true for the 2.5D case (surface ESP, see Sect. 7.5) but not true for the 2D case. In 2D, there exist exact algorithms for ESP problems, such as, for example, the MLP problem in 2D space.

9.10 Problems

Problem 9.1 Discuss differences between critical edge, critical line, and critical face, first-class simple cube-cube versus non-first-class simple cube-cube, and between end-angle, middle-angle, and inner-angle.

Problem 9.2 What has been changed in the original RBA compared to the revised RBA discussed in Sect. 9.5.

Problem 9.3 Discuss differences between (2,3)-cube-arc, maximal (2,3)-cube-arc, 2-cube-arc, maximal 2-cube-arc, 2-cube-arc unit, 3-cube-arc unit, regular cube-arc unit, and cube-arc unit.

Problem 9.4 Show that Procedure 22 can be computed in $\mathcal{O}(1)$.

Problem 9.5 Procedures 20 and 21 are both modified from Algorithm 7. Underline the modifications of Procedure 20 in Fig. 9.30, and of Procedure 21 in Fig. 9.31.

Problem 9.6 Can Algorithm 36 be simplified?

Problem 9.7 Algorithm 37 is also modified from Algorithm 7. Underline modifications of Algorithm 37 in Fig. 9.35.

Problem 9.8 Consider the last line of Algorithm 33. Would it be possible to replace the original RBA by Algorithm 7 by or Algorithm 8?

Problem 9.9 Consider Line 9 of Algorithm 33. Assume that the Dijkstra algorithm is implemented in a Fibonacci heap. This defines a smaller upper bound for the time complexity of Algorithm 33. Identify such a smaller upper bound.

Problem 9.10 Find a non-trivial lower bound for the length of an MLP, as calculated by Algorithm 33.

Problem 9.11 (Programming exercise) Design an "arc" version of an rubberband algorithm for computing MLPs for cube-curves with an end-angle, and compare its performance with that of Algorithms 34 in Fig. 9.12.

Problem 9.12 Prove that the approximation factor $\{1 + 4(m + 1) \cdot [r(\varepsilon_1) + \sqrt{2} \times \varepsilon_2]/L\}$ in Theorem 9.12 can be replaced by $\{1 + 2(m + 1) \cdot [r(\varepsilon_1) + \sqrt{2} \times \varepsilon_2]/L\}$.

Problem 9.13 (Programming exercise) Implement the generic RBA (i.e., Algorithm 11 for the 3D case) as described in Fig. 3.24, and compare its performance with those of Algorithms 33 (in Fig. 9.5) and 37 (in Fig. 9.35).

Problem 9.14 (Open problems)

(a) What is the smallest (say, in number of cubes or in number of critical edges— both are equivalent) simple cube-curve which does not have any end-angle?
(b) What is the smallest (say, in number of cubes or in number of critical edges— both are equivalent) simple cube-curve which does not have any of its MLP vertices at a grid point location?

Problem 9.15 (Research problem) Design an algorithm for generating random simple cube curves. Note that there cannot be any crossings or 'touchings'.

9.11 Notes

The difficulty of ESP problems can also be verified by combinatorial considerations; see [45, 46]. For the MLP problem, a combinatorial difficulty is characterised by the complexity of computing the step set (i.e., a subset of the set of all critical edges). such that each edge in this set contains (exactly) one vertex of the MLP.

The problem of length estimation in picture analysis dates back to the beginning of the 1970s. For example, [47] presented a method for extracting a smooth polygonal contour from a digitised image aiming at a minimum-length polygonal approximation. For later work, we cite [3, 4, 18, 20] as examples which were focused on the problem in 2D. Reference [33, 34] compared the DSS and MLP techniques[25] for measuring the length of a digital curve in 2D (in fact, both techniques allow generalisations to 3D and beyond) based on time-efficient (linear) algorithms for both 2D techniques.

For the 3D case, [1] presents two methods for estimating the length of digitised straight lines. Reference [13] gives four types of characterisation schemes for this problem. Reference [31] studies shortest paths between points on a digital 3D surface. Reference [30] considers local length estimators for curves in 3D space, derived from their chain codes. Reference [48] discusses the problem of approximating the length of a parametric curve $\gamma(t)$, with $\gamma : [0, 1] \to \mathbb{R}^n$, from sampled points $q_i = \gamma(t_i)$, where the parameters t_i are unknown. See also [14, 15, 50, 53].

The computation of the length of a simple cube-curve in 3D Euclidean space was a subject in [27], but the proposed method may produce errors for specific curves. Reference [11] presents the original rubberband algorithm (see Sect. 9.2) for computing an approximating MLP in the tube $\mathbf{g}$ of cube-curve g, with a measured runtime allowing to expect a general $\mathcal{O}(n)$ behaviour, where n is the number of grid cubes in the given cube-curve. However, no proof was given that this original rubberband algorithm actually converges toward the correct MLP, and also no analysis of its worst-case time complexity.

3D MLP calculations generalise MLP computations in 2D; see, for example, [29, 54] for theoretical results and [19, 59] for 2D robotics scenarios. Shortest curve calculations in image analysis also use graph metrics instead of the Euclidean metric; see, for example, [57].

Interest in 3D MLPs was also raised by the issue of multigrid-convergent length estimation for digitised curves. The length of a simple cube-curve in 3D Euclidean space can be defined by that of the MLP; see [33, 55, 56], which is there characterised to be a *global approach* toward length measurement. A *local approach* for 3D length estimation, allowing only weighted steps within a restricted neighbourhood, was considered in [26] and [27]. Alternatively to the MLP, the length of 3D digital curves can also be measured (within linear time in the number of grid points on the curve) based on DSS-approximations [16] (DSS = digital straight segment).

The computation of 3D MLPs was first published in [9–11, 32], proposing a 'rubberband' algorithm.[26] This iterative algorithm was experimentally tested and

---

[25]The *DSS technique* represents a digital curve, assumed to be a sequence of grid points, by subsequent digital straight segments (DSSs) of maximum length; the *MLP technique* considers a digital curve as a sequence of grid cells, calculating a minimum-length polygon (MLP) in the union of these cells.

[26]Not to be confused with a 2D image segmentation algorithm of the same name [44].

showed "linear run-time behavior" with respect to a pre-selected accuracy constant $\varepsilon > 0$. It proved to be correct for tested inputs, where correctness was possible to be tested manually. However, in [9–11, 32], no mathematical proof was given for linear run time or general convergence (in the sense of our definition of approximate algorithms) to the exact solution. Nevertheless, the algorithm has been used since 2002 (e.g., in DNA research).

This original RBA is also published in the book [33]. Applications of this algorithm are also in 3D medical imaging; see, for example, [21, 61].

The correctness and linearity problem of the original RBA was approached along the following steps:

Reference [38] only considered a very special class of simple cube-curves and developed a provable correct MLP algorithm for this class. The main idea was to decompose a cube-curve of that class into arcs at end angles (see Definition 3 in [38]), which means that the cube-curves have to have end-angles, and then the algorithm can be applied.

Reference [39] constructed an example of a simple cube-curve whose MLP does not have any of its vertices at a corner of a grid cube. It follows that any cube-curve with this property does not have any end angle, and this means that we cannot use the MLP algorithm as proposed in [38]. This was the basic importance of the result in [39]: it showed the existence of cube-curves which require further algorithmic studies.

Reference [42] showed that the original RBA requires a modification (in its Option 3) to guarantee that calculated curves are always contained in the tube **g**. This revised RBA achieves (as the original RBA) minimisation of length by moving vertices along critical edges.

Reference [41] (finally) extended the revised RBA into the edge-based RBA and showed that it is correct for *any* simple cube-curve. [41] also presented a totally new algorithm, the face-based RBA, and showed that it is also correct for any simple cube-curve. [43] proposed a very simple approximate face-based RBA. It was proved that the edge-based and the face-based RBAs have time complexity $\kappa(\varepsilon) \cdot \mathcal{O}(m)$, where m is the number of critical edges in the given simple cube-curve, and

$$\kappa(\varepsilon) = (L_0 - L)/\varepsilon \qquad (9.44)$$

where L_0 is the length of the initial path, L is the true (i.e., optimum) path. This chapter reported about those publications. We also recall from Chap. 1 that an algorithm is an $(1 + \varepsilon)$-*approximation algorithm* for a minimisation problem P iff, for each input instance I of P, the algorithm delivers a solution that is at most $(1 + \varepsilon)$ times the optimum solution [25].

The introduction of MLPs in simple cube curves follows [32, 55, 56].

Visibility graphs are common in computational geometry; see, for example, [2, 23, 28, 37, 49, 60].

For convex analysis, see, for example, [8, 51, 52]. For Theorem 9.11, see [52], Theorem 3.5. For Proposition 9.1, see [8], page 72. For Proposition 9.2, see [52], page 264.

Lemma 9.25 is the first part (i.e., in the case that n is even) of Lemma 8 in [6].

References

1. Amarunnishad, T.M., Das, P.P.: Estimation of length for digitized straight lines in three dimensions. Pattern Recognit. Lett. **11**, 207–213 (1990)
2. Asano, T., Asano, T., Guibas, L., Hershberger, J., Imai, H.: Visibility of disjoint polygons. Algorithmica **1**, 49–63 (1986)
3. Asano, T., Kawamura, Y., Klette, R., Obokata, K.: Minimum-length polygons in approximation sausages. In: Proc. Int. Workshop Visual Form. LNCS, vol. 2059, pp. 103–112. Springer, Berlin (2004)
4. Bailey, D.: An efficient Euclidean distance transform. In: Proc. Int. Workshop Combinatorial Image Analysis. LNCS, vol. 3322, pp. 394–408. Springer, Berlin (2004)
5. Bajaj, C.: The algebraic complexity of shortest paths in polyhedral spaces. In: Proc. Allerton Conf. Commum. Control Comput., pp. 510–517 (1985)
6. Bajaj, C.: The algebraic degree of geometric optimization problems. Discrete Comput. Geom. **3**, 177–191 (1988)
7. Berlekamp, E.R.: Factoring polynomials over large finite fields. Math. Comp. **24**, 713–735 (1970)
8. Boyd, S., Vandenberghe, L.: Convex Optimization. Cambridge University Press, Cambridge, UK (2004)
9. Bülow, T., Klette, R.: Rubber band algorithm for estimating the length of digitized spacecurves. In: Proc. Intern. Conf. Pattern Recognition, vol. 3, pp. 551–555 (2000)
10. Bülow, T., Klette, R.: Approximation of 3D shortest polygons in simple cube curves. In: Proc. Digital and Image Geometry. LNCS, vol. 2243, pp. 281–294. Springer, Berlin (2001)
11. Bülow, T., Klette, R.: Digital curves in 3D space and a linear-time length estimation algorithm. IEEE Trans. Pattern Anal. Mach. Intell. **24**, 962–970 (2002)
12. Burden, R.L., Faires, J.D.: Numerical Analysis, 7th edn. Brooks Cole, Pacific Grove (2000)
13. Chattopadhyay, S., Das, P.P.: Estimation of the original length of a straight line segment from its digitization in three dimensions. Pattern Recognit. **25**, 787–798 (1992)
14. Choi, J., Sellen, J., Yap, C.-K.: Approximate Euclidean shortest path in 3-space. In: ACM Conf. Computational Geometry, pp. 41–48. ACM Press, New York (1994)
15. Choi, J., Sellen, J., Yap, C.-K.: Precision-sensitive Euclidean shortest path in 3-space. In: Proc. Annu. ACM Sympos. Computational Geometry, pp. 350–359 (1995)
16. Coeurjolly, D., Debled-Rennesson, I., Teytaud, O.: Segmentation and length estimation of 3D discrete curves In: Proc. Digital and Image Geometry. LNCS, vol. 2243, pp. 299–317. Springer, Berlin (2001)
17. Cormen, T.H., Leiserson, C.E., Rivest, R.L., Stein, C.: Introduction to Algorithms. MIT Press, Cambridge (2001)
18. Dorst, L., Smeulders, A.W.M.: Length estimators for digitized contours. Comput. Vis. Graph. Image Process. **40**, 311–333 (1987)
19. Dror, M., Efrat, A., Lubiw, A., Mitchell, J.: Touring a sequence of polygons. In: Proc. STOC, pp. 473–482 (2003)
20. Ellis, T.J., Proffitt, D., Rosen, D., Rutkowski, W.: Measurement of the lengths of digitized curved lines. Comput. Graph. Image Process. **10**, 333–347 (1979)
21. Ficarra, E., Benini, L., Macii, E., Zuccheri, G.: Automated DNA fragments recognition and sizing through AFM image processing. IEEE Trans. Inf. Technol. Biomed. **9**, 508–517 (2005)
22. GAP—Groups, Algorithms, Programming—a system for computational discrete algebra. www-gap.mcs.st-and.ac.uk/gap.html (2011). Accessed July 2011
23. Ghosh, S.K., Mount, D.M.: An output sensitive algorithm for computing visibility graphs. SIAM J. Comput. **20**, 888–910 (1991)
24. Herstein, I.N.: Topics in Algebra, 2nd edn. Wiley, New York (1975)
25. Hochbaum, D.S. (ed.): Approximation Algorithms for NP-Hard Problems. PWS Pub. Co., Boston (1997)
26. Jonas, A., Kiryati, N.: Length estimation in 3-D using cube quantization. In: Proc. Vision Geometry. SPIE, vol. 2356, pp. 220–230 (1994)

27. Jonas, A., Kiryati, N.: Length estimation in 3-D using cube quantization. J. Math. Imaging Vis. **8**, 215–238 (1998)
28. Kapoor, S., Maheshwari, S.N.: Efficient algorithms for Euclidean shortest path and visibility problems with polygonal. In: Proc. Annu. ACM Sympos. on Computational Geometry, pp. 172–182 (1988)
29. Karavelas, M.I., Guibas, L.J.: Static and kinetic geometric spanners with applications. In: Proc. ACM–SIAM Symp. Discrete Algorithms, pp. 168–176 (2001)
30. Kiryati, N., Kubler, O.: On chain code probabilities and length estimators for digitized three-dimensional curves. Pattern Recognit. **28**, 361–372 (1995)
31. Kiryati, N., Szekely, G.: Estimating shortest paths and minimal distances on digitized three-dimensional surfaces. Pattern Recognit. **26**, 1623–1637 (1993)
32. Klette, R., Bülow, T.: Critical edges in simple cube-curves. In: Proc. Discrete Geometry Computational Imaging. LNCS, vol. 1953, pp. 467–478. Springer, Berlin (2000)
33. Klette, R., Rosenfeld, A.: Digital Geometry. Morgan Kaufmann, San Francisco (2004)
34. Klette, R., Yip, B.: The length of digital curves. Mach. Graph. Vis. **9**, 673–703 (2000)
35. Klette, R., Kovalevsky, V., Yip, B.: Length estimation of digital curves. In: Proc. Vision Geometry. SPIE, vol. 3811, pp. 117–129 (1999)
36. Knuth, D.E.: The Art of Computer Programming, vol. 2, 3rd edn. Addison-Wesley, Reading (1997)
37. Lee, D.T.: Proximity and reachability in the plane. Ph.D. thesis, University of Illinois at Urbana–Champaign, Urbana (1978)
38. Li, F., Klette, R.: Minimum-length polygon of a simple cube-curve in 3D space. In: Proc. Int. Workshop Combinatorial Image Analysis. LNCS, vol. 3322, pp. 502–511. Springer, Berlin (2004)
39. Li, F., Klette, R.: The class of simple cube-curves whose MLPs cannot have vertices at grid points. In: Proc. Discrete Geometry Computational Imaging. LNCS, vol. 3429, pp. 183–194. Springer, Berlin (2005)
40. Li, F., Klette, R.: Minimum-length polygons of first-class simple cube-curves. In: Proc. Computer Analysis Images Patterns. LNCS, vol. 3691, pp. 321–329. Springer, Berlin (2005)
41. Li, F., Klette, R.: Shortest paths in a cuboidal world. In: Proc. Int. Workshop Combinatorial Image Analysis. LNCS, vol. 4040, pp. 415–429. Springer, Berlin (2006)
42. Li, F., Klette, R.: Analysis of the rubberband algorithm. Image Vis. Comput. **25**(10), 1588–1598 (2007)
43. Li, F., Pan, X.: An approximation algorithm for computing minimum-length polygons in 3D images. In: Proc. The 10th Asian Conference on Computer Vision (ACCV 2010). LNCS, vol. 6495, pp. 641–652. Springer, Berlin (2011)
44. Luo, H., Eleftheriadis, A.: Rubberband: an improved graph search algorithm for interactive object segmentation. In: Proc. Int. Conf. Image Processing, vol. 1, pp. 101–104 (2002)
45. Mitchell, J.S.B.: Geometric shortest paths and network optimization. In: Sack, J.-R., Urrutia, J. (eds.) Handbook of Computational Geometry, pp. 633–701. Elsevier, Amsterdam (2000)
46. Mitchell, J.S.B., Sharir, M.: New results on shortest paths in three dimensions. In: Proc. SCG, pp. 124–133 (2004)
47. Montanari, U.: A note on minimal length polygonal approximations to a digitalized contour. Commun. ACM **13**, 41–47 (1970)
48. Noakes, L., Kozera, R., Klette, R.: Length estimation for curves with different samplings. In: Digital and Image Geometry. LNCS, vol. 2243, pp. 334–346. Springer, Berlin (2001)
49. Overmars, M.H., Welzl, E.: New methods for constructing visibility graphs. In: Proc. Annu. ACM Sympos. on Computational Geometry, pp. 164–171 (1988)
50. Papadimitriou, C.H.: An algorithm for shortest path motion in three dimensions. Inf. Process. Lett. **20**, 259–263 (1985)
51. Roberts, A.W., Varberg, V.D.: Convex Functions. Academic Press, New York (1973)
52. Rockafellar, R.T.: Convex Analysis. Princeton University Press, Princeton (1970)
53. Sharir, M., Schorr, A.: On shortest paths in polyhedral spaces. SIAM J. Comput. **15**, 193–215 (1986)

54. Sklansky, J., Kibler, D.F.: A theory of nonuniformly digitized binary pictures. IEEE Trans. Syst. Man Cybern. **6**, 637–647 (1976)
55. Sloboda, F., Zat'ko, B., Klette, R.: On the topology of grid continua. In: Proc. Vision Geometry. SPIE, vol. 3454, pp. 52–63 (1998)
56. Sloboda, F., Zat'ko, B., Stoer, J.: On approximation of planar one-dimensional grid continua. In: Klette, R., Rosenfeld, A., Sloboda, F. (eds.) Advances in Digital and Computational Geometry, pp. 113–160. Springer, Singapore (1998)
57. Sun, C., Pallottino, S.: Circular shortest path on regular grids. CMIS Report 01/76, CSIRO Math. Information Sciences, Australia (2001)
58. Sunday, D.: Algorithm 15: convex hull of a 2D simple polygonal path. www.softsurfer.com/Archive/algorithm_0203/ (2011). Accessed July 2011
59. Talbot, M.: A dynamical programming solution for shortest path itineraries in robotics. Electr. J. Undergrad. Math. **9**, 21–35 (2004)
60. Welzl, E.: Constructing the visibility graph for n line segments in $\mathcal{O}(n^2)$ time. Inf. Process. Lett. **20**, 167–171 (1985)
61. Wolber, R., Stäb, F., Max, H., Wehmeyer, A., Hadshiew, I., Wenck, H., Rippke, F., Wittern, K.: Alpha-Glucosylrutin: Ein hochwirksams Flavonoid zum Schutz vor oxidativem Stress. J. Dtsch. Dermatol. Ges. **2**, 580–587 (2004)
62. Wolfram Mathworld. Good Prime. mathworld.wolfram.com/GoodPrime.html (2011). Accessed July 2011

Part IV
Art Galleries

The image above shows a hallway in Bertel Thorvaldsen's Museum in Copenhagen. Museums often come with the challenge of optimising a walkway for visiting the offered exhibits within given time constraints.

This fourth part applies rubberband algorithms for solving shortest path problems such as the safari, zookeeper, or watchman route problem, which are all in the category of *art gallery algorithms*. This part also provides a κ-linear approximate solution for touring a finite number of simple polygons, which are pairwise disjoint and not necessarily convex, and it also provides an approximate solution to the unconstrained touring polygons problem which is known to be NP-hard.

Chapter 10
Touring Polygons

> *The more constraints one imposes, the more one frees one's self.*
> *And the arbitrariness of the constraint serves only to obtain*
> *precision of execution.*
>
> *Igor Stravinsky (1882–1971)*

Assume that two points p and q are given as well as a finite ordered set of simple polygons, all in the same plane; the basic version of a touring-a-sequence-of-polygons problem (TPP) is to find a shortest path such that it starts at p, then visits these polygons in the given order, and ends at q. This chapter describes four approximation algorithms for unconstrained versions of problems defined by touring an ordered set of polygons. It contributes to an approximate and partial answer to the previously open problem "What is the complexity of the touring-polygons problem for pairwise disjoint, simple, and not necessarily convex polygons?" by providing $\kappa(\varepsilon)\mathcal{O}(n)$ approximation algorithms for solving this problem, either for given start and end points p and q, or allowing to have those variable, where n is the total number of vertices of the given k simple and pairwise disjoint polygons; $\kappa(\varepsilon)$ defines the numerical accuracy depending on a selected $\varepsilon > 0$.

10.1 About TPP

The *touring-polygon problem* (TPP), as defined above, does have various applications such as for parts cutting in the industry (e.g., moving the head of a cutting robot to subsequent start positions of planar shapes) or for route planning in general when the task consists in visiting selected polygonal regions.

Let π be a plane. Consider simple, pairwise disjoint polygons $P_i \subset \pi$, where $i = 1, 2, \ldots, k$, and two points $p, q \in \pi \setminus \bigcup_{i=1}^{k} P_i$. As always in the book so far, a simple polygon P is a planar region whose frontier ∂P is represented by a polygonal path (i.e., a sequence of line segments whose endpoints define the vertices of P).

Let $\rho(p, p_1, p_2, \ldots, p_k, q)$ be a polygonal path in π that starts at $p_0 = p$, is then incident with points p_i in the given order, and ends at $p_{k+1} = q$, with $p_i \in \pi$, for

F. Li, R. Klette, *Euclidean Shortest Paths*,
DOI 10.1007/978-1-4471-2256-2_10, © Springer-Verlag London Limited 2011

$i = 1, 2, \ldots, k$. We denote such a path briefly by $\rho(p, q)$ if this does not cause any confusion (i.e., if intermediate points p_i are known by context or not important to be listed).

Definition 10.1 A path $\rho(p, q)$ *visits a polygon* P at point $r \in P$ if the path intersects P, and r is the first (i.e., along the path) point in P on this path.

Obviously, $r \in \partial P$. The (unconstrained) *fixed TPP* is defined as follows:

> Find a shortest path $\rho(p, p_1, p_2, \ldots, p_k, q)$ such that it visits each of the polygons P_i in the given order at point p_i, for $i = 1, 2, \ldots, k$.

The path may be further constrained by some predefined properties, and we discuss examples below.—If the start and end points of the path are not given then the problem becomes the (unconstrained) *floating TPP*:

> Let $P_0 = P_k$. Find a cyclic shortest path $\rho(p_0, p_1, p_2, \ldots, p_{k-1}, p_k)$, with $p_0 = p_k$, such that it visits each of the polygons P_i in the given order at point p_i, for $i = 0, 1, 2, \ldots, k$.

Obviously, in this case it does not matter at which polygon to start, and we could re-order the polygons modulo k.

We just mention an example of further constraints (without discussing it further in this chapter). Let $F \subset \pi$ be a simple, not necessarily convex polygon which contains the union of some polygons, all given in the plane π. Then F is called a *fence* with respect to those polygons.

In our case, we have polygons $P_1, \ldots, P_k$. Let $F_i \subset \pi$ be a fence for polygons P_i and P_{i+1}, for $i = 0, 1, 2, \ldots, k$, with degenerated polygons $P_0 = \{p = p_0\}$ and $P_{k+1} = \{q = p_{k+1}\}$. An example of a *constrained fixed TPP* is defined as follows:

> Find a shortest path $\rho(p, p_1, p_2, \ldots, p_k, q)$ such that it visits each of the polygons P_i in the given order at point p_i, for $i = 1, 2, \ldots, k$, also satisfying the condition that F_i contains the subpath from p_i to p_{i+1}, for $i = 0, 1, 2, \ldots, k$.

The subpath from p_i to p_{i+1}, for $i = 0, 1, 2, \ldots, k$, needs to be a shortest path within a given polygon F_i.

In a variety of industries, such as clothing, window manufacturing, or metal sheet processing, it is necessary to cut a set of parts (modelled by polygons) from large sheets of paper, cloth, glass, metal, and so forth. Motivated by such applications, we can consider the following three models of cutting scenarios:

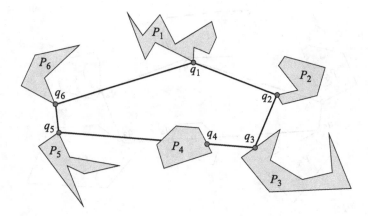

Fig. 10.1 The general ordered P-TSP. Not necessarily convex polygons are assumed to be given in a particular order. The cutting tool may travel across the given polygons

- *Continuous cutting.* The path of the cutting tool visits each object (i.e., polygon) to be cut just once. The tool can engage the object at any point on its frontier, but must cut the entire object before it travels to the next object. Accordingly, the same frontier point must be used for entry and departure from the object.
- *Endpoint cutting.* The tool can enter and exit the object only at some predefined frontier points; however, it may cut the object in sections (i.e., it may visit an object repeatedly).
- *Intermittent cutting.* This is the most general version of the problem in which the object can be cut in sections and there is no restriction on the frontier points that can be used for entry or exit.

The continuous cutting problem, where each object is a polygon, is also called the *plate-cutting travelling salesman problem* (P-TSP). The P-TSP is a generalisation of the *travelling salesman problem* (TSP). If each polygon degenerates into a single vertex then the P-TSP becomes the TSP which is known to be NP-hard. It follows that the P-TSP is NP-hard as well.

The *ordered P-TSP* is a simplified TSP by providing a predefined order of visits; see Fig. 10.1. The polygons may still be nonconvex as well. This simplifies further if polygons are all assumed to be convex; see Fig. 10.2. If, additionally, the start point is also given (see Fig. 10.3), then we have a 'fixed problem'. This fixed ordered P-TSP coincides with the fixed TPP as defined earlier; in this chapter we prefer to use the naming based on touring polygons rather than on a travelling salesman.

10.2 Contributions in This Chapter

In this chapter, we focus on the unconstrained *fixed TPP* (i.e., given start and end point of the path) and *floating TPP* (i.e., no given start or end point) under the

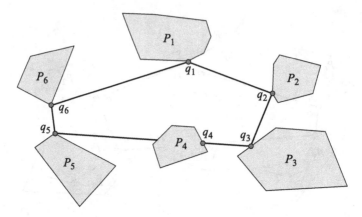

Fig. 10.2 Illustration for the ordered P-TSP also assuming that all polygons are convex

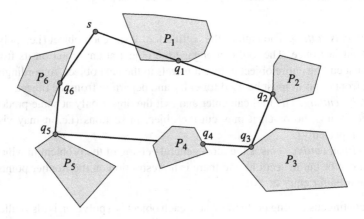

Fig. 10.3 The P-TSP, now also with a given start point s

condition that the convex hulls of the input polygons P_i are pairwise disjoint, but the polygons P_i itself may be nonconvex.

Algorithm 39 in Sect. 10.3 is for solving the fixed TPP by providing an approximation algorithm running in time $\kappa(\varepsilon) \cdot \mathcal{O}(n)$, where n is the total number of vertices of all polygons. Previously proposed solution techniques[1] can only handle the fixed TPP, the fixed safari problem, and the fixed watchman route problem, all for convex polygons only. Our solution technique is suitable for solving both the fixed and the floating TPP with the same time complexity, also allowing nonconvex polygons P_i with pairwise disjoint convex hulls. (Our method might also be useful for solving the floating watchman route, the floating safari problem, and the floating zookeeper problem; but this is not yet a subject in this chapter; see Chaps. 11 and 12.)

---

[1]See [2].

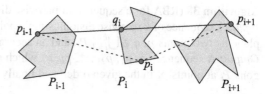

Fig. 10.4 Repositioning in Line 5 of Algorithm 38. Point p_i moves into a new position q_i

Our approximate algorithms are based (again) on the idea of a rubberband algorithm. We recall that such an algorithm starts with an *initial path* through a provided sequence of *steps* (here polygons P_i), and runs then in iterations through those (possibly "adjusted") sequence again while reducing (compared to the previous run) the length of the current path in each run. The important issue is to guarantee that the resulting Cauchy sequence of lengths is actually converging to the minimum length (i.e., the *global minimum*).

We recall (see Chap. 1) that a Euclidean path is a δ-*approximation (Euclidean) path* for an ESP problem iff its length is at most δ times the optimum solution.

We will also refer to a convex hull algorithm. We provided different options in Chap. 4, such as Algorithm 15 or Algorithm 14. Such an algorithm reads an ordered sequence of vertices of a planar simple polygonal curve ρ and outputs an ordered sequence of vertices of the convex hull of ρ; its running time is $\mathcal{O}(|V(\rho)|)$.

The chapter is structured as follows: Sect. 10.3 provides approximation algorithms for the case of the fixed TPP, with either convex or not necessarily convex input polygons, and then some modifications of those two algorithms for solving the floating TPP, with either convex or not necessarily convex input polygons, in the approximate sense. Section 10.4 reports about experiments, and Sect. 10.5 concludes.

10.3 The Algorithms

In this section, we do not only deal with the fixed TPP, we also discuss an approximate solution for the floating TPP.

First, we provide Algorithm 38 which is only guaranteed to find a fixed TPP solution as a *local minimum*, and not necessarily as the intended global minimum. However, if the input polygons P_i are all convex, then this simple algorithm already outputs an approximate fixed TPP solution (with adjustable accuracy) in the global sense.

Line 5 of Algorithm 38, the local optimisation step, is illustrated in Fig. 10.4.

Analogously to the proof of Theorem 3.3, we analyse the approximation factor of Algorithm 38 (see Fig. 10.5) as follows: This algorithm calculates a $(1 + 2k \times r(\varepsilon)/L)$-approximate solution for the fixed TPP, and it is not necessarily only restricted to convex polygons P_i, where L is the length of a shortest path (i.e., the intended *global minimum*), $r(\varepsilon)$ the upper error bound for distances between p_i and its corresponding optimal vertex p_i' (i.e., $d_e(p_i, p_i') \leq r(\varepsilon)$, for $i = 1, \ldots, k$), and we recall that d_e is the Euclidean metric.

Algorithm 38 (RBA for a sequence of pairwise disjoint simple polygons)
Input: A sequence of k pairwise disjoint simple polygons $P_1, P_2, \ldots, P_k$ in the same
plane π; two points $p, q \notin \bigcup_{i=1}^{k} P_i$, and an accuracy constant $\varepsilon > 0$.
Output: A sequence $\langle p, p_1, p_2, \ldots, p_k, q \rangle$ which starts at $p = p_0$, then visits polygons P_i at points p_i in the given order, and finally ends at $q = p_{k+1}$.

1: For each $i \in \{1, 2, \ldots, k\}$, let initial vertex p_i be a vertex of P_i.
2: Let $L_0 = \infty$. Calculate $L_1 = \sum_{i=0}^{k} d_e(p_i, p_{i+1})$, where $p_0 = p$ and $p_{k+1} = q$.
3: **while** $L_0 - L_1 \geq \varepsilon$ **do**
4:     **for** $i = 1, 2, \ldots, k$ **do**
5:         Compute a point $q_i \in \partial P_i$ such that $d_e(p_{i-1}, q_i) + d_e(q_i, p_{i+1}) =$
             $\min\{d_e(p_{i-1}, p) + d_e(p, p_{i+1}) : p \in \partial P_i\}$.
6:         Update the path $\langle p, p_1, p_2, \ldots, p_k, q \rangle$ by replacing p_i by q_i.
7:     **end for**
8:     Let $L_0 = L_1$ and calculate $L_1 = \sum_{i=0}^{k} d_e(p_i, p_{i+1})$.
9: **end while**
10: Return $\langle p, p_1, p_2, \ldots, p_k, q \rangle$.

Fig. 10.5 RBA for solving the fixed TPP problem for a sequence of pairwise disjoint simple
polygons

This approximation factor follows because, for each $i \in \{1, 2, \ldots, k\}$, the error of
the difference between $d_e(p_i, p_{i+1})$ and $d_e(p'_i, p'_{i+1})$ is at most $2 \times r(\varepsilon)$ because of
$d_e(p_i, p'_i) \leq r(\varepsilon)$. We obtain that

$$L \leq \sum_{i=0}^{k} d_e(p_i, p_{i+1}) \leq \sum_{i=0}^{k} [d_e(p'_i, p'_{i+1}) + 2 \times r(\varepsilon)]$$
$$= L + 2k \times r(\varepsilon).$$

Thus, the output path is a $\{1 + 2k \times r(\varepsilon)/L\}$-approximation path.

Recall the analysis of $\kappa(\varepsilon)$ in Sect. 8.3, where $\kappa(\varepsilon) = \frac{L_0 - L}{\varepsilon}$ is the usual function
which only depends upon the difference between the length L_0 of an initial path
and L of the optimum path, and the accuracy constant ε. Let L_m be the length
of the mth updated path, for $m = 0, 1, 2, \ldots$, with $L_m - L_{m+1} \geq \varepsilon$ (otherwise the
algorithm stops). It follows that

$$\kappa(\varepsilon) = \frac{L_0 - L}{\varepsilon} \geq 1 + \frac{L_1 - L}{\varepsilon} \geq \cdots \geq m + \frac{L_m - L}{\varepsilon}. \tag{10.1}$$

The sequence $\{m + \frac{L_m - L}{\varepsilon}\}$ is monotonously decreasing, lower bounded by 0, and
stops at the first m_0 where $L_{m_0} - L_{m_0+1} < \varepsilon$. This defines a *local minimum* in this
approximation process. However, it is still possible that L_{m_0+1} is not yet "close"
to L in the case of nonconvex polygons.

For convex input polygons, we apply the following

Lemma 10.1 *For the unconstrained TPP, if all input polygons are pairwise disjoint
and convex, then local optimality is equivalent to global optimality.*

Algorithm 39 (Algorithm for the fixed TPP; polygons may be nonconvex)

Input: A sequence of k simple polygons $P_1, P_2, \ldots, P_k$ such that the convex hulls $C(P_1), C(P_2), \ldots, C(P_k)$ are pairwise disjoint; two points $p, q \notin \bigcup_{i=1}^{k} C(P_i)$, and an accuracy constant $\varepsilon > 0$.

Output: A sequence $\langle p, p_1, p_2, \ldots, p_k, q \rangle$ which starts at p, then visits polygon P_i at p_i in the given order, and finally ends at q.

1: For $i \in \{1, 2, \ldots, k\}$, apply a linear-time convex hull algorithm for computing the convex hull $C(P_i)$.
2: Let $C(P_1), C(P_2), \ldots, C(P_k)$, p, and q be the input of Algorithm 38 for computing an approximate shortest route $\langle p, p_1, \ldots, p_k, q \rangle$.
3: For $i = 1, 2, \ldots, k - 1$, find a point $q_i \in \partial P_i$ such that $d_e(p_{i-1}, q_i) + d_e(q_i, p_{i+1}) = \min\{d_e(p_{i-1}, p) + d_e(p, p_{i+1}) : p \in \partial P_i\}$.
4: Update the path for each i by $p_i = q_i$.
5: Let $P_1, P_2, \ldots, P_k$, p and q be the input of Algorithm 38, and points p_i as obtained in Line 3 are the initial vertices p_i in Line 1 of Algorithm 38. Continue with running Algorithm 38.
6: Return $\langle p, p_1, \ldots, p_{k-1}, p_k, q \rangle$ as provided in Line 4.

Fig. 10.6 An improved version of RBA for solving the fixed TPP problem for a sequence of pairwise disjoint simple polygons

Thus, we immediately obtain

Corollary 10.1 *If all input polygons are convex then Algorithm* 38 *outputs an approximate global solution for the fixed TPP.*

Obviously, $\lim_{\varepsilon \to 0} r(\varepsilon) = 0$. Thus, Algorithms 38 may be "tuned" by a very small $\varepsilon > 0$ to be of very high accuracy. The time complexity of the algorithm is discussed later below.

Now we provide a second (heuristic) algorithm which applies Algorithm 38 on the convex hulls $C(P)$ of the input polygons P in order to obtain an 'improved' initial path whose vertices are located on the frontier of the convex hulls (see Fig. 10.6); then we transform this path in the algorithm into another one such that its vertices are on the frontier of the input polygons; finally, the algorithm applies Algorithm 38 again on the input polygons to find a further improved solution to the fixed TPP.

Line 2 iterates through the convex hulls. The iteration through step sets P_i only occurs in Line 4 (i.e., when applying Algorithm 38 for the second time, using the same ε). Algorithm 39 provides a $(1 + (L_2 - L_1)/L)$-approximate global solution for the floating TPP, where L is the length of an optimal path, L_1 is the length of the path obtained in Line 2, and L_2 the length of the final path obtained in Line 5. Note that $L_2 \geq L_1$, and $L_2 = L_1$ if all polygons P_i are convex.

Theorem 10.1 *Algorithms* 38 *and* 39 *may be computed in time* $\kappa(\varepsilon)\mathcal{O}(n)$, *where* n *is the total number of vertices of the involved* k *polygons* P_i.

Algorithm 40 (The "floating version" of Algorithm 38)
Input: A sequence of k pairwise disjoint simple polygons $P_0, P_1, \ldots, P_{k-1}$ in a
plane π; an accuracy constant $\varepsilon > 0$.
Output: A sequence $\langle p_0, p_1, p_2, \ldots, p_k \rangle$ (where $p_k = p_0$) which visits polygon P_i at p_i in the given order $i = 0, 1, 2, \ldots, k - 1$, and finally $i = 0$
again.

1: For each $i \in \{0, 1, \ldots, k - 1\}$, let initial p_i be a vertex of P_i.
2: Let $L_0 = \infty$. Calculate $L_1 = \sum_{i=0}^{k-1} d_e(p_i, p_{i+1})$.
3: **while** $L_0 - L_1 \geq \varepsilon$ **do**
4:     **for** $i = 0, 1, \ldots, k - 1$ **do**
5:         Compute a point $q_i \in \partial P_i$ such that $d_e(p_{i-1}, q_i) + d_e(q_i, p_{i+1}) = \min\{d_e(p_{i-1}, p) + d_e(p, p_{i+1}) : p \in \partial P_i\}$.
6:         Update the route $\langle p_0, p_1, \ldots, p_{k-1} \rangle$ by replacing p_i by q_i.
7:     **end for**
8:     Let L_0 be L_1 and calculate $L_1 = \sum_{i=0}^{k-1} d_e(p_i, p_{i+1})$.
9: **end while**
10: Return $\langle p_0, p_1, \ldots, p_{k-1}, p_k \rangle$.

Fig. 10.7 RBA for solving the floating TPP problem for a sequence of pairwise disjoint simple
polygons

Proof In Line 5 of Algorithm 38, each locally optimal point q_i can be computed
using on the order of $|V(P_i)|$ operations, where $V(P_i)$ is the set of vertices of P_i.
Thus, each iteration of the for loop takes $\mathcal{O}(n)$ operations at most. Furthermore,
the number of runs through the outer while-loop is upper bounded by $\kappa(\varepsilon)$; see
Eq. (10.1). Thus, Algorithm 38 runs in time $\kappa(\varepsilon)\mathcal{O}(n)$.   □

Extensive experiments showed that $\kappa(\varepsilon)$ was always much too large to estimate
the actual number of runs through the outer while-loop. Obviously, $\kappa(\varepsilon)$ also depends on the selection of the initial path. By taking fixed initial points p_i (e.g., the
uppermost, leftmost vertex of P_i), the function $\kappa(\varepsilon)$ would only depend on ε and
the configuration of the input polygons P_i.

Now we propose two algorithms for the floating TPP. They are derived in a
straightforward way from the two algorithms described above.

Algorithm 40 (see Fig. 10.7) calculates a $[1 + 2k \cdot r(\varepsilon)/L]$-approximate solution
for the floating TPP and convex polygons P_i, where L is the length of an optimal
path, and $r(\varepsilon)$ the upper error bound for distances between p_i and a corresponding
optimal vertex p'_i.[2]

Algorithm 41 (see Fig. 10.8) provides a $(1 + (L_2 - L_1)/L)$-approximate solution
for the floating TPP with not necessarily only convex polygons P_i, where L is the

---

[2]Regarding a proof of correctness for Algorithm 40 and convex polygons P_i, it is analogous to the
proof of Theorem 3 in [11]. This proof is actually theoretically challenging, but it is too long for
this chapter. The method of [2], as cited before for the fixed TPP and convex polygons P_i, is not
applicable here for showing correctness for the case of convex input polygons.

Algorithm 41 (The "floating version" of Algorithm 39)

Input: A sequence of k simple polygons $P_0, P_1, \ldots, P_{k-1}$ such that convex hulls $C(P_0), C(P_1), \ldots, C(P_{k-1})$ are pairwise disjoint and an accuracy constant $\varepsilon > 0$.

Output: A sequence $\langle p_0, p_1, p_2, \ldots, p_{k-1}, p_k \rangle$ (where $p_k = p_0$) which visits polygon P_i at p_i in the given order $i = 0, 1, 2, \ldots, k - 1$, and then $i = 0$ again.

1: For $i \in \{0, 1, \ldots, k - 1\}$, apply a linear-time convex hull algorithm for computing $C(P_i)$.
2: Let $C(P_0), C(P_1), \ldots, C(P_{k-1})$ be the input of Algorithm 40 for computing an approximate shortest route $\langle p_0, p_1, \ldots, p_k \rangle$.
3: For $i = 0, 1, 2, \ldots, k - 1$, find a point $q_i \in \partial P_i$ such that $d_e(p_{i-1}, q_i) + d_e(q_i, p_{i+1}) = \min\{d_e(p_{i-1}, p) + d_e(p, p_{i+1}) : p \in \partial P_i\}$.
4: Update p_i by letting p_i be q_i.
5: Let $P_0, P_1, \ldots, P_{k-1}$ be the input of Algorithm 40, and points p_i as obtained from Line 3 be the initial points in Line 1 of Algorithm 40 for computing an approximate shortest route $\langle p_0, p_1, \ldots, p_{k-1}, p_k \rangle$.
6: Return $\langle p_0, p_1, \ldots, p_{k-1}, p_k \rangle$.

Fig. 10.8 An improved version of RBA for solving the floating TPP problem for a sequence of pairwise disjoint simple polygons

length of an optimal path, L_1 is the length of the path obtained in Line 2, and L_2 the length of the final path in Line 5.

Regarding the time complexity of Algorithms 40 and 41, it is obvious that they are the same as that of Algorithms 38 and 39, that is, running in time $\kappa(\varepsilon)\mathcal{O}(n)$, where n is the total number of vertices of the k involved polygons P_i, and $\kappa(\varepsilon)$ is the function as defined before.

10.4 Experimental Results

Both Algorithms 39 and 41 were implemented in Java. On the left (on the right) in Fig. 10.9, the red route is obtained by Line 2 of Algorithm 39 (of Algorithm 41), the blue one is the initial route in Line 4 of Algorithm 39 (of Algorithm 41), and the green one is the final route of Algorithm 39 (of Algorithm 41) when applying $\varepsilon = 10^{-10}$. Routes follow the predefined order of those polygons.

In Table 10.1, L_1, L_2, and L are as defined in the output of Algorithm 39, while L_1', L_2', and L' are the corresponding values defined in the output of Algorithm 41. Values δ and δ' are defined by $(1 + (L_2 - L_1)/L)$ and $(1 + (L_2' - L_1')/L')$, respectively. Note their closeness to 1.0. Both L and L' are approximate because they are found by running Algorithms 38 and 39 for 100,000 times (each time with a randomly selected initial path), and then selecting the minimum. Note that $L_2 \geq L$ $(L_2' \geq L')$ follows from being approximate solutions for the fixed or the floating TPP. For more examples of measured run times, see Fig. 10.10.

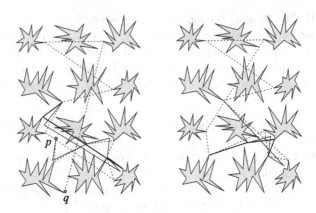

Fig. 10.9 Routes as calculated in Lines 2, 4, and 5 of Algorithms 39 and 41

Table 10.1 Resulting data obtained from Algorithms 39 and 41 on the input example shown in Fig. 10.9

L_1	L_2	L	δ
2,867.069	2,888.999	2,887.736	1.0076
L'_1	L'_2	L'	δ'
2,521.294	2,532.700	2,532.700	1.000

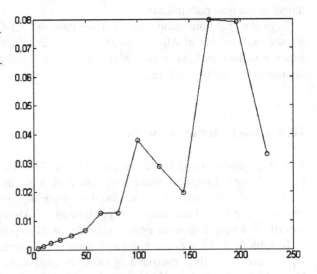

Fig. 10.10 Running time (in seconds) of Algorithm 41, for numbers of randomly generated polygons up to 230. Implementation: in Java on a PC with Pentium Dual-Core CPU E5200 2.50 GHz, 1.99 GB memory; all input polygons with 14 vertices. (Obviously, this implementation was not optimised for speed, but indicates an approximate linear increase for this small set of samples.)

10.5 Concluding Remarks and Future Work

In this chapter, we present $(1 + (L_2 - L_1)/L)$-approximation algorithms for finding approximate solutions for both the fixed and the floating TPP, where simple input polygons have to satisfy the condition that their convex hulls are pairwise disjoint.

For extensive experimental results, showing the practical appearance of the theoretical upper time bound $\kappa(\varepsilon)\mathcal{O}(n)$, see Sects. 3.8 and 8.3.

The presented algorithms appear to be suitable for solving both the fixed and the floating TPP, and they have identical theoretical time complexity.[3] The RBA method could also be useful for solving the floating watchman route problem, the floating safari problem, and the floating zookeeper problem, and we will discuss those problems in the next two chapters.

It also remains an open problem to handle cases where the convex hulls of the polygons are not necessarily pairwise disjoint.

10.6 Problems

Problem 10.1 Discuss the differences between the fixed TPP, the floating TPP, and the constrained fixed TPP.

Problem 10.2 What are the differences between continuous cutting, endpoint cutting, and intermittent cutting.

Problem 10.3 Discuss the difference between P-TSP, TSP, and TPP.

Problem 10.4 Algorithm 38 is also modified from Algorithm 7. Underline the modifications of Algorithm 38 in Fig. 10.5.

Problem 10.5 Prove Lemma 10.1.

Problem 10.6 Algorithm 39 is modified from Algorithm 31. Underline the modifications of Algorithm 39 in Fig. 10.6.

Problem 10.7 Algorithm 40 is also modified from Algorithm 7. Underline the modifications of Algorithm 40 in Fig. 10.7.

Problem 10.8 Algorithm 41 is also modified from Algorithm 31. Underline the modifications of Algorithm 41 in Fig. 10.8.

Problem 10.9 Consider the input of Algorithm 39. Why is it required that the convex hulls $C(P_1), C(P_2), \ldots, C(P_k)$ be pairwise disjoint?

Problem 10.10 (Programming exercise) Implement Algorithms 38, 39, 40, and 41, and compare their performance.

---

[3]While the method of [2] is only suitable for the fixed TPP and convex polygons. Moreover, RBAs appear to be simpler and easier to understand and to implement than the algorithm in [2].

Problem 10.11 Consider input polygons in 3D space (not necessarily in parallel planes) for Algorithms 38, 39, 40, and 41. Would those algorithms still solve the TPP defined by those input polygons?

Problem 10.12 Read reference [2]. Conclude why the method presented there cannot be generalised to deal with the floating TPP.

Problem 10.13 (Research problem) Generalise the generic RBA (i.e., Algorithm 10 described in Fig. 3.22) for solving the TPP for a sequence of pairwise disjoint simple polygons. Compare its performance with that of Algorithm 38.

10.7 Notes

Notations in this chapter follow [2]; Lemma 10.1 is Lemma 1 in this paper. Our approximate algorithms apply RBAs; see Chap. 3 for the basic design of an RBA, and see also [10], for variants of RBAs.

The TPP already has a history of publications addressing its various applications; see, for example, [1, 2, 7, 10, 13]. The given three models of cutting scenarios have been defined in [7]; this paper discusses the general ordered P-TSP as illustrated in Fig. 10.1. In [1], it was assumed that all the polygons in an ordered P-TSP are convex.

Reference [7] focused on solving the continuous cutting problem where each object is a polygon [i.e., the plate-cutting travelling salesman problem (P-TSP)]. The P-TSP is a generalisation of the well-known travelling salesman problem (TSP) [9]. The P-TSP is more general than the *generalised TSP* (GTSP) as discussed in [8, 13]. If each polygon degenerates into a single vertex then the P-TSP becomes the TSP which is known to be NP-hard [3]. It follows that the P-TSP is NP-hard as well. Figure 10.3 shows the P-TSP as considered in [2]. See also [6].

Reference [7] solved the ordered P-TSP by a heuristic approach based on a Lagrange relaxation method as discussed in [4, 5], without providing a time complexity analysis for this proposed approach. As a follow-up of this work, [1] proved that a further simplified ordered P-TSP, where polygons are all assumed to be convex (see Fig. 10.2), is solvable in polynomial time. If, additionally, the start point is also given (see Fig. 10.3), then the authors of [2] claim that they can solve this fixed problem in time $\mathcal{O}(kn \log(n/k))$, where n is the total number of vertices of polygons $P_i \subset \pi$, for $i = 1, 2, \ldots, k$. According to [2], "*one of the most intriguing open problems*" identified by their results "*is to determine the complexity of the fixed TPP for pairwise disjoint nonconvex simple polygons*".

Assuming that Chazelle's triangulation method actually defines a linear-time algorithm (see our discussion in Chap. 5), a shortest path, connecting two given points within a simple polygon, and fully contained in this polygon, can then be constructed in time linear in the number of vertices of this polygon [12].

For a convex hull algorithm, see, Chap. 4.

References

1. Dror, M.: Polygon plate-cutting with a given order. IIE Trans. **31**, 271–274 (1999)
2. Dror, M., Efrat, A., Lubiw, A., Mitchell, J.: Touring a sequence of polygons. In: Proc. STOC, pp. 473–482 (2003)
3. Garey, M.R., Graham, R.L., Johnson, D.S.: Some NP-complete geometric problems. In: Proc. ACM Sympos. Theory Computing, pp. 10–22 (1976)
4. Geoffrion, A.M.: Lagrangian relaxation and its uses in integer programming. In: Mathematical Programming Study, vol. 2, pp. 82–114. North-Holland, Amsterdam (1974)
5. Guignard, M., Kim, S.: Lagrangian decomposition: a model yielding stronger Lagrangian bounds. Math. Program. **39**, 215–228 (1987)
6. Hochbaum, D.S. (ed.): Approximation Algorithms for NP-Hard Problems. PWS Pub. Co., Boston (1997)
7. Hoeft, J., Palekar, U.S.: Heuristics for the plate-cutting traveling salesman problem. IIE Trans. **29**, 719–731 (1997)
8. Laporte, G., Mercure, H., Nobert, Y.: Generalized traveling salesman problem through n clusters. Discrete Appl. Math. **18**, 185–197 (1987)
9. Lawler, E., Lenstra, J., Rinnooy Kan, A., Shmoys, D.: The Traveling Salesman Problem. A Guided Tour of Combinatorial Optimization. Wiley, New York (1985)
10. Li, F., Klette, R.: Rubberband algorithms for solving various 2D or 3D shortest path problems (invited talk). In: IEEE Proc. Computing: Theory and Applications, pp. 9–18. The Indian Statistical Institute, Kolkata (2007)
11. Li, F., Klette, R.: Watchman route in a simple polygon with a rubberband algorithm (with downloadable source of algorithm). Mi-tech report-51. www.mi.auckland.ac.nz/ index.php?option=com_content&view=article&id=127&Itemid=113
12. Mitchell, J.S.B.: Geometric shortest paths and network optimization. In: Sack, J.-R., Urrutia, J. (eds.) Handbook of Computational Geometry, pp. 633–701. Elsevier, Amsterdam (2000)
13. Noon, C.E., Bean, J.C.: An efficient transformation of the generalized traveling salesman problem. Inf. Syst. Oper. Res. **31**, 39–44 (1993)

Chapter 11
Watchman Routes

Sometimes the path you are on is not as important as the direction you are heading. For, no matter if you take the Holland Tunnel or the George Washington Bridge you get to New Jersey, so long as you are heading west. The procedure involved in your path to Nirvana may meander, but a road worth travelling will have its twists and turns.

Kevin Patrick Smith (born 1970)

So far, the best result in running time for solving the floating watchman route problem (i.e., shortest path for viewing any point in a simple polygon with given start point) is $\mathcal{O}(n^4 \log n)$, published in 2003 by M. Dror, A. Efrat, A. Lubiw, and J. Mitchell. This chapter provides an algorithm with $\kappa(\varepsilon) \cdot \mathcal{O}(kn) + \mathcal{O}(k^2 n)$ runtime, where n is the number of vertices of the given simple polygon P, and k the number of essential cuts; $\kappa(\varepsilon)$ defines the numerical accuracy depending on a selected constant $\varepsilon > 0$. Moreover, the presented RBA appears to be significantly simpler, easier to understand and to be implemented than previous ones for solving the fixed watchman route problem.

11.1 Essential Cuts

Let P be a planar, simple, topologically closed polygon (i.e., $P = P^{\bullet}$) with n vertices, and let ∂P be its frontier. A point $p \in P$ is *visible* from point $q \in P$ iff $pq \subset P$. The *(floating) watchman route problem* (WRP) of computational geometry is defined as follows:

Compute a shortest route $\rho \subset P$ such that any point $p \in P$ is visible from at least one point on ρ.

F. Li, R. Klette, *Euclidean Shortest Paths*,
DOI 10.1007/978-1-4471-2256-2_11, © Springer-Verlag London Limited 2011

Fig. 11.1 A watchman route.
Note that the positions of the
vertices on the *dashed lines*
need to be optimised for
obtaining an ESP

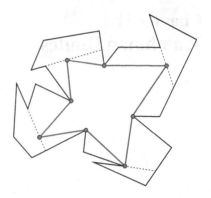

This is actually equivalent to the requirement that all points $p \in P$ are visible just from the vertices of the path ρ, that is, for any $p \in P$ there is a vertex q on ρ such that $pq \subset P$. See Fig. 11.1. We also cite an important theorem from this area:

Theorem 11.1 *There is a unique (floating) watchman route in a simple polygon, except for those cases where there is an infinite number of different shortest routes, all of equal length.*

If the start point of the route is given, then this refined problem is known as being the *fixed* WRP. In the rest of this chapter, let s be the start point of a fixed WRP.

Given the large time complexity of published exact algorithms for solving the WRP, finding efficient approximation algorithms became an interesting subject in recent years.

Recall the definition of a δ-approximation (see Definition 2.2). In case of the WRP, the optimum solution is defined by the length of the shortest path. We also recall that a Euclidean path is a δ-approximation path for an ESP problem iff its length is at most δ times the optimum solution.

A vertex v of P is called *reflex* or *concave* if v's internal angle is greater than $180°$. Let u be a vertex of P which is adjacent to a reflex vertex v. Assume that the straight line uv intersects an edge of P at v'. Then the segment $C = vv'$ partitions P into two parts.

Definition 11.1 Such a segment C is called a *cut* of P if C makes a convex vertex at v in the part containing the starting point s; v is called the *defining* vertex of C, and v' is called the *hit* point of C.

For the floating WRP, each reflex vertex defines two cuts because there does not exist a starting point. That part of P which contains s is called an *essential* part of C and is denoted by $P(C)$. See Fig. 11.2. The other part of P is called the *pocket* induced by cut C, and C is the *associated* cut of the pocket.

Fig. 11.2 Example of cuts
and essential cuts. There are
eight cuts C_1, C_2, C_3, C_4,
C_5, C_6, C_8, and C_{12}. There
are five essential cuts C_1, C_3,
C_5, C_8, and C_{12}, where v_6 is
the starting point

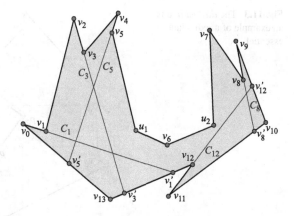

Definition 11.2 A cut C *dominates* a cut C' iff $P(C)$ contains $P(C')$. A cut is
called *essential* if it is not dominated by another cut. A pocket is called *essential* if
it does not contain any other pocket.

A pocket is essential iff its associated cut is essential.

If two points u and v are on two different edges of P, such that the segment
uv partitions P into two parts, then we say that uv is a *general cut* of P. We may
arbitrarily select one of both endpoints of the segment uv to be its *start point*. In the
rest of this chapter, for an essential cut C of P, we identify the defining vertex of C
with its start point.

If $C_0, C_1, \ldots, C_{k-1}$ are all the essential cuts of P such that their start points are
ordered clockwise around on the frontier ∂P of P, then we say that $C_0, C_1, \ldots,$
C_{k-1}, and P *satisfy the condition of the fixed (or floating) watchman route problem
(WRP)*.

Example 11.1 In Fig. 11.2, polygon P and the essential cuts C_1, C_3, C_5, C_8, and
C_{12} satisfy the condition of the fixed WRP, where v_6 is the starting point.   □

Let e be an edge of P, and u and v be the endpoints of e. We say that the *direction
of edge e* is from u to v if this is the same direction as defined by a counterclockwise
scan of ∂P. The *direction of an essential cut* is the same direction as the one of that
edge where it is collinear with.

We also say that a point *lies to the right (left)* of an essential cut if the point
lies locally to the right (left) in the subpolygon separated by the essential cut (i.e.,
possibly also on that essential cut).

Definition 11.3 An essential cut C is called *redundant* if there is a watchman route
ρ such that ρ is completely contained in the pocket $P(C)$; cut C is called *non-
redundant* otherwise.

Definition 11.4 Let C be a cut, v the defining vertex of C, and v' the hit point of C.
Let u be a vertex of P such that u, v and v' are collinear. Let w be a vertex of P

Fig. 11.3 The *dashed line* is
an example of a redundant
essential cut

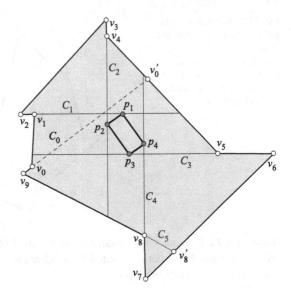

such that w is incident with u and $w \neq v$. If w is a reflex vertex, then the straight
line uw intersects an edge of P at w'. The segment ww' is called the *associated cut*
of C. The part containing both C and its associated cut is called an *associated part*
of C, denoted by $P'(C)$.

Example 11.2 Figure 11.3 shows one redundant essential cut. The figure also shows
a shortest watchman route, and the watchman route is not uniquely defined for this
example. C_5 is an associated cut of C_0. The polygon $v_0 v_9 v_8 v_8' v_6 v_5 v_0' v_0$ is the associated part of C_0.   □

There exist only three categories of contacts between a watchman route ρ and an
essential cut C:

- A *reflection contact* if ρ and C have exactly one point in common,
- a *crossing contact* if both have two points in common (i.e., C is redundant in this
  case), and
- a *tangential contact* if ρ and C share a line segment of nonzero length.

We say that ρ *contacts* an essential cut if ρ makes a reflection, crossing or tangential
contact with such a cut. An essential cut C is *active* if ρ makes a reflection contact
with C.[1]

Figure 11.4 shows a watchman route $\rho = (p_0, p_1, p_2, v_6, p_4, p_3, p_4, v_6, p_0)$
which makes reflection contacts with C_1, C_5, and C_8 at p_0, p_2, and p_3, respectively. Path ρ makes a crossing contact with C_3 at p_1 and p_1'; ρ makes a crossing

---

[1] We do not have to consider any non-essential cuts when defining active cuts. Otherwise the route
would be longer.

Fig. 11.4 Reflection and crossing contacts between a watchman route ρ and essential cuts

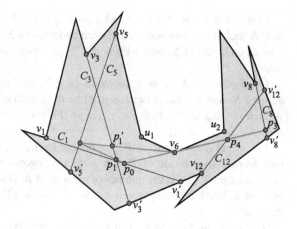

Fig. 11.5 Three types of contacts between a watchman route ρ and essential cuts

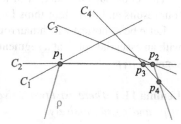

contact with C_{12} at p_4 (both incident points are identical to p_4); ρ does not make a tangential contact with any of the cuts.

In Fig. 11.5, ρ makes reflection contacts with C_1 and C_3 at p_1 and p_2, respectively, a crossing contact with C_4 at p_3 and p_4, and a tangential contact with C_2. Path ρ and cut C_2 share a line segment $p_1 p_2$. Cuts C_1 and C_3 are active cuts.

Now let $\{C_0, C_1, \ldots, C_{k-1}\}$ be the sequence of all essential cuts of P.

Definition 11.5 $\{C_0, C_1, \ldots, C_{k-1}\}$ is called *in good order* iff $v_0, v_1, \ldots, v_{k-1}$ are located around ∂P in order; v_i is the defining vertex of C_i, for $i = 0, 1, \ldots, k-1$.

Now let $\{C_{a_0}, C_{a_1}, \ldots, C_{a_{m-1}}\}$ be the sequence of all active cuts.

Definition 11.6 $\{C_{a_0}, C_{a_1}, \ldots, C_{a_{m-1}}\}$ is called *in good order* iff

$$0 \leq a_0 < a_1 < \cdots < a_{m-1} \leq k-1.$$

Consider two points $p, q \in P$. If the segment $pq \subset P$ then we say that q *can see* p (with respect to P), and p is a *visible point* of q. Let $q \in P$ and consider a segment $S \subset P$. If, for each point $p \in S$, q can see p, then we say that q *can see* S.

Let q be a point in P, $S \subset P$ a segment, $p \in S$, and point p is not an endpoint of S. If q can see p, but for any sufficiently small $\varepsilon > 0$, q cannot see p', where $p' \in S$ and Euclidean distance $d_e(p, p') = \varepsilon$, then we say that p is a *visible extreme point* of q (with respect to S and P).

Consider a segment $S \subset P$ and a point $q \in P \setminus S$. If there exists a subsegment $S' \subseteq S$ such that q can see S', and each endpoint of S' is a visible extreme point of q or an endpoint of S, then we say that S' is a *maximal visible segment* of q (with respect to P).

Let $S_0, S_1, \ldots, S_{k-1}$ be k segments ($k \geq 2$) in 3-dimensional Euclidean space, and $p \in S_0$ and $q \in S_{k-1}$. Let $L_S(p, q)$ be the length of the shortest path, starting at p, then visiting segments $S_1, \ldots, S_{k-2}$ in this order, and finally ending at q, where $S = \langle S_0, S_1, \ldots, S_{k-1} \rangle$.

Let p and q be points in P. We denote by $L_P(p, q)$ the length of the shortest path from p to q inside of P. Let ρ be a polygonal path and $V(\rho)$ the set of all vertices of ρ; $|V(\rho)|$ is the number of vertices of ρ. Denote by $C(S)$ the convex hull of a set S. For k non-empty sets $S_0, S_1, \ldots, S_{k-1}$, $\prod_{i=0}^{k-1} S_i$ is the cross product of those sets.

This ends our introduction of technical terms. We also recall here in one place four results and four algorithms for later use:

Let ρ be a shortest watchman route, and assume that ρ makes a tangential contact with an essential cut C at a segment $p_1 p_2$, where p_1 and p_2 are the endpoints of the segment $p_1 p_2$.

Lemma 11.1 *There exist two active cuts C_1 and C_2 such that ρ contacts C_1 and C_2 at p_1 and p_2 successively.*

Lemma 11.2 *There is a shortest watchman route that contacts the sequence of all active cuts in good order.*

Lemma 11.3 *A closed curve is a watchman route if and only if the curve has at least one point to the left of (or on) each essential cut.*

Theorem 11.2 *Given a simple polygon P, the set C of all essential cuts for the watchman route in P can be computed in $\mathcal{O}(n)$ time.*

We also refer to an algorithm for the 2D ESP problem (see Chap. 6 for options) that has as input a simple polygon P and two points $p, q \in P$ and computes a shortest path from p to q inside of P as follows: first a decomposition into triangles or trapezoids (see Chap. 5), and then the shortest path on those triangles or trapezoids, using, e.g., some adaptation of the Dijkstra algorithm. For the running time of such an algorithm we use $\mathcal{O}(|V(\partial P)|)$ in this chapter (i.e., assuming that triangulation can actually be done in linear time). However, there are other (slower, but already implemented) ways for solving the problem 'ESP in a simple polygon', as Chap. 6 discusses.

We also use a *tangent calculation*[2] which has as input an ordered sequence of vertices of a planar convex polygonal curve ∂P and a point $p \notin P$; its output are

---

[2]See [29].

Fig. 11.6 The floating watchman route algorithm and the call-structure of procedures used in the algorithm

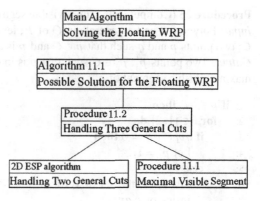

two points $t_i \notin P$ such that pt_i are tangents to ∂P, for $i = 1, 2$; the running time is $\mathcal{O}(\log |V(\partial P)|)$.

Finally, we also use a *winding number inclusion*[3] which has as input a polygon P and a point p; its output tells us whether p is inside of P or not; the running time is $\mathcal{O}(|V(\partial P)|)$.

The remainder of this chapter is organised as follows: Sect. 11.2 proposes the main algorithm of this chapter. Section 11.3 discusses its correctness and time complexity.

11.2 Algorithms

In this section, we describe and discuss the algorithm for solving the floating watchman route problem, thus (finally) arriving at the problem this chapter is aiming at: solving this problem with reasonable run times.

The call-structure of used procedures is shown in Fig. 11.6. The main algorithm of this chapter is based on Procedure 26 in Fig. 11.7, which applies the 2D ESP algorithm and Procedure 25. We present the used procedures first and the main algorithm at the end.

We apply again the following '3-point local optimisation strategy' of an RBA: In each iteration, we update (by finding a local minimum or optimal vertex) the second vertex p_i for every 3-subsequent-vertices subsequence p_{i-1}, p_i, p_{i+1}, constrained by the step set $\{S_1, S_2, \ldots, S_k\}$. The first procedure below computes the maximal visible segment, which is actually an element of the step set of the used RBA. The second procedure is used for updating the vertices.

For Procedure 25, see the pseudocode in Fig. 11.7:

Case 1: p is not an endpoint of C. For $i \in \{1, 2\}$, if q can see v_i (see left of Fig. 11.8), let p_i' be v_i; otherwise, let V_i be the set of vertices in $V(\partial P)$ such that each vertex in V_i is in $\triangle qpv_i$. Apply the convex hull algorithm to compute $C(V_i)$.

---

[3] See [30].

Procedure 25 (Compute a maximal visible segment)
Input: Polygon P and a general cut C of P; let v_1 and v_2 be the two endpoints of C; two points p and q such that $p \in C$ and p is a visible point of $q \in \partial P \setminus C$.
Output: Two points $p'_1, p'_2 \in C$ such that p is in the segment $p'_1 p'_2$, and $p'_1 p'_2$ is the maximal visible segment of q.

1: **if** $p \notin \partial P$ **then**
2:     **for** $i \in \{1, 2\}$ **do**
3:         **if** $qv_i \cap \partial P = \emptyset$ **then**
4:             Let $p'_i = v_i$.
5:         **else**
6:             Let V_i be the set of vertices in $V(\partial P)$ such that each vertex in V_i is inside of $\triangle qpv_i$.
7:             Apply a linear-time convex hull algorithm to compute $C(V_i)$.
8:             Apply the tangent algorithm to find a point $p'_i \in C$ such that qp'_i is a tangent to $C(V_i)$.
9:         **end if**
10:     **end for**
11: **else**
12:     Without loss of generality, assume that $p = v_1$. Let $p'_1 = p$.
13:     Proceed analogously as in Lines 4–8, but now for $i = 2$.
14: **end if**
15: Output p'_i, for $i = 1, 2$, and Stop.

Fig. 11.7 Compute a maximal visible segment of a general cut

Fig. 11.8 Illustration for Procedure 25

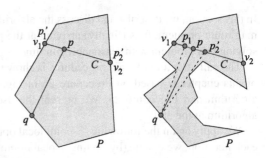

Apply the tangent algorithm to find a point $p'_i \in C$ such that qp'_i is a tangent to $C(V_i)$ (see on the right of Fig. 11.8).

Case 2: p is an endpoint of C. Without loss of generality, assume that $p = v_1$. Let p'_1 be p. Let V_2 be the set of vertices in $V(\partial P)$ such that each vertex in V_2 is in $\triangle qpv_i$. Apply the convex hull algorithm to compute $C(V_2)$. Apply the tangent algorithm to find a point $p'_2 \in C$ such that qp'_2 is a tangent to $C(V_2)$.

In Line 1 of Procedure 25, p is not an endpoint of C. In Line 3, q can see v_i, see left of Fig. 11.8. Line 8 is illustrated on the right of Fig. 11.8. In Line 11, p is an endpoint of C.

Fig. 11.9 Illustration for
Line 2 of Procedure 26

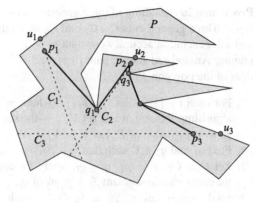

Fig. 11.10 Illustration for
Line 4 of Procedure 26

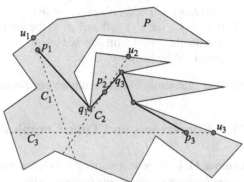

Now we turn to Procedure 26. In Line 2 of this procedure, q_1, p_2, q_3 appear consecutively in V. See Fig. 11.9.

We call the line segment S_2 in Line 4 of Procedure 26 (in Figs. 11.10 and 11.11) *associated to* the updated (optimal) point p_2.

The pseudocode in Fig. 11.12 defines now Algorithm 42. In this algorithm, all subscripts are taken modulo k. In Line 1 (see Fig. 11.13), $p_k = s$ and $p_{k+1} = p_0$. In Line 5, $C_k = p_k = s$ and $C_{k+1} = C_0$. In Line 7, sequence V_1 is the updated sequence V_0, after inserting U_i. In Line 9, we use the updated original sequence V instead of V_1 for the next iteration.

In order to describe the main algorithm of this chapter conveniently, we introduce some symbols for the input and output of Algorithm 42. Let S_C be the sequence $\{C_0, C_1, \ldots, C_{k-1}\}$. Let the route obtained by Algorithm 42 be $\rho(S_C, V, \varepsilon, \varepsilon_1)$ on input S_C, V, ε, and ε_1, where V is defined as in Line 9 of Algorithm 42.

The 'key idea' of Algorithm 43 in Fig. 11.14 is to maintain a watchman route by shrinking it properly so as to compute the set of all non-redundant essential cuts. When the loop ends, there does not exist any redundant essential cut in S_C, and the program terminates.

We choose accuracy value $\varepsilon_0 = 1$ to speed up in Lines 1, 3, and 5 during maintaining the watchman route. Finally, we choose the accuracy constant ε to be sufficiently small, such as 10^{-15} in Line 8 to obtain a high accuracy.

Procedure 26 (Handling of three general cuts)
Input: Three general cuts C_1, C_2 and C_3 of P, three points $p_i \in C_i$, for $i = 1, 2, 3$,
and a degeneration accuracy constant $\varepsilon_1 > 0$.
Output: An updated shorter path $\rho(p_1, \ldots, p_2, \ldots, p_3)$ that might also contain vertices of the polygon P.

1: For each $i \in \{1, 2\}$, let $\{p_i, p_{i+1}\}$ (where $p_i \in C_i$) be the input for the 2D ESP algorithm; the output is a set V_{ii+1}—the set of vertices of a shortest path from p_i to p_{i+1} inside of P. Let V be $V_{12} \cup V_{23}$.
2: Find q_1 and $q_3 \in V$ such that $\langle q_1, p_2, q_3 \rangle$ is a subsequence of V.
3: Let $C = C_2$, $p = p_2$, $q = q_i$, apply Procedure 25 (in Fig. 11.7) to find the maximal visible segment $S_i = p'_1 p'_2$ of q_i, $i = 1, 3$.
4: Find vertex $p'_2 \in S_2 = S_1 \cap S_3$ such that $d_e(q_1, p'_2) + d_e(p'_2, q_3) = \min\{d_e(q_1, p') + d_e(p', q_3) : p' \in S_2\}$.
5: If $C_2 \cap C_1$ (or C_3) $\neq \emptyset$ and p'_2 is the intersection point, then ε_1-transform p'_2 into another point (still denoted by p'_2) in C_2.
6: Update V by letting p_2 be p'_2.

Fig. 11.11 Calculation of a shorter subpath based on three subsequent general cuts

11.3 Correctness and Time Complexity

Lemma 11.4 *There is a shortest watchman route that contacts the sequence of all non-redundant essential cuts in good order.*

Proof Let $C_{a_0}, C_{a_1}, \ldots, C_{a_{m-1}}$ be all the active cuts (where "a" is for "active") such that

$$0 \leq a_0 < a_1 < \cdots < a_{m-1} \leq k - 1.$$

By Lemma 11.2, path ρ contacts $C_{a_0}, C_{a_1}, \ldots, C_{a_{m-1}}$ in this order. Let ρ contact active cut C_{a_i} at point p_{a_i}, where $i = 0, 1, \ldots, m - 1$; $\rho(p_{a_j}, p_{a_{j+1}})$ the section of ρ from p_{a_j} to $p_{a_{j+1}}$ inside of P, where $j = 0, 1, \ldots, m - 1$ (all subscripts are taken modulo m); see Fig. 11.15.

It is sufficient to prove that for any $j \in \{0, 1, \ldots, m - 1\}$, if $a_j + 1 < a_{j+1}$, then for any integer $n \in [a_j + 1, a_{j+1} - 1]$, essential cut C_n is visited (i.e., contacted) by a section of ρ, that is, the polyline $\rho(p_{a_j}, p_{a_{j+1}})$. By definition of C_{a_i} $(i = 0, 1, \ldots, m - 1)$, C_n is not an active cut. Thus, ρ contacts C_n by either a crossing contact or a tangential contact.

Case 1: If ρ contacts C_n by a tangential contact, then by Lemma 11.1, C_n must intersect with two successive active cuts which must be C_{a_j} and $C_{a_{j+1}}$ at points p_{a_j} and $p_{a_{j+1}}$ where ρ contacts C_{a_j} and $C_{a_{j+1}}$. Thus, C_n is contacted by the polyline $\rho(p_{a_j}, p_{a_{j+1}})$. See Fig. 11.16.

Case 2: ρ contacts C_n by a crossing contact. Then C_n must be contacted by the polyline $\rho(p_{a_j}, p_{a_{j+1}})$. [In Fig. 11.4, C_3 is contacted by the polyline $\rho(p_0, p_2) = $

Algorithm 42 (Approximate solution for the WRP with start point)

Input: A starting point s, k essential cuts $C_0, C_1, \ldots, C_{k-1}$, and P, which satisfy the condition of the fixed WRP, points $p_i \in C_i$, where $i = 0, 1, 2, \ldots, k-1$, an accuracy constant $\varepsilon > 0$, and a degeneration constant $\varepsilon_1 > 0$.

Output: An updated closed $\{1 + 2k[r(\varepsilon) + \varepsilon_1]/L\}$-approximation path $\rho(s, p_0, \ldots, p_1, \ldots, p_{k-1}, s)$, which may also contain vertices of P, where L is the length of an optimal path, $r(\varepsilon)$ the upper error bound for distances between p_i and the corresponding optimal vertex p_i': $d_e(p_i, p_i') \le r(\varepsilon)$, for $i = 0, 1, \ldots, k-1$.

1: For $i \in \{0, 1, \ldots, k-1\}$, let p_i be an arbitrary point of C_i.
2: Let V_0 and V be a sequence of points $\langle p_0, p_1, \ldots, p_{k-1} \rangle$, L_1 be $\sum_{i=0}^{k} L_P(p_i, p_{i+1})$, and L_0 be ∞.
3: **while** $L_0 - L_1 \ge \varepsilon$ **do**
4:    **for** each $i \in \{0, 1, \ldots, k-1\}$ **do**
5:       Let $C_{i-1}, C_i, C_{i+1}, p_{i-1}, p_i, p_{i+1}$ and P be the input for Procedure 26, which updates p_i in V_0.
6:       Let U_i be the sequence of vertices of the path $\rho(p_{i-1}, \ldots, p_i, \ldots, p_{i+1})$ with respect to C_{i-1}, C_i and C_{i+1} (inside of P); let U_i be $\langle q_1, q_2, \ldots, q_m \rangle$.
7:       Insert (after p_{i-1}) the points of sequence U_i (in the given order) into V_0, producing $V_1 = \langle p_0, p_1, \ldots, p_{i-1}, q_1, q_2, \ldots, q_m, p_{i+1}, \ldots, p_{k-1} \rangle$.
8:    **end for**
9:    Let L_0 be L_1 and V_0 be V.
10:   Compute the perimeter L_1 of the polygon, given by the sequence V_1 of vertices.
11: **end while**
12: Output sequence V_1, and the desired length equal to L_1.

Fig. 11.12 Calculation of an approximate solution for the fixed WRP

(p_0, p_1, p_2).] Otherwise, C_n must be contacted by polyline $\rho \setminus \rho(p_{a_j}, p_{a_{j+1}})$. As the defining vertices of C_{a_j}, C_n, and $C_{a_{j+1}}$ are located in order around ∂P, and C_n is not contacted by the polyline $\rho(p_{a_j}, p_{a_{j+1}})$, the polyline $\rho \setminus \rho(p_{a_j}, p_{a_{j+1}})$ must cross $\rho(p_{a_j}, p_{a_{j+1}})$ at least twice so as to contact C_n by a crossing contact. Let two points $p, q \in \rho(p_{a_j}, p_{a_{j+1}})$ be given such that $\rho \setminus \rho(p_{a_j}, p_{a_{j+1}})$ enters $\rho(p_{a_j}, p_{a_{j+1}})$ at point p to contact C_n at $r \in C_n$, and finally leaves $\rho(p_{a_j}, p_{a_{j+1}})$ at point q. Note that ρ contacts C_n by a crossing contact. Thus, polyline $\rho(p, r, q)$ is the shortest path from p to q inside of the simple polygon P. Also note that the polyline from p to q along with the polyline $\rho(p_{a_j}, p_{a_{j+1}})$ is a shortest path inside of the simple polygon P. Therefore, we obtain two different shortest paths from p to q inside of the simple polygon P. See Fig. 11.17. This is a contradiction. $\qquad\square$

Lemma 11.5 *The final set S_C in Algorithm 43 is the set of all non-redundant essential cuts.*

Fig. 11.13 Illustration for
Line 1 of Algorithm 42

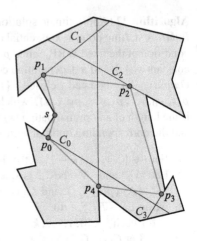

Proof Let S'_C be the set of all non-redundant essential cuts. In Algorithm 43, let S^i_C be S_C obtained after the ith while-loop. Let S^N_C be the final S_C, where $i = 0, 1, 2, \ldots, N$, and N is a non-negative integer. By Lemma 11.3, we have that

$$S'_C \subseteq S^i_C$$

where $i = 0, 1, 2, \ldots, N$. On the other hand, we have that

$$S^N_C \subset S^{N-1}_C \subset S^{N-2}_C \subset \cdots \subset S^1_C \subset S^0_C.$$

Thus, $S'_C = S^N_C$. This proves the lemma. $\square$

We cite a few basic results of convex analysis which are used in the following (also partially already used in Chap. 9):

Algorithm 43 (Algorithm for solving the floating WRP)
Input and output: the same as for Algorithm 42 except that we do not have to select a start point s.

1: Apply Algorithm 42 on the input S_C, V, ε_0, and ε_1 for finding a watchman route $\rho(S_C, V, \varepsilon_0, \varepsilon_1)$.
2: **for** $i \in \{0, 1, \ldots, k-1\}$ **do**
3:    **if** $C_i \in S_C$ is such that the pocket $P(C_i)$ completely contains the route $\rho(S_C, V, \varepsilon_0, \varepsilon_1)$ **then**
4:        Update S_C and V by removing C_i and p_i from S_C and V, respectively, where p_i is the vertex of ρ in C_i.
5:        Apply Algorithm 42 on updated S_C, V (obtained from last step), ε_0, and ε_1 for finding a watchman route $\rho(S_C, V, \varepsilon_0, \varepsilon_1)$.
6:    **end if**
7: **end for**
8: Apply Algorithm 42 on S_C, V, ε, and ε_1 for finding a watchman route R.

Fig. 11.14 Main algorithm for calculating an approximate solution of the floating WRP

Fig. 11.15 Illustration for
Lemma 11.4: $\rho(p_{a_j}, p_{a_{j+1}})$,
a section of ρ

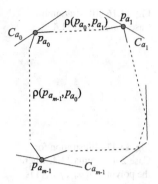

Theorem 11.3 *Let S_1 and S_2 be convex sets in $\mathbb{R}^m$ and $\mathbb{R}^n$, respectively. Then, $S_1 \times S_2$ is a convex set in $\mathbb{R}^{m+n}$, where $m, n \in \mathbb{N}$.*

Proposition 11.1 *Each line segment is a convex set.*

Proposition 11.2 *Each norm on $\mathbb{R}^n$ is a convex function.*

Proposition 11.3 *A non-negative weighted sum of convex functions is a convex function.*

Proposition 11.4 *Let f be a convex function. If x is a point where f has a finite local minimum, then x is a point where f has its global minimum.*

By Theorem 11.3 and Propositions 11.1–11.3, we have the following

Corollary 11.1 $L_S(p, q): S_0 \times S_{k-1} \to \mathbb{R}$ *is a convex function.*

Let C_1, C_2, and P satisfy the condition of the floating WRP. By Corollary 11.1, we also have the following

Corollary 11.2 $L_P(p, q): C_1 \times C_2 \to \mathbb{R}$ *is a convex function.*

Let $S_i \subseteq C_i$ be the line segment associated with the final updated point $p_i \in C_i$ in Algorithm 43, where $i = 0, 1, 2, \ldots, k-1$. Analogously to Theorem 3.3, we have the following

Theorem 11.4 *If the chosen accuracy constant ε is sufficiently small, then Algorithm 43 outputs a $\{1 + 2k \cdot [r(\varepsilon) + \varepsilon_1]/L\}$-approximation path (in fact, a loop) with respect to the step set $\langle S_0, S_1, \ldots, S_{k-1}, S_0 \rangle$, for any initial path.*

We sketch the proof of Theorem 11.4. Algorithm 7 is called an *arc version* of an RBA because we do not return from q to p. If we use points p and q for fixing two points on the ESP, but we find a shortest path which is a loop and passes through

Fig. 11.16 Illustration for
Case 1 in Lemma 11.4

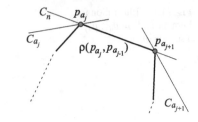

Fig. 11.17 Illustration for
Case 2 in Lemma 11.4, where
the polyline $\rho(p_{a_j}, p_{a_{j+1}})$ is
in *red colour*,
$\rho \setminus \rho(p_{a_j}, p_{a_{j+1}})$ *green*, and
$\rho(p, r, q)$ is *dark green*. The
shortest path from p to q,
being a section of the polyline
$\rho(p_{a_j}, p_{a_{j+1}})$, is in *dark red*
(colour refers to the e-copy of
the book)

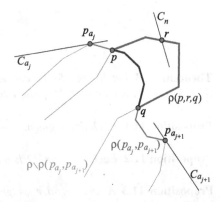

line segments $\langle S_1, S_2, \ldots, S_k, S_1 \rangle$ in this order, then we obtain a *loop version* of
Algorithm 7.

Basically, following the same way as demonstrated with the proof of Theorem 3.3, we can prove that the loop version of Algorithm 7 outputs a $\{1 + 2k \cdot [r(\varepsilon) + \varepsilon_1]/L\}$-approximation path which is a loop. In this sense, Algorithm 43
defines a $\{1 + 2k \cdot [r(\varepsilon) + \varepsilon_1]/L\}$-approximation path (a loop) for the step set
$\langle S_0, S_1, \ldots, S_{k-1}, S_0 \rangle$. We do not provide details of the proof here because of similarities with previous proofs.

Theorem 11.4 says that Algorithm 43 outputs an approximate local minimal solution to the floating WRP. We also have the following

Theorem 11.5 *Algorithm 43 outputs a* $\{1 + 2k \cdot [r(\varepsilon) + \varepsilon_1]/L\}$-*approximation solution to the floating WRP.*

Proof By Corollary 11.2,

$$\sum_{i=0}^{k-1} L_P(p_i, p_{i+1}) : \prod_{i=0}^{k-1} C_i \to \mathbb{R}$$

is a convex function, where

$$L_P(p_i, p_{i+1})$$

is defined as in Line 2 of Algorithm 42. By Proposition 11.4 and Theorem 11.4, we
have shown the theorem. □

Regarding the time complexity of our solution to the floating WRP, we first state two corollaries without proofs:

Corollary 11.3 *Procedure* 25 *can be computed in time* $\mathcal{O}(|V(\partial P)|)$.

Corollary 11.4 *Procedure* 26 *can be computed in time* $\mathcal{O}(|V(\partial P)|)$.

Furthermore, note that the main computation is in two stacked loops. The while-loop takes $\kappa(\varepsilon)$ iterations. By Corollary 11.4, the for-loop can be computed in time $\mathcal{O}(k \cdot |V(\partial P)|)$. Thus, we have the following

Corollary 11.5 *Algorithm* 42 *can be computed in time* $\kappa(\varepsilon) \cdot \mathcal{O}(k \cdot |V(\partial P)|)$.

This finally allows us to summarise the complexity results for the main algorithm of this chapter:

Theorem 11.6 *Algorithm* 43 *can be computed in time*

$$\kappa(\varepsilon) \cdot \mathcal{O}\big(k \cdot \big|V(\partial P)\big|\big) + \kappa(\varepsilon_0) \cdot \mathcal{O}\big(k^2 \cdot \big|V(\partial P)\big|\big).$$

Proof By Corollary 11.5, Line 1 can be computed in time $\kappa(\varepsilon_0) \cdot \mathcal{O}(k \cdot |V(\partial P)|)$. Inside of the for-loop, Line 3 can be computed in time $\mathcal{O}(k \cdot |V(\partial P)|)$; Line 4 can be computed in time $\mathcal{O}(k)$. The main computation of the for-loop occurs in Line 5. By Corollary 11.5, it can be performed in time $\kappa(\varepsilon_0) \cdot \mathcal{O}(k \cdot |V(\partial P)|)$.

As the computation inside of the for-loop can happen at most k times, the for-loop can be computed in time $\kappa(\varepsilon_0) \cdot \mathcal{O}(k^2 \cdot |V(\partial P)|)$. By Corollary 11.5, Line 8 can be executed in time $\kappa(\varepsilon) \cdot \mathcal{O}(k \cdot |V(\partial P)|)$. Thus, Algorithm 43 requires

$$\kappa(\varepsilon) \cdot \mathcal{O}\big(k \cdot \big|V(\partial P)\big|\big) + \kappa(\varepsilon_0) \cdot \mathcal{O}\big(k^2 \cdot \big|V(\partial P)\big|\big)$$

computation time.   □

By Lemmas 11.4 and 11.5, and Theorems 11.2 and 11.6, we conclude:

Theorem 11.7 *This chapter provided a* $\{1 + 2k \cdot [r(\varepsilon) + \varepsilon_1]/L\}$-*approximation solution to the floating WRP, having time complexity* $\kappa(\varepsilon) \cdot \mathcal{O}(k \cdot |V(\partial P)|) + \kappa(\varepsilon_0) \cdot \mathcal{O}(k^2 \cdot |V(\partial P)|)$, *where* k *is the number of essential cuts, and* L *is the length of an optimal watchman route.*

The main algorithm of this chapter (Algorithm 43) is just another example for following the general RBA methodology. In some simple polygons, we find the exact solution for the floating WRP, in others we 'just' converge to the correct solution.

A large number of experiments also indicates that $\kappa(\varepsilon)$ does not depend on k, where k is the number of essential cuts. It remains an open problem to prove a smallest upper bound for $\kappa(\varepsilon)$.

11.4 Problems

Problem 11.1 Describe simple polygons P such that there exists a single point q_1 in P for which every point q in P is visible from q_1. In this case, what is the shortest watchman route for P?

Problem 11.2 Does there always exist at least one cut or general cut for any simple polygon P? Given a cut C of P, does there always exist an associated cut of C?

Problem 11.3 Consider the floating WRP. How to define a cut, essential part of that cut, and the pocket induced by the cut? When is a cut active?

Problem 11.4 Consider the fixed WRP. Is there any difference for the performance of Algorithm 42 if the starting point s is identical to a vertex of P, on ∂P (i.e., the frontier of P), or in $P°$ (i.e., the interior of P)?

Problem 11.5 Consider the following problem: Compute a shortest route $\rho \subset P$ such that any frontier point $p \in \partial P$ is visible from at least one point on ρ. Would that be equivalent to the floating WRP?

Problem 11.6 Show that there exists a shortest watchman route ρ for any simple polygon P such that there exists a tangential contact between ρ and at least one essential cut of P.

Problem 11.7 How do cuts differ when addressing either the fixed WRP or the floating WRP?

Problem 11.8 Show that in the case of the floating WRP, there is the possibility of an existence of an infinite number of shortest watchman routes. Provide an example of such a polygon.

Problem 11.9 Consider the case of the fixed WRP. Prove that each essential cut is also a non-redundant cut.

Problem 11.10 (Research problem) Discuss the difference in computational time complexity between the fixed WRP and the floating WRP by providing 'sharp' upper or lower bounds.

Problem 11.11 (Programming exercise) Implement the Algorithm 43 and discuss its performance for inputs of varying complexity.

Problem 11.12 (Research problem) Specify the generic RBA (see Sect. 3.10) for the particular case of solving the fixed WRP.

11.5 Notes

The (floating) watchman route problem (WRP) of computational geometry is discussed in [4]. Definitions in this chapter, related to the general WRP, follow [13, 34], but some definitions also follow [9, 11]. In particular, see page 379 in [9]. Figure 11.5 is a modification of Fig. 2 in [11]. For Lemma 11.1, see page 14 in [11]. See [9] for Lemma 11.2 (Lemma 3.2 in the source) and Lemma 11.3 (Lemma 2.1). Theorem 11.2 is Theorem 1 in [34]. For Algorithm 2D ESP, see [20], pages 639–641. For linear-time convex hull algorithms, see Algorithm 15 or Algorithm 14 in Chap. 4. An expanded version of this chapter is MI-tech report no. 51 at www.mi.auckland.ac.nz/.

For convex analysis, see, for example, [5, 27, 28]. For example, Theorem 11.3 is Theorem 3.5 in [28], for Proposition 11.2 see page 72 in [5], and for Proposition 11.4, see page 264 in [28].

A simplified WRP of finding a shortest route in a simple isothetic polygon was solved in 1988 in [10] by presenting an $\mathcal{O}(n \log \log n)$ algorithm. In 1991, [11] claimed to have presented an $\mathcal{O}(n^4)$ algorithm, solving the fixed WRP. In 1993, [36] obtained an $\mathcal{O}(n^3)$ solution for the fixed WRP. In the same year, this was further improved to a quadratic time algorithm [35]. However, four years later, in 1997, [17] pointed out that the algorithms in both [11] and [36] were flawed, but presented a solution for fixing those errors. Interestingly, two years later, in 1999, [37] found that the solution given by [17] was also flawed. By modifying the (flawed) algorithm presented in [36, 37] gave an $\mathcal{O}(n^4)$ runtime algorithm for the fixed WRP.

In 1995 and 1999, [22] and [9] each gave an $\mathcal{O}(n^6)$ algorithm for the WRP. This was improved in 2001 by an $\mathcal{O}(n^5)$ algorithm in [31]; this paper also proved Theorem 11.1.

So far the best known result for the fixed WRP is due to [13] by presenting in 2003 an $\mathcal{O}(n^3 \log n)$ runtime algorithm.

In 1995, [19] published an $\mathcal{O}(\log n)$-approximation algorithm for solving the WRP. In 1997, [8] gave a 99.98-approximation algorithm with time complexity $\mathcal{O}(n \log n)$ for the WRP. In 2001, [32] presented a linear-time algorithm for an approximate solution of the fixed WRP such that the length of the calculated watchman route is at most twice of that of the shortest watchman route. The coefficient of accuracy was improved to $\sqrt{2}$ in [33] in 2004. Most recently, [34] presented a linear-time algorithm for the WRP for computing an approximate watchman route of length at most twice of that of the shortest watchman route.

There are several generalisations and variations of watchman route problems; see, for example, [6, 7, 9, 12, 14–16, 18, 21–26]. [1–3] show that some of these problems are NP-hard, and the authors solve them by approximation algorithms.

References

1. Alsuwaiyel, M.H., Lee, D.T.: Minimal link visibility paths inside a simple polygon. Comput. Geom. **3**(1), 1–25 (1993)

2. Alsuwaiyel, M.H., Lee, D.T.: Finding an approximate minimum-link visibility path inside a simple polygon. Inf. Process. Lett. **55**, 75–79 (1995)
3. Arkin, E.M., Mitchell, J.S.B., Piatko, C.: Minimum-link watchman tours. Report, University at Stony Brook (1994)
4. Asano, T., Ghosh, S.K., Shermer, T.C.: Visibility in the plane. In: Sack, J.-R., Urrutia, J. (eds.) Handbook of Computational Geometry, pp. 829–876. Elsevier, Amsterdam (2000)
5. Boyd, S., Vandenberghe, L.: Convex Optimization. Cambridge University Press, Cambridge, UK (2004)
6. Carlsson, S., Jonsson, H., Nilsson, B.J.: Optimum guard covers and m-watchmen routes for restricted polygons. In: Proc. Workshop Algorithms Data Struct. LNCS, vol. 519, pp. 367–378. Springer, Berlin (1991)
7. Carlsson, S., Jonsson, H., Nilsson, B.J.: Optimum guard covers and m-watchmen routes for restricted polygons. Int. J. Comput. Geom. Appl. **3**, 85–105 (1993)
8. Carlsson, S., Jonsson, H., Nilsson, B.J.: Approximating the shortest watchman route in a simple polygon. Technical report, Lund University, Sweden (1997)
9. Carlsson, S., Jonsson, H., Nilsson, B.J.: Finding the shortest watchman route in a simple polygon. Discrete Comput. Geom. **22**, 377–402 (1999)
10. Chin, W., Ntafos, S.: Optimum watchman routes. Inf. Process. Lett. **28**, 39–44 (1988)
11. Chin, W.-P., Ntafos, S.: Shortest watchman routes in simple polygons. Discrete Comput. Geom. **6**, 9–31 (1991)
12. Czyzowicz, J., Egyed, P., Everett, H., Lenhart, W., Lyons, K., Rappaport, D., Shermer, T., Souvaine, D., Toussaint, G., Urrutia, J., Whitesides, S.: The aquarium keeper's problem. In: Proc. ACM–SIAM Sympos. Data Structures Algorithms, pp. 459–464 (1991)
13. Dror, M., Efrat, A., Lubiw, A., Mitchell, J.: Touring a sequence of polygons. In: Proc. STOC, pp. 473–482 (2003)
14. Gewali, L.P., Lombardo, R.: Watchman Routes for a Pair of Convex Polygons. Lecture Notes in Pure Appl. Math., vol. 144 (1993)
15. Gewali, L.P., Ntafos, S.: Watchman routes in the presence of a pair of convex polygons. In: Proc. Canad. Conf. Comput. Geom., pp. 127–132 (1995)
16. Gewali, L.P., Meng, A., Mitchell, J.S.B., Ntafos, S.: Path planning in 0/1/infinity weighted regions with applications. ORSA J. Comput. **2**, 253–272 (1990)
17. Hammar, M., Nilsson, B.J.: Concerning the time bounds of existing shortest watchman routes. In: Proc. FCT'97. LNCS, vol. 1279, pp. 210–221 (1997)
18. Kumar, P., Veni Madhavan, C.: Shortest watchman tours in weak visibility polygons. In: Proc. Canad. Conf. Comput. Geom., pp. 91–96 (1995)
19. Mata, C., Mitchell, J.S.B.: Approximation algorithms for geometric tour and network design problems. In: Proc. Ann. ACM Symp. Computational Geometry, pp. 360–369 (1995)
20. Mitchell, J.S.B.: Geometric shortest paths and network optimization. In: Sack, J.-R., Urrutia, J. (eds.) Handbook of Computational Geometry, pp. 633–701. Elsevier, Amsterdam (2000)
21. Mitchell, J.S.B., Wynters, E.L.: Watchman routes for multiple guards. In: Proc. Canad. Conf. Comput. Geom., pp. 126–129 (1991)
22. Nilsson, B.J.: Guarding art galleries: methods for mobile guards. Ph.D. thesis, Lund University, Sweden (1995)
23. Nilsson, B.J., Wood, D.: Optimum watchmen routes in spiral polygons. In: Proc. Canad. Conf. Comput. Geom., pp. 269–272 (1990)
24. Ntafos, S.: The robber route problem. Inf. Process. Lett. **34**, 59–63 (1990)
25. Ntafos, S.: Watchman routes under limited visibility. Comput. Geom. **1**, 149–170 (1992)
26. Ntafos, S., Gewali, L.: External watchman routes. Vis. Comput. **10**, 474–483 (1994)
27. Roberts, A.W., Varberg, V.D.: Convex Functions. Academic Press, New York (1973)
28. Rockafellar, R.T.: Convex Analysis. Princeton University Press, Princeton (1970)
29. Sunday, D.: Algorithm 14: tangents to and between polygons. http://softsurfer.com/Archive/algorithm_0201/ (2011). Accessed July 2011
30. Sunday, D.: Algorithm 3: fast winding number inclusion of a point in a polygon. http://softsurfer.com/Archive/algorithm_0103/ (2011). Accessed July 2011

31. Tan, X.: Fast computation of shortest watchman routes in simple polygons. Inf. Process. Lett. **77**, 27–33 (2001)
32. Tan, X.: Approximation algorithms for the watchman route and zookeeper's problems. In: Proc. Computing and Combinatorics. LNCS, vol. 2108, pp. 201–206. Springer, Berlin (2005)
33. Tan, X.: Approximation algorithms for the watchman route and zookeeper's problems. Discrete Appl. Math. **136**, 363–376 (2004)
34. Tan, X.: A linear-time 2-approximation algorithm for the watchman route problem for simple polygons. Theor. Comput. Sci. **384**, 92–103 (2007)
35. Tan, X., Hirata, T.: Constructing shortest watchman routes by divide-and-conquer. In: Proc. ISAAC. LNCS, vol. 762, pp. 68–77 (1993)
36. Tan, X., Hirata, T., Inagaki, Y.: An incremental algorithm for constructing shortest watchman route algorithms. Int. J. Comput. Geom. Appl. **3**, 351–365 (1993)
37. Tan, X., Hirata, T., Inagaki, Y.: Corrigendum to 'An incremental algorithm for constructing shortest watchman routes'. Int. J. Comput. Geom. Appl. **9**, 319–323 (1999)

Chapter 12
Safari and Zookeeper Problems

The obstacle is the path.

Zen Proverb

So far, the best result in running time for solving the fixed safari route problem (SRP) is $\mathcal{O}(n^2 \log n)$ published in 2003 by M. Dror, A. Efrat, A. Lubiw, and J. Mitchell. The best result in running time for solving the floating zookeeper route problem (ZRP) is $\mathcal{O}(n^2)$ published in 2001 by X. Tan. This chapter provides an algorithm for the "floating" SRP with $\kappa(\varepsilon) \cdot \mathcal{O}(kn + m_k)$ runtime, where n is the number of vertices of the given search space or domain D (a simple polygon), k the number of convex polygons P_i in D, and m_k is the total number of vertices of all polygons P_i. This chapter also provides an algorithm for the floating ZRP with $\kappa(\varepsilon) \cdot \mathcal{O}(kn)$ runtime, where n is the number of vertices of all polygons involved, and k the number of the "cages". Extensions of the presented algorithms can solve more general SRPs and ZRPs if each convex polygon is replaced by a convex region such as convex polybeziers (beziergons, or parabolic splines) or ellipses.

12.1 Fixed and Floating Problems; Dilations

Let a search space or domain D contain k pairwise-disjoint convex polygons P_i $(i = 0, 1, \ldots, k - 1)$ such that exactly one edge of each of these polygons P_i is incident with the frontier of D. We are interested in a route with vertices $p_i \in \partial P_i$, for $i = 1, 2, \ldots, k$. We consider indices modulo k, and identify index k with index 0 this way. The start point p and end point q are assumed to be identical; that means we have $p = q = p_0 \in \partial P_0$. Recall that P_i° denotes the interior of P_i. The *safari route problem* (SRP) is defined as follows:

F. Li, R. Klette, *Euclidean Shortest Paths*,
DOI 10.1007/978-1-4471-2256-2_12, © Springer-Verlag London Limited 2011

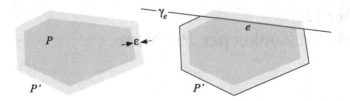

Fig. 12.1 *Left*: ε_d-outer polygon of a convex polygon P. *Right*: ε_d polygon of P, where $\varepsilon_d = \varepsilon_3$

Compute a shortest route $\rho \subset D$ inside of D such that ρ visits each P_i at a point p_i on the frontier of P_i, allowing that straight segments $p_i p_{i+1} \pmod{k}$ intersect P_i° in one or several segments, for $i = 0, 1, \ldots, k-1$.

If the start point of the route is given then this refined safari problem is known as the *fixed* SRP. Otherwise, the problem is known as the *floating* SRP.

This chapter provides an algorithm for the floating SRP with $\kappa(\varepsilon) \cdot \mathcal{O}(kn + m_k)$ runtime, where n is the number of vertices of the given simple polygon D, k the number of convex polygons P_i, and m_k is the total number of vertices of all polygons P_i; $\kappa(\varepsilon)$ defines the numerical accuracy depending upon a selected parameter $\varepsilon > 0$.

The *zookeeper route problem* (ZRP) is defined as follows (informally speaking, "the zookeeper is not supposed to enter any of the cages on his path, but to visit all"):

Compute a shortest route ρ inside of D such that ρ visits each "cage" P_i (a convex polygon) at a point p_i on its frontier, and such that this path is not intersecting any of the interiors P_i°, for $i = 0, 1, \ldots, k-1$.

Analogously to the SRP, if the start point of the route is given then this refined zookeeper route problem is known as the *fixed* ZRP. Otherwise, the problem is known as the *floating* ZRP.

This chapter provides an algorithm for the floating ZRP with $\kappa(\varepsilon) \cdot \mathcal{O}(kn)$ runtime, where n is the number of vertices of the given simple polygon D and all the convex polygons P_i, k is the number of convex polygons P_i, and $\kappa(\varepsilon)$ is defined as before in this book.

Let P be a convex polygon such that exactly one edge of P, denoted by e, is incident with ∂D (i.e., the frontier of D). Let P' be a convex polygon such that P' completely contains P, and such that P' represents a *Minkowski addition* (also known as *dilation*) of P such that the distance between each edge of P' and its corresponding edge in P is $\varepsilon_d > 0$ (where d is for dilation), then P' is called an ε_d-*outer polygon* of P^1 (see Fig. 12.1 left). Let l_e be the straight line which contains

---

[1]The ε_d-outer polygon of P is a "shell" around P of uniform thickness $\varepsilon_d > 0$.

Fig. 12.2 An example of
three polygons which satisfy
the SRP (or ZRP) condition.
The *light shaded* simple
polygon P is the domain that
contains two convex polygons
P_1 and P_2. The figure also
shows a shortest path
connecting a point p_1 in P_1
with a point p_2 in P_2

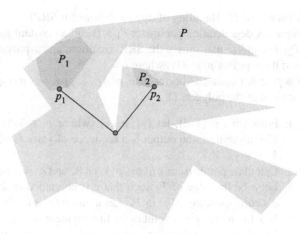

the edge e. Line l_e cuts P' into two polygons. One of them completely contains P.
We call this polygon the ε_d-*polygon* of P (with respect to e). See Fig. 12.1, right.

Let $P_0, P_1, \ldots, P_{k-1}$, and D be k convex polygons and one simple polygon such
that P_i and P_{i+1} are pairwise-disjoint; any P_i is completely contained in D, and
exactly one edge e_i of P_i is incident with the frontier ∂D of the simple polygon D,
where $i = 0, 1, 2, \ldots, k - 1$. Note that D is not required to be convex.

Definition 12.1 If those edges $e_0, e_1, \ldots, e_{k-1}$ are ordered clockwise around ∂D,
then we say that convex polygons $P_0, P_1, \ldots, P_{k-1}$, and domain D *satisfy the SRP*
(or *ZRP*) *condition*.

For example, in Fig. 12.2, P_1, P_2, and P satisfy the SRP (or ZRP) condition.

Let polygons P_i and D satisfy this condition; $P_i \cap D = e_i$ (i.e., e_i is an edge of
P_i which is incident with ∂D). If two points $p_1^i, p_2^i \in \partial P_i \setminus e_i$, and p_1^i, p_2^i (p_1^i and
p_2^i may be identical) are ordered counterclockwise around ∂P_i, then we say that
(p_1^i, p_2^i) is a *ZRP pair*.

Let D' be the simple polygon obtained by modifying D by replacing e_i by
$\partial P_i \setminus e_i$, where $i = 0, 1, 2, \ldots, k - 1$.

For generalisations at the end of the chapter, also recall that a *simple curve* is a
curve in a plane such that each point of it can be expressed as $(x(t), y(t))$, where
t varies in an interval $[a, b]$, and both $x(t)$ and $y(t)$ are differentiable in $[a, b]$.
A *simple region* is a planar, convex, topologically-closed region such that its frontier consists of finite simple curves. Such simple regions will generalise the convex
polygons P_i at the end of this chapter.

This ends the introduction of technical terms for this chapter. We also recall here
in one place two results and two algorithms for later use:

Lemma 12.1 *A solution to the safari problem must visit polygons P_i in the same
order as it meets ∂D.*

Procedure 27 (Handling of three polygons for SRP)

Input: A degeneration parameter $\varepsilon_1 > 0$ and a constant $\varepsilon_d \geq \varepsilon_1$, four polygons P_1, P_2, P_3, and D that satisfy the SRP condition, an ε_d-polygon of P_2, denoted by P_2', and three points $p_i \in \partial P_i$, where $i = 1, 2, 3$.

Output: An updated shorter path $\rho(p_1, \ldots, p_2, \ldots, p_3)$ which might also contain vertices of the polygon D.

1: For each $i \in \{1, 2\}$, let $\{p_i, p_{i+1}\}$ (where $p_i \in \partial P_i$) be the input for the 2D ESP algorithm; the output is a sequence of vertices V_{ii+1} of $\rho_D(p_i, p_{i+1})$. Let $V = V_{12} \cup V_{23}$.

2: Calculate points $q_1 = \rho_D(p_1, p_2) \cap \partial P_2'$ and $q_3 = \rho_D(p_2, p_3) \cap \partial P_2'$.

3: Let e be the edge of P_2 such that e is contained in the frontier of D; let e' be e's corresponding edge in P_2'. Let u and v (u' and v') be the two endpoints of e (e'). Let u'' (v'') be a point in the line segment uu' (vv') such that $d_e(u'', u) = \varepsilon_1$ ($d_e(v'', v) = \varepsilon_1$). If q_1 or q_3 is located on segment uu'' or vv'', then reset it to be u'' or v''.

4: **if** $q_1 q_3 \cap P_2' \neq \emptyset$ **then**

5:    Let p_2' be the point in segment $q_1 q_3 \cap P_2'$ such that $d_e(q_1, p_2') = \min\{d_e(q_1, p') : p' \in q_1 q_3 \cap P_2'\}$.

6: **else**

7:    Find vertex $p_2' \in \partial P_2'$ such that $d_e(q_1, p_2') + d_e(p_2', q_3) = \min\{d_e(q_1, p') + d_e(p', q_3) : p_2' \in \partial P_2'\}$.

8: **end if**

9: Update V by letting $p_2 = p_2'$.

Fig. 12.3 Handling of three polygons for the SRP case

Lemma 12.2 *A solution to the zookeeper problem must visit polygons P_i in the same order as it meets ∂D.*

We again assume (as in Chap. 11) an algorithm for calculating a 2D ESP (as discussed in Chap. 6) that has as input a simple polygon D and two points $p, q \in D$ and which calculates a set of vertices of a shortest path from p to q inside of D, and we assume that this algorithm is in linear time $\mathcal{O}(|V(\partial D)|)$ for simplicity.

We also use again *tangent calculation* which has as input an ordered sequence of vertices of a planar convex polygonal curve ∂D and a point $p \notin D$; its output are two points $t_i \notin D$ such that pt_i are tangents to ∂D, for $i = 1, 2$; the running time is in $\mathcal{O}(\log |V(\partial D)|)$.

12.2 Solving the Safari Route Problem

The algorithm for solving the SRP is modified from Algorithm 42. All modifications are underlined. We start presenting Procedure 27 in Fig. 12.3 which is frequently used by the Algorithm 44, which is later specified in Fig. 12.7.

Fig. 12.4 Illustration for
Procedure 27. Parts of P_2 and
P_2' (indicated by the
rectangle) are shown
magnified in Fig. 12.5

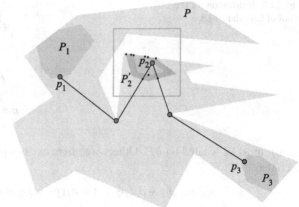

Fig. 12.5 Magnified sections
of P_2 and P_2' (the *rectangle* in
Fig. 12.4)

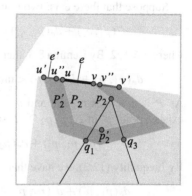

In Line 2 of Procedure 27, q_1 (q_3) is the intersection point between path
$\rho_D(p_1, p_2)$ (path $\rho_D(p_2, p_3)$) and P_2' (ε_d-polygon of P_2). See Fig. 12.4. Line 3
is also illustrated in Fig. 12.4.

In Line 7 of Algorithm 44, sequence V_1 is the updated sequence V_0, after inserting U_i. In Line 9, we use the updated original sequence V instead of V_1 for the next
iteration.

Let P be a convex region, ∂P is the boundary of P, p_1 and p_2 are two points
outside of P such that $P \cap p_1 p_2 = \emptyset$, where $p_1 p_2$ is the straight line passing through
points p_1 and p_2. Then elementary geometry (we include this for completeness)
shows the following

Corollary 12.1 *There is a unique point q in ∂P such that*

$$d_e(p_1, q) + d_e(p_2, q) = \min\{d_e(p_1, q) + d_e(p_2, q) : q \in P\}.$$

Proof If p is a point in P such that

$$d_e(p_1, p) + d_e(p_2, p) = \min\{d_e(p_1, q) + d_e(p_2, q) : q \in P\} \qquad (12.1)$$

Fig. 12.6 Illustration for the
proof of Corollary 12.1

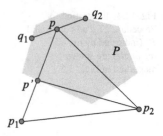

then p must be located on ∂P. Otherwise, there exists a point p' in $\triangle pp_1p_2 \cap \partial P$
such that

$$d_e(p_1, p') + d_e(p_2, p') < d_e(p_1, p) + d_e(p_2, p).$$

This is a contradiction to Eq. (12.1). See Fig. 12.6.

Suppose that there exist two points q_i in ∂P such that

$$d_e(p_1, q_i) + d_e(p_2, q_i) = \min\{d_e(p_1, q) + d_e(p_2, q) : q \in P\}$$

where $i = 1, 2$. By Lemma 8.3, there exists a point p in-between q_1 and q_2 such that

$$d_e(p_1, p) + d_e(p_2, p) < \min\{d_e(p_1, q) + d_e(p_2, q) : q \in P\}. \tag{12.2}$$

As P is a convex region, and q_1, q_2 are in ∂P, p must be in $P\setminus\partial P$. Then there
exists a point p' in $\triangle pp_1p_2 \cap \partial P$ such that

$$d_e(p_1, p') + d_e(p_2, p') < d_e(p_1, p) + d_e(p_2, p).$$

By Inequality (12.2), we have that

$$d_e(p_1, p') + d_e(p_2, p') < \min\{d_e(p_1, q) + d_e(p_2, q) : q \in P\}.$$

This is a contradiction.   □

Theorem 12.1 *Algorithm 44 outputs a $\{1 + 2k \cdot [r(\varepsilon) + \varepsilon_1]/L\}$-approximation solution to the SRP.*

Proof Algorithm 44 (see Fig. 12.7) defines a function f mapping from $\prod_{i=0}^{k-1} \partial P_i$
to $\mathbb{R}$. Analogously to the proof of Corollary 3.5, the total number of different values
of f is finite. By Corollary 12.1, f is a continuous function. Thus, the range of f
must be a singleton. By Lemma 12.1, this proves the theorem.   □

Regarding the time complexity of this solution to the SRP, note that   the main
computation is in the two stacked loops. The while-loop takes $\kappa(\varepsilon)$ iterations. Pro-
cedure 27 can be computed in time $\mathcal{O}(|V(\partial D)| + |V(\partial P_i)|)$. Thus, the for-loop can
be computed in time $\mathcal{O}(k \cdot |V(\partial D)| + \sum_{i=0}^{k-1} |V(\partial P_i)|)$. We obtain

Corollary 12.2 *Algorithm 44 requires $\kappa(\varepsilon) \cdot \mathcal{O}(k \cdot |V(\partial D)| + \sum_{i=0}^{k-1} |V(\partial P_i)|)$ time.*

By Lemma 12.1, Theorem 12.1, and Corollary 12.2, we have the following

Algorithm 44 (Algorithm for solving the <u>SRP</u>)
Input: An accuracy constant $\varepsilon > 0$, a degeneration parameter $\varepsilon_1 > 0$ and a constant $\varepsilon_d \geq \varepsilon_1$; k <u>convex</u> polygons $P_0, P_1, \ldots, P_{k-1}$, and a domain <u>$D$</u>, which satisfy the condition of the <u>SRP</u>, and points $p_i \in \partial P_i$, where $i = 0, 1, 2, \ldots, k-1$.
Output: An updated closed $\{1 + 2k \cdot [r(\varepsilon) + \varepsilon_1]/L\}$-approximation path (i.e., the route) $\rho(p_0, \ldots, p_1, \ldots, p_{k-1})$ which may also contain vertices of <u>polygon D</u>.

1: For each $i \in \{0, 1, \ldots, k-1\}$, let p_i be a point in ∂P_i.
2: Let $V_0 = V = \langle p_0, p_1, \ldots, p_{k-1} \rangle$, let L_1 be <u>$\sum_{i=0}^{k-1} L_D(p_i, p_{i+1})$</u>, and let L_0 be ∞.
3: **while** $L_0 - L_1 \geq \varepsilon$ **do**
4:    **for** each $i \in \{0, 1, \ldots, k-1\}$ **do**
5:        Let P_{i-1}, P_i, P_{i+1}, p_{i-1}, p_i, p_{i+1} and <u>D</u> be the input for Procedure <u>27</u>, which updates p_i in V_0.
6:        Let U_i be the sequence of vertices of the path $\rho(p_{i-1}, \ldots, p_i, \ldots, p_{i+1})$ with respect to <u>P_{i-1}, P_i and P_{i+1} (inside of D)</u>; let $U_i = \langle q_1, q_2, \ldots, q_m \rangle$.
7:        Insert (after p_{i-1}) the points of sequence U_i (in the given order) into V_0, i.e., we have $V_1 = \langle p_0, p_1, \ldots, p_{i-1}, q_1, q_2, \ldots, q_m, p_{i+1}, \ldots, p_{k-1} \rangle$.
8:    **end for**
9:    Let $L_0 = L_1$ and $V_0 = V$.
10:   Compute the perimeter L_1 of the polygon, given by the sequence V_1 of vertices.
11: **end while**
12: Output sequence V_1, and the length value of interest equals L_1.

Fig. 12.7 Algorithm for solving the <u>SRP</u>. The algorithm is modified from Algorithm 42, and all modifications are *underlined*

Theorem 12.2 *Algorithm* 44 *is a* $\{1 + 2k \cdot [r(\varepsilon) + \varepsilon_1]/L\}$-*approximation solution to the SRP, having time complexity* $\kappa(\varepsilon) \cdot \mathcal{O}(k|V(\partial D)| + \sum_{i=0}^{k-1} |V(\partial P_i)|)$, *where* k *is the number of convex polygons, and* L *is the length of an optimal safari route.*

12.3   Solving the Zookeeper Route Problem

In this section, we modify the algorithms and theorems of Sect. 12.2 for solving the ZRP. All modifications are underlined in the algorithms.

In Line 2 of Procedure 28 (see Figs. 12.8, 12.9, 12.10), q_1 (q_3) is the intersection point between the path $\rho_{D'}(p_1, p_2)$ (path $\rho_{D'}(p_2, p_3)$) and P_2' (ε_d-polygon of P_2). Recall that D' is the simple polygon obtained by modifying D by replacing e_i by $\partial P_i \backslash e_i$, where $i = 0, 1, 2, \ldots, k-1$. Line 3 is also illustrated in Fig. 12.8.

In Line 7 of Algorithm 45 (see Fig. 12.11), sequence V_1 is the updated sequence V_0, after inserting U_i. In Line 9, we use the updated original sequence V instead

Fig. 12.8 Illustration for Procedure 28, where $p_i = p_1^i = p_2^i$ ($i = 1, 2, 3$), and $p'_2 = p'^2_1 = p'^2_2$. Magnified sections of P_2 and P'_2 (the *rectangle*) are shown in Fig. 12.9

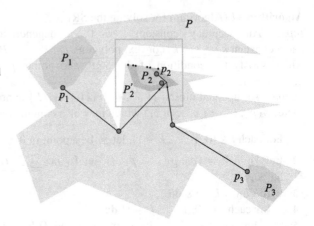

Fig. 12.9 Magnified sections of P_2 and P'_2 (the *rectangle* in Fig. 12.8)

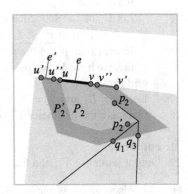

of V_1 for the next iteration. Note: sequence V_1 is the updated sequence V_0, after inserting U_i.

Analogously to the proof of Theorem 12.1, we obtain

Theorem 12.3 *Algorithm 45 outputs a $\{1 + 2k \cdot [r(\varepsilon) + \varepsilon_1]/L\}$-approximation solution to the ZRP.*

Analogously to the proof of Theorem 12.2, we can prove that

Theorem 12.4 *This chapter provided a $\{1 + 2k \cdot [r(\varepsilon) + \varepsilon_1]/L\}$-approximation solution to the ZRP, having time complexity $\kappa(\varepsilon) \cdot \mathcal{O}(k(|V(\partial D)| + \sum_{i=0}^{k-1} |V(\partial P_i)|))$, where k is the number of convex polygons, and L is the length of an optimal zookeeper route.*

By underlining text, we illustrate the *adaptivity* of the RBA approach to a range of varying route problems.

Procedure 28 (Handling of three polygons for ZRP)
Input: A degeneration parameter $\varepsilon_1 > 0$ and a constant $\varepsilon_d \geq \varepsilon_1$; four polygons P_1, P_2, P_3, and D satisfying the <u>ZRP</u> condition; the ε_d-polygon of P_2, denoted by P_2'; three ZRP pairs $(p_1^i, p_2^i) \in \partial P_i$, where $i = 1, 2, 3$.
Output: An updated shorter path $\rho(p_1^1, \ldots, p_2^1, \ldots, p_1^2, \ldots, p_2^2, \ldots, p_1^3, \ldots, p_2^3)$, which might also contain vertices of the polygon D.

1: For each $i \in \{1, 2\}$, let $\{p_2^i, p_1^{i+1}\}$ (where $p_2^i \in \partial P_i$, $p_1^{i+1} \in \partial P_{i+1}$) be the input for the 2D ESP algorithm; the output is a sequence of vertices V_{ii+1} of $\rho_{D'}(p_2^i, p_1^{i+1})$. Let V be $V_{12} \cup V_{23}$.
2: Calculate points $q_1 = \rho_{D'}(p_2^1, p_1^2) \cap \partial P_2'$ and $q_3 = \rho_{D'}(p_2^2, p_1^3) \cap \partial P_2'$.
3: Let e be the edge of P_2 such that e is contained in the frontier of D; let e' be e's corresponding edge in P_2'. Let u and v (u' and v') be the two endpoints of e (e'). Let u'' (v'') be a point in the line segment uu' (vv') such that $d_e(u'', u) = \varepsilon_1$ ($d_e(v'', v) = \varepsilon_1$). If q_1 or q_3 are located on segment uu'' or vv'', then reset it to be u'' or v''.
4: **if** $q_1q_3 \cap P_2' \neq \emptyset$ **then**
5:     Apply the tangent algorithm to find two points $p_1'^2, p_2'^2 \in \partial P_2$ such that $q_1 p_1'^2 (q_3 p_2'^2)$ is a tangent to P_2, and $p_1'^2, p_2'^2$ are ordered counterclockwise around ∂P_2.
6: **else**
7:     Find vertex $p_2' \in \partial P_2'$ such that $d_e(q_1, p_2') + d_e(p_2', q_3) = \min\{d_e(q_1, p') + d_e(p', q_3) : p_2' \in \partial P_2'\}$, let $p_1'^2$ and $p_2'^2$ be p_2'.
8: **end if**
9: Update V by letting $(p_1^2, p_2^2) = (p_1'^2, p_2'^2)$.

Fig. 12.10 Handling of three polygons for solving the ZRP

12.4 Some Generalisations

So far, algorithms for solving various Euclidean shortest path problems typically only consider an environment containing geometric objects which have straight edges. For example, the search space or obstacles in ESPs are modelled by polygons in 2D, polyhedrons in 3D, or the surface of polyhedrons in 2.5D. However, obstacles with smooth surfaces started to be investigated in computational geometry. In this section, we discuss some generalisation of the SRP and ZRP, where each convex polygon is replaced by a simple region as defined earlier in this chapter.

In the SRP, if the simple polygon D is ignored (i.e., $D = \mathbb{R}^2$) and instead of convex polygons P_i we consider arbitrary simple polygons, then this SRP is known as the touring polygons problem (TPP); see Chap. 10.

In this section, we modify (or generalise) the SRP so that the simple polygon D is ignored, and all the convex polygons P_i are replaced by ellipses. See Fig. 12.12. We call this the *touring ellipses problem (TEP)*. By E_i we denote the k pairwise-

Algorithm 45 (Algorithm for solving the <u>ZRP</u>)

Input: An accuracy constant $\varepsilon > 0$, a degeneration parameter $\varepsilon_1 > 0$, a constant $\varepsilon_d \geq \varepsilon_1$; k convex polygons $P_0, P_1, \ldots, P_{k-1}$, and D, which satisfy the condition of the <u>ZRP</u>, and ZRP pair $(p_1^i, p_2^i) \in \partial P_i$, where $i = 0, 1, 2, \ldots, k-1$.

Output: An <u>updated closed $\{1 + 2k \cdot [r(\varepsilon) + \varepsilon_1]/L\}$-approximation path</u> (i.e., the "route") $\rho(p_1^0, \ldots, p_2^0, \ldots, p_1^1, \ldots, p_2^1, \ldots, p_1^{k-1}, \ldots, p_2^{k-1})$, which may also con-tain <u>vertices of polygon D</u>.

1: For each $i \in \{0, 1, \ldots, k-1\}$, let (p_1^i, p_2^i) be a ZRP pair in ∂P_i.

2: Let $V_0 = V = \langle p_1^0, \ldots, p_2^0, \ldots, p_1^1, \ldots, p_2^1, \ldots, p_1^{k-1}, \ldots, p_2^{k-1} \rangle$; let L_1 be $\sum_{i=0}^{k-1} L_D(p_1^i, p_2^{i+1})$; and let L_0 be ∞.

3: **while** $L_0 - L_1 \geq \varepsilon$ **do**

4:     **for** each $i \in \{0, 1, \ldots, k-1\}$ **do**

5:         Let $P_{i-1}, P_i, P_{i+1}, (p_1^{i-1}, p_2^{i-1}), (p_1^i, p_2^i), (p_1^{i+1}, p_2^{i+1})$, and D be the in-put for Procedure <u>28</u>, which updates (p_1^i, p_2^i) in V_0.

6:         Let U_i be the sequence of vertices of the path $\rho(p_1^{i-1}, \ldots, p_2^{i-1}, \ldots, p_1^i, \ldots, p_2^i, \ldots, p_1^{i+1}, \ldots, p_2^{i+1})$ with respect to P_{i-1}, P_i, and P_{i+1} (inside of D); let $U_i = \langle q_1, q_2, \ldots, q_m \rangle$.

7:         Insert (after p_2^{i-1}) the points of sequence U_i (in the given order) into V_0, i.e., we have $V_1 = \langle p_1^0, \ldots, p_2^0, \ldots, p_1^1, \ldots, p_2^1, \ldots, p_1^{i-1}, \ldots, p_2^{i-1}, q_1, q_2, \ldots, q_m, p_1^{i+1}, \ldots, p_2^{i+1}, \ldots, p_1^{k-1}, \ldots, p_2^{k-1} \rangle$.

8:     **end for**

9:     Let $L_0 = L_1$ and $V_0 = V$ (note: we use the updated original sequence V instead of V_1 for the next iteration).

10:    Compute the perimeter L_1 of the polygon, given by the sequence V_1 of ver-tices.

11: **end while**

12: Output sequence V_1, and the desired length equals to L_1.

Fig. 12.11 Main algorithm for solving the ZRP. This is a modification of the provided SRP algo-rithm in the previous section. All modifications are *underlined*

disjoint ellipses, where $i = 0, 1, \ldots, k-1$; E_i° is the interior of E_i and ∂E_i the frontier of E_i.

> Compute a shortest route ρ such that ρ visits each ellipse E_i at a point p_i on ∂E_i (the frontier of E_i), allowing that straight segments $p_i p_{i+1}$ (mod k) intersect E_i° in one or several segments, for $i = 0, 1, \ldots, k-1$.

We describe Algorithm 46 (see Fig. 12.13) for solving the TEP. It is derived from Algorithm 7. In this algorithm, all subscripts are taken modulo k.

Fig. 12.12 An example of touring a sequence of *ellipses*

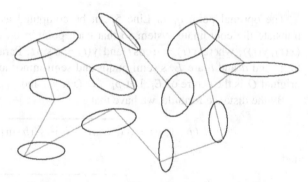

The optimal point q_i in Line 5 can be computed as follows: Each point p on ∂E_i can be expressed as $(x(t), y(t))$, where

$$x(t) = x_c + a \cos t \cos \theta - b \sin t \sin \theta,$$

$$y(t) = y_c + a \cos t \sin \theta + b \sin t \cos \theta,$$

the parameter t varies from $-\pi$ to π, (x_c, y_c) is the centre of the ellipse E_i, a and b are E_i's semi-major and semi-minor axes, respectively, and θ is the angle between the X-axis and the major axis of E_i. Therefore, the coordinates of p have the form

$$(a_1 \cos t - b_1 \sin t + c_1, a_2 \cos t + b_2 \sin t + c_2)$$

where $t \in [-\pi, \pi]$, a_i, b_i, and c_i $(i = 1, 2)$ are functions of x_c, y_c, a, b, and θ. Let $p_{i-1} = (x_1, y_1)$ and $p_{i+1} = (x_2, y_2)$.

Algorithm 46 (RBA for the fixed ESP problem of pairwise-disjoint ellipses in 2D space)

Input: A sequence of k pairwise-disjoint ellipses $E_0, E_1, \ldots, E_{k-1}$ and an accuracy constant $\varepsilon > 0$.

Output: A sequence $\langle p_0, p_1, p_2, \ldots, p_{k-1}, p_0 \rangle$ of a $[1 + 2(k + 1)r(\varepsilon)/L]$-approximation path which starts at p_0, then visits ∂E_i at p_i in the given order, and finally ends at p_0.

1: For each $j \in \{0, 1, \ldots, k - 1\}$, let p_j be a point ∂E_j.
2: $L_{\text{current}} = \sum_{j=0}^{k-1} d_e(p_j, p_{j+1})$, where $p_0 = p_k$; and let $L_{\text{previous}} = \infty$.
3: **while** $L_{\text{previous}} - L_{\text{current}} \geq \varepsilon$ **do**
4:     **for** each $j \in \{0, 1, 2, \ldots, k - 1\}$ **do**
5:         Compute a point $q_j \in S_j$ such that $d_e(p_{j-1}, q_j) + d_e(q_j, p_{j+1}) = \min\{d_e(p_{j-1}, p) + d_e(p, p_{j+1}) : p \in \partial E_j\}$.
6:         Update the path $\langle p_0, p_1, \ldots, p_{k-1}, p_0 \rangle$ by replacing p_j by q_j.
7:     **end for**
8:     Let $L_{\text{previous}} = L_{\text{current}}$ and $L_{\text{current}} = \sum_{j=0}^{k-1} d_e(p_j, p_{j+1})$.
9: **end while**
10: Return $\{p_0, p_1, \ldots, p_{k-1}, p_0\}$.

Fig. 12.13 RBA for the fixed ESP problem of pairwise-disjoint ellipses in 2D space

The optimal point q_i in Line 5 can be computed as follows: We rotate and translate the coordinate system so that each point p on ∂E_i can be expressed as $(x(t), y(t))$, where $x(t) = a \cos t$ and $y(t) = b \sin t$. Parameter t varies from $-\pi$ to π, and a and b are E_i's semi-major and semi-minor axes, respectively, and the original O is the centre of E_i. Let $p_{i-1} = (x_1, y_1)$ and $p_{i+1} = (x_2, y_2)$.

By the distance formula, we have that

$$d_e(p_{i-1}, p) = \sqrt{(a \cos t - x_1)^2 + (b \sin t - y_1)^2}$$

and

$$d_e(p, p_{i+1}) = \sqrt{(a \cos t - x_2)^2 + (b \sin t - y_2)^2}.$$

Let p'_{i-1} (or p'_{i+1}) be the intersection point between line segment $p_{i-1}O$ (or $p_{i+1}O$) and E_i. Let t_1 (or t_2) in $[-\pi, \pi]$ such that $p'_{i-1} = (x(t_1), y(t_1))$ [or $p'_{i+1} = (x(t_2), y(t_2))$]. Then, by Lemma 8.1 and Fig. 8.3, we obtain an interval $[t_1, t_2]$ for an optimal point $q_i = (x(t_i), y(t_i))$ where t_i is in $[t_1, t_2]$.

Analogously to Corollary 12.1, the optimal value of t is unique. It can be computed using binary search as follows: Rewrite $[t_1, t_2]$ as $[a_1, b_1]$. Let

$$\begin{aligned} d(t) &= d_e(p_{i-1}, p) + d_e(p, p_{i+1}) \\ &= \sqrt{(a \cos t - x_1)^2 + (b \sin t - y_1)^2} \\ &\quad + \sqrt{(a \cos t - x_2)^2 + (b \sin t - y_2)^2}. \end{aligned}$$

Let $d'(t)$ be the derivative of $d(t)$ with respect to t. Since $d(t)$ has a unique minimum $d(t_i)$ when t is in $[a_1, b_1]$, it follows that $d'(a_1) < 0$ and $d'(b_1) > 0$. Let $t_0 = (a_1 + b_1)/2$. If $d'(t_0) = 0$, then let $t_i = t_0$, and we are done. Otherwise, if $d'(t_0) < 0$ (or $d'(t_0) > 0$), then let $a_2 = t_0$ and $b_2 = b_1$ (or $a_2 = a_1$ and $b_2 = t_0$). We shrink the search interval $[a_1, b_1]$ by half to be $[a_2, b_2]$. Repeat this procedure until we obtain an interval $[a_n, b_n]$ such that $b_n - a_n$ is sufficiently small. Then, we let $t_i = (a_n + b_n)/2$. Since $b_n - a_n = \frac{(b_1 - a_1)}{2^{n-1}}$ $(n = 1, 2, \ldots)$ converges rapidly to 0, and because binary search is of $\mathcal{O}(\log n)$ time complexity, the optimal value of t can be found in constant time.

Thus, analogously to the analysis in Sect. 3.5 and to the proof of Theorem 12.1, we have the following

Theorem 12.5 *Algorithm 46 outputs a $\{1 + 2k \cdot [r(\varepsilon)]/L\}$-approximation solution to the TEP in time $\kappa(\varepsilon)\mathcal{O}(k)$.*

In the SRP (or ZRP) case, if each convex polygon is replaced by a simple region, then this results in a *general* SRP (or ZRP). For example, Fig. 12.14 (or Fig. 12.15) shows an initial route of the general SRP (ZRP). It is clear that Algorithm 44 (or Algorithm 45) can be modified to solve such a SRP (or ZRP) without increasing the asymptotic time complexity.

Fig. 12.14 An example of an initial route of a SRP defined by *ellipses*

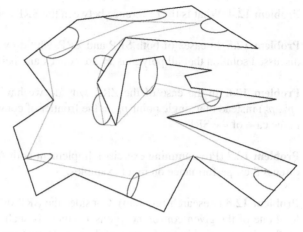

Fig. 12.15 An example of an initial route of a ZRP defined by *ellipses*

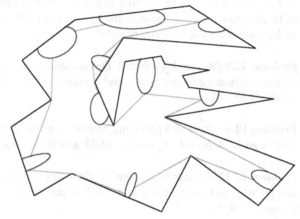

12.5 Problems

Problem 12.1 Each point p on the frontier ∂E of the ellipse E can be expressed as $(x(t), y(t))$, where $x(t) = 3\cos t$ and $y(t) = 2\sin t$. Parameter t varies from $-\pi$ to π. Let $p_1 = (-4, -3)$ and $p_2 = (5, -4)$. Compute an optimal point q in ∂E such that $d_e(p_1, q) + d_e(q, p_2)$ is minimal.

Problem 12.2 In Problem 12.1, let $d(t) = d_e(p_1, p) + d_e(p, p_2)$. Derive an explicit formula for $d(t)$. Solve the equation $d'(t) = 0$ without converting the trigonometric equation into a polynomial equation; rather solve the trigonometric equation numerically for t. *Hint:* We can start with the finite interval $[0, 2\pi)$ which contains all real roots, apply binary search to generate an interval containing at least one of the roots, and then apply the Newton–Raphson method to compute that root at high accuracy.

Problem 12.3 What is the difference between the fixed SRP (ZRP) and the floating SRP (ZRP)?

Problem 12.4 What is the difference between the SRP and the ZRP?

Problem 12.5 In cases of both SRP and ZRP, why do we have to assume for the discussed solution that all polygons P_i are convex and pairwise-disjoint?

Problem 12.6 In the case of the ZRP, why do we have to consider a ZRP pair (p_1^i, p_2^i) instead of a single point p_i on the frontier of convex polygon P_i, as we did in the case of the SRP?

Problem 12.7 (Programming exercise) Implement both Algorithms 44 and 45 and discuss their performance on input examples.

Problem 12.8 (Research problem) Consider the SRP or ZRP. Assume that for at least one of the given convex polygons P_i, there is such an endpoint of an edge e of P_i, e is contained in the frontier of domain D, that this endpoint is also a vertex of D. How to define and compute the ε_d-polygon of P_i? How does the existence of such a case affect the performance of the algorithm?

Problem 12.9 (Research problem) Discuss how variations (i.e., smaller or larger) of degeneration parameter $\varepsilon_1 > 0$ and constant $\varepsilon_d \geq \varepsilon_1$ affect the run-time of Algorithms 44 and 45.

Problem 12.10 (Research problem) Starting with the generic RBA (see Sect. 3.10), specify an RBA for solving the fixed SRP and the fixed ZRP.

Problem 12.11 (Research problem) Is there any simpler way to compute the optimal point q_j in Line 5 of Algorithm 46? In particular, consider the case where all ellipses are disks.

Problem 12.12 (Research problem) Generalise Algorithm 46 for handling a sequence of k pairwise-disjoint ellipsoids (or, in particular, balls) $E_0, E_1, \ldots, E_{k-1}$ in 3D space. Does this change the time complexity of the algorithm compared to the provided algorithm for the 2D case?

12.6 Notes

The safari route problem (SRP) was introduced in [9]. Reference [9] claimed to have an $\mathcal{O}(kn^2)$ time algorithm for solving this problem, where n is the total number of vertices of polygon D and of all polygons P_i, for $i = 1, 2, \ldots, k$. In 1994, [14] improved the result to an $\mathcal{O}(n^2)$ time algorithm for the floating SRP, not using anymore the restriction used in [9] of forcing the route through a specific point. In 2003, [15] showed that there is an error in the algorithm proposed in [9], and presented an $\mathcal{O}(n^3)$ time algorithm for the fixed SRP, where n is again the total number of vertices of D and all polygons P_i, for $i = 1, 2, \ldots, k$. The algorithm runs in $\mathcal{O}(n^4)$

time for the floating SRP. In the same year, the result was improved by [4] with an algorithm running in time $\mathcal{O}(kn \log(n/k))$ for the fixed SRP.

The zookeeper route problem (ZRP) was introduced in [3]. Both [2] (in 1987) and [3] (in 1992) present an $\mathcal{O}(n^2)$ algorithm for the fixed ZRP, where n is the total number of vertices of D and P_i, and $i = 1, 2, \ldots, k$. In 1994, [5] improved this to $\mathcal{O}(n \log^2 n)$. In 2003, [1] improved this further to $\mathcal{O}(n \log n)$; the algorithm is still for the fixed ZRP, and n is the input size as defined above. Reference [13] (in 2001) gave an $\mathcal{O}(n^2)$ algorithm for the floating ZRP.

Lemma 12.1 is Lemma 2 in [9]. Lemma 12.2 is Lemma 2 in [3]. For Algorithm 2D ESP, see [7], pages 639–641. For the tangent calculation, see [12].

The touring polygons problem (TPP) was discussed in [4]; see also Chap. 10. Obstacles with smooth surfaces start to be discussed in computational geometry; see, for example, [8, 10, 11].

For binary search algorithms, see, for example, [6].

References

1. Bespamyatnikh, S.: An $\mathcal{O}(n \log n)$ algorithm for the zoo-keepers problem. Comput. Geom. **24**, 63–74 (2003)
2. Chin, W.-P., Ntafos, S.: Optimum zookeeper routes. Congr. Numer. **58**, 257–266 (1987)
3. Chin, W.-P., Ntafos, S.: The zookeeper route problem. Inf. Sci. **63**, 245–259 (1992)
4. Dror, M.A., Efrat, A., Lubiw, A., Mitchell, J.: Touring a sequence of polygons. In: Proc. STOC, pp. 473–482 (2003)
5. Hershberger, J., Snoeyink, J.: An efficient solution to the zookeeper's problem. In: Proc. Canad. Conf. Comput. Geom, pp. 104–109 (1994)
6. Knuth, D.E.: The Art of Computer Programming: Volume 3, 3rd edn., pp. 409–426. Addison-Wesley, Reading (1997)
7. Mitchell, J.S.B.: Geometric shortest paths and network optimization. In: Sack, J.-R., Urrutia, J. (eds.) Handbook of Computational Geometry, pp. 633–701. Elsevier, Amsterdam (2000)
8. Mitchell, J.S.B., Sharir, M.: New results on shortest paths in three dimensions. In: Proc. SCG, pp. 124–133 (2004)
9. Ntafos, S.: Watchman routes under limited visibility. Comput. Geom. **1**, 149–170 (1992)
10. Pocchiola, M., Vegter, G.: Computing the visibility graph via pseudo-triangulations. In: Proc. ACM-SIAM Sympos. Discrete Algorithms, pp. 248–257 (1995)
11. Pocchiola, M., Vegter, G.: Minimal tangent visibility graphs. Comput. Geom. **6**, 303–314 (1996)
12. Sunday, D.: Algorithm 14: tangents to and between polygons. http://softsurfer.com/Archive/algorithm_0201/. Accessed July 2011
13. Tan, X.: Shortest zookeeper routes in simple polygons. Inf. Process. Lett. **77**, 23–26 (2001)
14. Tan, X., Hirata, T.: Shortest safari routes in simple polygons. In: LNCS, vol. 834, pp. 523–531 (1994)
15. Tan, X., Hirata, T.: Finding shortest safari routes in simple polygons. Inf. Process. Lett. **87**, 179–186 (2003)

Appendix
Mathematical Details

All's well that ends well.

A.1 Derivatives for Example 9.6

We provide a complete list of all $\frac{\partial d_i}{\partial t_i}$ (for $i = 0, 1, \ldots, 19$) for the cube-curve g used in Example 9.6 and shown in Fig. 9.17:

$$d_{t_0} = \frac{t_0}{\sqrt{t_0^2 + t_1^2 + 4}} + \frac{t_0 - t_{19}}{\sqrt{(t_0 - t_{19})^2 + 4}},$$

$$d_{t_1} = \frac{t_1}{\sqrt{t_0^2 + t_1^2 + 4}} + \frac{t_1 - t_2}{\sqrt{(t_1 - t_2)^2 + 5}},$$

$$d_{t_2} = \frac{t_2 - t_1}{\sqrt{(t_2 - t_1)^2 + 5}} + \frac{t_2 - 1}{\sqrt{(t_2 - 1)^2 + (t_3 - 1)^2 + 4}},$$

$$d_{t_3} = \frac{t_3 - 1}{\sqrt{(t_2 - 1)^2 + (t_3 - 1)^2 + 4}} + \frac{t_3}{\sqrt{t_3^2 + t_4^2 + 4}},$$

$$d_{t_4} = \frac{t_4}{\sqrt{t_3^2 + t_4^2 + 4}} + \frac{t_4 - 1}{\sqrt{(t_4 - 1)^2 + t_5^2 + 4}},$$

$$d_{t_5} = \frac{t_5}{\sqrt{(t_4 - 1)^2 + t_5^2 + 4}} + \frac{t_5 - t_6}{\sqrt{(t_5 - t_6)^2 + 4}},$$

$$d_{t_6} = \frac{t_6 - t_5}{\sqrt{(t_6 - t_5)^2 + 4}} + \frac{t_6 - 1}{\sqrt{(t_6 - 1)^2 + t_7^2 + 4}},$$

$$d_{t_7} = \frac{t_7}{\sqrt{(t_6 - 1)^2 + t_7^2 + 4}} + \frac{t_7 - 1}{\sqrt{(t_7 - 1)^2 + t_8^2 + 4}},$$

F. Li, R. Klette, *Euclidean Shortest Paths*,
DOI 10.1007/978-1-4471-2256-2, © Springer-Verlag London Limited 2011

$$d_{t_8} = \frac{t_8}{\sqrt{(t_7 - 1)^2 + t_8^2 + 4}} + \frac{t_8 - t_9}{\sqrt{(t_8 - t_9)^2 + 4}},$$

$$d_{t_9} = \frac{t_9 - t_8}{\sqrt{(t_9 - t_8)^2 + 4}} + \frac{t_9 - 1}{\sqrt{(t_9 - 1)^2 + t_{10}^2 + 4}},$$

$$d_{t_{10}} = \frac{t_{10}}{\sqrt{(t_9 - 1)^2 + t_{10}^2 + 4}} + \frac{t_{10} - 1}{\sqrt{(t_{10} - 1)^2 + (t_{11} - 1)^2 + 4}},$$

$$d_{t_{11}} = \frac{t_{11} - 1}{\sqrt{(t_{11} - 1)^2 + (t_{10} - 1)^2 + 4}} + \frac{t_{11}}{\sqrt{t_{11}^2 + t_{12}^2 + 1}},$$

$$d_{t_{12}} = \frac{t_{12}}{\sqrt{t_{11}^2 + t_{12}^2 + 1}} + \frac{t_{12} - t_{13}}{\sqrt{(t_{12} - t_{13})^2 + 4}},$$

$$d_{t_{13}} = \frac{t_{13} - t_{12}}{\sqrt{(t_{13} - t_{12})^2 + 4}} + \frac{t_{13} - 1}{\sqrt{(t_{13} - 1)^2 + (t_{14} - 1)^2 + 4}},$$

$$d_{t_{14}} = \frac{t_{14} - 1}{\sqrt{(t_{13} - 1)^2 + (t_{14} - 1)^2 + 4}} + \frac{t_{14}}{\sqrt{t_{14}^2 + (t_{15} - 1)^2 + 4}},$$

$$d_{t_{15}} = \frac{t_{15} - 1}{\sqrt{t_{14}^2 + (t_{15} - 1)^2 + 4}} + \frac{t_{15} - t_{16}}{\sqrt{(t_{15} - t_{16})^2 + 16}},$$

$$d_{t_{16}} = \frac{t_{16} - t_{15}}{\sqrt{(t_{16} - t_{15})^2 + 16}} + \frac{t_{16} - t_{17}}{\sqrt{(t_{16} - t_{17})^2 + 4}},$$

$$d_{t_{17}} = \frac{t_{17} - t_{16}}{\sqrt{(t_{17} - t_{16})^2 + 4}} + \frac{t_{17}}{\sqrt{t_{17}^2 + (t_{18} - 1)^2 + 1}},$$

$$d_{t_{18}} = \frac{t_{18} - 1}{\sqrt{t_{17}^2 + (t_{18} - 1)^2 + 1}} + \frac{t_{18} - t_{19}}{\sqrt{(t_{18} - t_{19})^2 + 101}},$$

$$d_{t_{19}} = \frac{t_{19} - t_{18}}{\sqrt{(t_{19} - t_{18})^2 + 101}} + \frac{t_{19} - t_0}{\sqrt{(t_{19} - t_0)^2 + 4}}.$$

A.2 GAP Inputs and Outputs

1. To Compute the Factors of the Determinant of Polynomial f(x) = 84*x^6-228*x^5+361*x^4+20*x^3+210*x^2-200*x+25.

1.1 Create the rows of a (2n-1)x(2n-1) matrix, where n is 6.

```
r1:=[1,-228,361,20,210,-200,25,0,0,0,0];
r2:=[0,1,-228,361,20,210,-200,25,0,0,0];
r3:=[0,0,1,-228,361,20,210,-200,25,0,0];
r4:=[0,0,0,1,-228,361,20,210,-200,25,0];
r5:=[0,0,0,0,1,-228,361,20,210,-200,25];
```

```
r6:= [6*1,-5*228,4*361,3*20,2*210,-200,0,0,0,0,0];
r7:= [0,6*1,-5*228,4*361,3*20,2*210,-200,0,0,0,0];
r8:= [0,0,6*1,-5*228,4*361,3*20,2*210,-200,0,0,0];
r9:= [0,0,0,6*1,-5*228,4*361,3*20,2*210,-200,0,0];
r10:=[0,0,0,0,6*1,-5*228,4*361,3*20,2*210,-200,0];
r11:=[0,0,0,0,0,6*1,-5*228,4*361,3*20,2*210,-200];

m:=[r1,r2,r3,r4,r5,r6,r7,r8,r9,r10,r11];
```

1.2 Compute the Determinant.

```
gap> d:=DeterminantMatDestructive(m); 313645192522810211250000000
```

```
gap> d:=84*d; 2634619617191605774500000000
```

1.3 Compute the Factors of the Determinant.

```
gap> FactorsInt(d); [ 2, 2, 2, 2, 2, 2, 2, 2, 3, 3, 3, 3, 5, 5,
5, 5, 5, 5, 5, 5, 5, 7, 11,
  19249, 204797, 214309 ]
```

2. Factorising f(x) mod 13.

```
gap> F:=GaloisField(13); GF(13) gap> e:=Elements(F); [ 0*Z(13),
Z(13)^0, Z(13), Z(13)^2, Z(13)^3, Z(13)^4, Z(13)^5, Z(13)^6,
  Z(13)^7, Z(13)^8, Z(13)^9, Z(13)^10, Z(13)^11 ]
gap> x:= X(F,"x"); x gap>
f:=84*x^6-228*x^5+361*x^4+20*x^3+210*x^2-200*x+25;;
gap> Factors(f); [ Z(13)^5*x+Z(13)^10, x^2+Z(13)^5,
x^3+Z(13)^3*x^2+Z(13)^2*x+Z(13)^3 ]
```

3. Factorising f(x) mod 19.

```
gap> F:=GaloisField(19); GF(19) gap> e:=Elements(F); [ 0*Z(19),
Z(19)^0, Z(19), Z(19)^2, Z(19)^3, Z(19)^4, Z(19)^5, Z(19)^6,
  Z(19)^7, Z(19)^8, Z(19)^9, Z(19)^10, Z(19)^11, Z(19)^12,
  Z(19)^13, Z(19)^14, Z(19)^15, Z(19)^16, Z(19)^17 ]
gap> x:= X(F,"x"); x gap>
f:=84*x^6-228*x^5+361*x^4+20*x^3+210*x^2-200*x+25;;
gap> Factors(f); [ Z(19)^3*x^6+x^3+x^2+Z(19)^8*x+Z(19)^14 ]
```

4. Factorising f(x) mod 37.

```
gap> F:=GaloisField(37); GF(37) gap> e:=Elements(F); [ 0*Z(37),
Z(37)^0, Z(37), Z(37)^2, Z(37)^3, Z(37)^4, Z(37)^5, Z(37)^6,
  Z(37)^7, Z(37)^8, Z(37)^9, Z(37)^10, Z(37)^11, Z(37)^12,
  Z(37)^13, Z(37)^14, Z(37)^15, Z(37)^16, Z(37)^17, Z(37)^18,
  Z(37)^19, Z(37)^20, Z(37)^21, Z(37)^22, Z(37)^23, Z(37)^24,
  Z(37)^25, Z(37)^26, Z(37)^27, Z(37)^28, Z(37)^29, Z(37)^30,
  Z(37)^31, Z(37)^32, Z(37)^33, Z(37)^34, Z(37)^35 ]
gap> x:= X(F,"x"); x gap>
f:=84*x^6-228*x^5+361*x^4+20*x^3+210*x^2-200*x+25;;
gap> Factors(f); [ Z(37)^24*x+Z(37)^0,
  x^5+Z(37)^26*x^4+Z(37)^22*x^3+Z(37)^30*x^2+Z(37)^9*x+Z(37)^10 ]
```

A.3 Matrices Q for Sect. 9.9

Table A.1 Computation of fourth (Part a), fifth (Part b), and sixth (Part c) row of matrix Q for proving that function $f_1^1(x)$ is irreducible

Part a

k	$a_{k,5}$	$a_{k,4}$	$a_{k,3}$	$a_{k,2}$	$a_{k,1}$	$a_{k,0}$
0	0	13	2	6	5	2
1	13	2	6	5	2	0
2	2	6	1	17	2	14
3	6	1	12	16	7	8
4	1	12	1	11	6	5
5	12	1	18	13	11	4
6	1	18	2	0	0	10
7	18	2	7	7	16	4
8	2	7	0	9	17	15
9	7	0	4	12	8	8
10	0	4	4	0	12	9
11	4	4	0	12	9	0
12	4	0	2	18	5	16
13	0	2	8	14	2	16
14	2	8	14	2	16	0
15	8	14	16	11	12	8
16	14	16	10	11	18	13
17	16	10	14	2	2	18
18	10	14	0	0	0	7
19	14	0	13	13	10	2

Part b

k	$a_{k,5}$	$a_{k,4}$	$a_{k,3}$	$a_{k,2}$	$a_{k,1}$	$a_{k,0}$
0	14	0	13	13	10	2
1	0	13	16	13	10	18
2	13	16	13	10	18	0
3	16	13	6	14	2	14
4	13	6	12	0	15	7
5	6	12	15	11	9	14
6	12	15	15	13	12	5
7	15	15	2	1	1	10
8	15	2	11	11	5	3
9	2	11	2	15	17	3
10	11	2	10	12	15	8
11	2	10	13	16	17	6
12	10	13	11	12	18	8
13	13	11	6	12	11	2
14	11	6	8	7	4	14
15	6	8	8	5	4	6
16	8	8	9	8	4	5
17	8	9	7	3	15	13
18	9	7	2	14	4	13
19	7	2	1	10	10	17

Part c

k	$a_{k,5}$	$a_{k,4}$	$a_{k,3}$	$a_{k,2}$	$a_{k,1}$	$a_{k,0}$
0	7	2	1	10	10	17
1	2	1	2	2	2	9
2	1	2	16	16	2	8
3	2	16	4	9	14	4
4	16	4	4	9	16	8
5	4	4	7	14	9	7
6	4	7	4	18	12	16
7	7	4	8	2	2	16
8	4	8	13	13	1	9
9	8	13	3	10	14	16
10	13	3	9	13	7	13
11	3	9	9	3	15	14
12	9	9	5	17	13	12
13	9	5	4	0	9	17
14	5	4	6	15	14	17
15	4	6	12	11	9	1
16	6	12	1	18	6	16
17	12	1	3	10	14	5
18	1	3	18	3	1	10
19	3	18	10	8	16	4

Table A.2 Computation of the second (Part a) and third (Part b) row of matrix Q for proving that function $f_2^2(x)$ is irreducible

k	$a_{k,4}$	$a_{k,3}$	$a_{k,2}$	$a_{k,1}$	$a_{k,0}$	k	$a_{k,4}$	$a_{k,3}$	$a_{k,2}$	$a_{k,1}$	$a_{k,0}$
Part a						Part b					
0	0	0	0	0	1	0	36	21	11	1	3
1	0	0	0	1	0	1	24	32	12	34	25
2	0	0	1	0	0	2	34	26	29	21	29
3	0	1	0	0	0	3	35	18	17	11	1
4	1	0	0	0	0	4	24	22	33	26	13
5	34	16	26	6	12	5	24	10	21	9	29
6	25	15	2	31	1	6	12	35	4	25	29
7	14	32	15	3	4	7	36	11	4	27	33
8	27	17	34	14	20	8	14	25	1	27	25
9	10	22	13	34	28	9	20	3	21	35	20
10	29	25	35	14	9	10	17	8	0	29	18
11	12	18	28	35	15	11	31	13	27	9	19
12	19	35	14	13	33	12	31	5	1	20	2
13	15	22	26	36	6	13	23	16	12	3	2
14	14	7	19	22	32	14	21	10	9	29	17
15	2	21	16	5	20	15	21	12	20	32	30
16	15	11	20	32	24	16	23	23	23	8	30
17	3	1	15	3	32	17	28	21	14	20	17
18	29	26	7	13	36	18	11	18	8	0	3
19	13	27	27	25	15	19	22	36	27	32	21
20	25	13	30	19	8	20	7	9	12	5	5
21	12	23	3	10	4	21	25	13	2	10	10
22	24	10	26	2	33	22	12	32	31	12	4
23	12	3	34	29	29	23	33	1	28	2	33
24	4	4	8	27	33	24	13	1	9	9	26
25	29	35	20	20	11	25	36	32	14	30	8
26	22	3	34	0	15	26	35	35	4	2	25
27	11	16	17	36	5	27	4	9	24	13	13
28	20	8	26	34	21	28	34	14	6	0	11
29	22	13	36	30	18	29	23	32	33	30	1
30	21	18	10	2	5	30	0	31	36	28	17
31	29	13	30	20	30	31	31	36	28	17	0
32	0	13	34	19	15	32	17	6	9	1	2
33	13	34	19	15	0	33	29	22	36	30	19
34	32	5	20	4	8	34	9	19	7	8	15
35	20	14	22	15	14	35	29	3	20	32	34
36	28	9	17	23	18	36	27	3	9	23	15
37	36	21	11	1	3	37	33	34	22	29	28

Table A.3 Computation of the fourth (Part a) and fifth (Part b) row of matrix Q for proving that function $f_2^2(x)$ is irreducible

k	$a_{k,4}$	$a_{k,3}$	$a_{k,2}$	$a_{k,1}$	$a_{k,0}$	k	$a_{k,4}$	$a_{k,3}$	$a_{k,2}$	$a_{k,1}$	$a_{k,0}$
Part a						Part b					
0	33	34	22	29	28	0	31	22	1	7	23
1	9	32	36	4	26	1	3	16	36	24	2
2	5	32	16	6	34	2	7	10	28	20	36
3	17	22	25	27	23	3	26	29	17	4	10
4	8	1	25	14	19	4	25	26	14	18	16
5	14	5	0	30	22	5	25	7	2	18	4
6	0	2	24	32	20	6	6	32	2	6	4
7	2	24	32	20	0	7	14	24	14	3	35
8	18	27	35	12	24	8	19	16	34	8	20
9	10	27	36	21	31	9	33	5	21	23	6
10	34	11	22	17	9	10	17	31	30	19	26
11	20	11	13	28	1	11	17	6	17	17	19
12	25	0	30	10	18	12	29	30	15	10	19
13	36	23	31	20	4	13	17	35	24	8	15
14	26	15	31	35	25	14	21	0	6	6	19
15	11	3	8	33	16	15	11	9	34	34	30
16	7	36	23	8	21	16	13	25	24	22	21
17	15	24	5	26	10	17	23	10	27	25	8
18	16	23	9	26	32	18	15	25	31	35	17
19	12	6	35	17	7	19	17	12	18	33	32
20	7	5	33	5	33	20	35	31	31	23	19
21	21	34	2	1	10	21	0	36	8	7	13
22	8	5	29	25	30	22	36	8	7	13	0
23	18	9	11	4	22	23	11	28	24	31	25
24	29	3	28	19	31	24	32	15	21	17	21
25	27	11	33	20	15	25	30	15	35	28	14
26	4	21	19	29	28	26	36	34	31	9	27
27	9	9	22	15	11	27	0	15	20	21	25
28	19	18	27	28	34	28	15	20	21	25	0
29	35	35	4	0	6	29	12	2	8	16	32
30	4	9	22	31	13	30	3	15	32	30	33
31	34	12	24	0	11	31	6	6	34	14	36
32	21	13	33	30	1	32	25	19	22	35	35
33	24	36	21	16	30	33	18	15	19	0	4
34	1	35	11	26	29	34	35	11	24	1	31
35	32	27	15	35	12	35	17	29	23	19	13
36	5	9	16	19	14	36	15	36	17	4	19
37	31	22	1	7	23	37	28	35	24	35	32

Table A.4 Computation of the second row of matrix Q for proving that function $f_2^3(x)$ is irreducible

k	$a_{k,1}$	$a_{k,0}$
0	0	1
1	1	0
2	0	7
3	7	0
4	0	10
5	10	0
6	0	5
7	5	0
8	0	9
9	9	0
10	0	11
11	11	0
12	0	12
13	12	0

Table A.5 Computation of the second (Part a) and third (Part b) row of matrix Q for proving that function $f_3^3(x)$ is irreducible

k	$a_{k,2}$	$a_{k,1}$	$a_{k,0}$	k	$a_{k,2}$	$a_{k,1}$	$a_{k,0}$
Part a				Part b			
0	0	0	1	0	0	3	1
1	0	1	0	1	3	1	0
2	1	0	0	2	3	1	2
3	5	9	5	3	3	3	2
4	8	11	12	4	5	3	2
5	12	6	1	5	2	8	12
6	1	5	8	6	5	4	10
7	10	4	5	7	3	3	12
8	2	4	11	8	5	0	2
9	1	3	10	9	12	8	12
10	8	6	5	10	3	3	8
11	7	12	1	11	5	9	2
12	8	12	9	12	8	8	12
13	0	3	1	13	9	6	1

Index